AN ENCYCLOPAEDIA OF
WORLD
BRIDGES

AN ENCYCLOPAEDIA OF
WORLD
BRIDGES

DAVID McFETRICH

PEN & SWORD
TRANSPORT

AN IMPRINT OF PEN & SWORD BOOKS LTD.
YORKSHIRE – PHILADELPHIA

First published in Great Britain in 2022 by
Pen and Sword Transport
An imprint of
Pen & Sword Books Ltd.
Yorkshire - Philadelphia

ISBN 978 1 52679 446 8

Typeset in 10/12 Goudy Old Style
by SJmagic DESIGN SERVICES, India.

Printed and bound in India by Replika Press Pvt. Ltd.

Pen & Sword Books Ltd incorporates the imprints of Pen & Sword Books Archaeology, Atlas, Aviation, Battleground, Discovery, Family History, History, Maritime, Military, Naval, Politics, Railways, Select, Transport, True Crime, Fiction, Frontline Books, Leo Cooper, Praetorian Press, Seaforth Publishing, Wharncliffe and White Owl.

For a complete list of Pen & Sword titles please contact

PEN & SWORD BOOKS LIMITED
47 Church Street, Barnsley, South Yorkshire, S70 2AS, England
E-mail: enquiries@pen-and-sword.co.uk
Website: www.pen-and-sword.co.uk

or

PEN AND SWORD BOOKS
1950 Lawrence Rd, Havertown, PA 19083, USA
E-mail: Uspen-and-sword@casematepublishers.com
Website: www.penandswordbooks.com

Contents

Introduction

My interest in bridges began when I was given a copy of the book *The World's Great Bridges* by Hubert Shirley Smith when I was about 15, a gift that helped me decide to become a civil engineer. During my career after leaving university, I both designed and built bridges and, although I changed career after eleven years, I have maintained that interest ever since.

The aim is for this volume to give brief descriptions of some of the world's most famous bridges that everyone knows, but also to include a wide selection of less well-known bridges that are interesting because they are historic or unusual in some way or have connections with important events or people in history. I have also added several bridges which, though not particularly interesting as structures, are tourist attractions that help to encourage people to visit developing countries which need this economic help. The bridges described in this book are not necessarily the longest, highest or oldest. Some of them indeed are, but all are included simply because I find them interesting.

There are nearly 1,000 main entries in the book but, occasionally, the details of some bridges that are located near each other and share some common theme are integrated into a single entry, for example the River Neva bridges in St Petersburg and the Ancient Roman Bridges still standing in Rome. Furthermore, brief descriptions of predecessor bridges at some sites where relevant information is available are also included. This means that, in total, more than 1,200 differently-named structures are covered. The heading to each entry generally gives the name of the bridge, its location and country. Where bridges cross between countries the heading gives the name of the nearest large town and the country in which it lies.

To help readers find examples of bridges with particular features or historic connections, I have included a section headed Bridge Lists that gives the names of some of the best examples of bridges in the book that have these characteristics. It should be noted, however, that this section is intended to be more a quick guide rather than a fully comprehensive index which would take up too much space.

In terms of countries covered by the book, there are entries for bridges in 174 of the 195 separate countries of the world that are listed by the United Nations, as shown in the table that follows directly after the Geographic Index. However, there are only three entries for Great Britain with one representative entry from each of England, Wales and Scotland: London Bridge, Menai Straits Bridges and Firth of Forth Bridges. These three entries, which include details of predecessor structures on these sites, are therefore also representative of bridges from four major epochs of British bridge-building: medieval, Industrial Revolution, Victorian and twenty-first century. (Bridges in the Northern Ireland part of the United Kingdom are included in this volume under Northern Ireland.) A total of 1,350 entries for British mainland bridges, covering more than 1,600 different structures altogether, are described separately in the sister volume *An Encyclopaedia of British Bridges* and, where appropriate, occasional mentions in this book about these British bridges are given in the form '(see *AEBB*)'.

It is perhaps worth pointing out that the seemingly arbitrary balance of bridges in favour of France and the USA may simply be because these are the countries best represented by books on foreign bridges that are most easily available in

British bookshops. However, the main criterion for selection is simply that the bridges chosen are ones that appealed to the author. So, if you think some of your favourite bridges are unfairly excluded, try to make the most of the unusual, sometimes strange, ones that have taken their place. Many of the source books I have used contain much more information than I have space for in this volume so I have included an extensive bibliography.

Where bridges are reasonably well known, their names are given in anglicised form but others are shown using the usual local form of spelling (including diacritics) but in Roman type. Dimensions relating to each bridge are normally given in the units used in that country or as quoted in the majority of the books in which the bridge is described.

How Bridges Work

There are five different types of force that can affect a bridge and each of its constituent parts. These forces, shown in the accompanying diagram, are:

Compression – in which an axial load presses on a structural member

Tension – in which an axial load pulls on it

Torsion – in which it is subject to twisting along its longitudinal axis

Shear – in which two opposing forces, working like the blades of a pair of scissors, are cutting into the member

Bending – in which one side of the member is being compressed and the opposite side stretched.

There are five main types of bridge:

Beam bridge – this supports its load by its resistance to bending; in truss bridges the bending is resisted by tension and compression forces in a triangular framework

Cantilever bridge – this also supports its load by its resistance to bending

Arch bridge – this supports its load by its resistance to compression

Suspension bridge – here, the main columns are in compression and the suspension cables and hangers are in tension

Stay bridge – the main columns are also in compression and the diagonal stays are in tension.

It should be noted that, except in beam bridges where the top part of the beam can form the traffic-bearing deck, in bridges that are principally arch, cable stay or suspension structures the

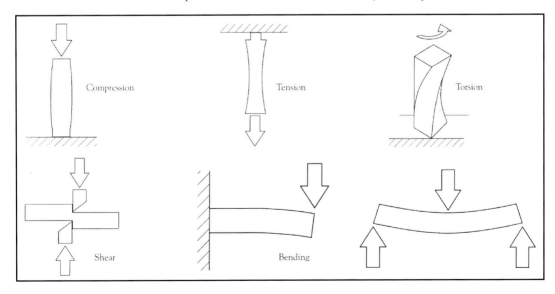

Forces on structures

deck itself will usually be a subsidiary structure of some kind, between support points provided by the main structure. On the Forth Bridge, for example, which is a huge cantilever bridge, this main cantilever framework supports an internal viaduct on which the trains run. Elsewhere, this viaduct would be a major beam type of structure in its own right.

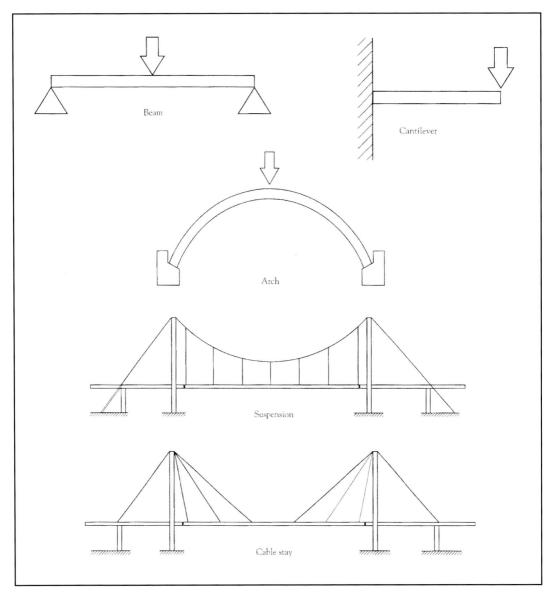

Types of bridge

Glossary

Abutment The support at the ends of either a single-span bridge or a series of bridge spans. See also Pier.

Aedicule A door or window opening flanked by classical columns and supporting an entablature and pediment, all in the manner of a shrine.

Anthemion A Greek type of carved ornament based on flowers.

Aqueduct An elevated bridge-like structure carrying an artificial watercourse.

Arcade A row of free-standing arches supported on columns or piers or, in a blind arcade, placed decoratively against a wall and usually standing on pilasters.

Arch A vertically curved structural member that supports loads across a gap by resistance to axial compression along its length. The shape of an arch in a bridge gives its distinctive character. The most common bridge arch shapes are semicircular, pointed, segmental and semi-elliptical, although there are also parabolic steel-framed and concrete arches. Arches exert a lateral thrust against their abutments, this thrust increasing as the arch becomes flatter. In multi-arch bridges, the thrust from one arch is counter-balanced by the massiveness of the intermediate piers or by the thrust from the neighbouring arch.

Architrave The moulded lowest part of a lintel over a door, window or other rectangular opening.

Archivolt The projecting stone edging, often moulded, to an arch ring.

Arch ring The exposed edge of a masonry arch barrel that is visible on the side elevation of a bridge and is composed of stone voussoirs or vertically curved courses of brickwork.

Arris The corner edge along the line formed by the intersection of two surfaces.

Ashlar High-quality masonry consisting of smoothly dressed blocks of approximately square-faced stonework laid in regular courses with very thin joints.

Back span A span in a cable stay or suspension bridge between a main pier and the end of the bridge which, for a cable stay bridge, fully or partially balances the dead loads from the main span.

Bailey bridge A prefabricated portable bridging system invented by Sir Donald Bailey in 1940 for military use in the Second World War.

Balanced cantilever A structure consisting of a central pier or column with equal-length projecting arms on each side. In balanced cantilever construction, work proceeds by simultaneously building outwards on each side of a pier, connecting together the two completed parts of a span when they meet at mid-span.

Balustrade An ornamental parapet consisting of a series of short vertical pillars (balusters), often moulded, standing on a base and supporting a handrail.

Banded (of a column) The encirclement of a column by a projecting strip, sometimes with a finish different from that of the column itself.

Barrel (of an arch) The curved, tunnel-like interior surface of an arched bridge comprising its working structure. The barrel of an arch is often constructed of a different (cheaper) material from that which can be seen on the elevation of the arch.

Bascule bridge A type of drawbridge in which the opening span pivots on a horizontal axis and is balanced by a shorter backspan, counterweight or both (from the French word *bascule* meaning seesaw). In castles such a structure is sometimes called a turning bridge.

Basket-handled A colloquial name sometimes given to a semi-elliptical or three-centred bridge arch.

Batter The slope of the face of a wall from the vertical.

Battlement A parapet of short alternating low and high sections, as in a castle.

Beam A straight structural member that spans, usually horizontally, across a gap to support loads through its resistance to bending.

Bearing A mechanism built into a bridge to transfer loads between structural sections, such as between deck and pier. One cause of the high cost of refurbishing or upgrading bridges is the difficulty of replacing worn, failed or undersized bearings. Where bearings have failed, often as a result of inadequate maintenance, additional stresses can be transferred into parts of the structure that were not designed to resist them, thereby endangering the structure.

Bent A two-dimensional frame of two or more legs placed at right angles to the line of the bridge it supports.

Blind arch A decorative arch placed against a wall.

Boom The top or bottom member of a truss or girder (sometimes called chord).

Bowstring bridge A bridge with a vertically curved upper girder that meets a horizontal tie at the bridge's support points, thus creating a profile similar to that of an archer's bow and bowstring. The horizontal tie counteracts the outward lateral force from the ends of the arch, meaning that the bridge piers and abutments have to support a vertical load only.

Box girder A fabricated beam with a hollow square or rectangular cross section.

Bracing Secondary structural members, usually diagonal, which take tension or compression forces in order to stiffen a structure, for example against deformation from wind loading.

Bucket bridge A crossing that consists of a large wooden box, which can hold several people, supported by pulley wheels running along one or a pair of suspension cables spanning a river, that is then pulled across the gap by hand or a simple winch.

Butterfly bridge A bridge in which the deck is suspended below twin arches, with the plane of each arch and its filament of deck support hangers being inclined outwards from the vertical to form a V-shaped cross section.

Buttress A projecting support to a wall, usually at right angles to it, that helps the wall to resist lateral thrust.

Cable stay bridge A bridge in which the intermediate supports to a continuous beam are not from either piers beneath the beam or vertical hangers from a suspension cable above the beam but from diagonal ties, made from high-strength steel cable, connected directly between the point of support and a main pier or pylon that extends to some distance above the beam. The diagonal inclination of these stays imparts a compressive horizontal force into the bridge deck. In some early bridges of this type the stays were rods rather than cables.

Caisson A prefabricated circular or rectangular structure that is placed in water where a pier is to be constructed, the water then being pumped out allowing construction work to proceed in the dry. A caisson is incorporated in the foundations and remains a permanent part of the structure (cf. Cofferdam).

Camelback truss A truss in which the top boom is not horizontal but consists of two sloping members each side, the ones nearer the support points being at a steeper angle from the horizontal.

Cantilever A beam that is firmly fixed at one end and unsupported at the other in the manner of a bracket.

Capital The top of a column or pilaster.

Castellated The shape given to the top of a parapet wall, in imitation of the battlements on a castle, where alternate sections are higher and lower than each other.

Cartouche A scroll-shaped ornamental tablet which, on a bridge, can contain information such as the bridge's name, designer and date of construction.

Cast iron Iron, smelted from iron ore and still containing carbon, that has been poured into moulds when in a molten state. The presence of the carbon makes cast iron brittle and unable to withstand tensile forces. It was replaced as a structural material by the introduction of wrought iron in the mid-nineteenth century.

Catenary The shape assumed by a cable of uniform self-weight when suspended freely between two points. It is thus the shape of a suspension bridge's main cables before the deck has been hung from them. Although the cables in the completed bridge are sometimes still called the catenary, the actual final shape is a compound curve between that of a catenary and a parabola (q.v.).

Cement A fine powder that, after mixing with water, soon dries into a hard state and is the active ingredient in concrete and grout.

Centring The temporary structure, traditionally made of timber and sometimes called formwork, built to provide the support and necessary shape for an arch during its construction.

Chamfer The face resulting from the removal of an edge or arris between two surfaces at an angle to each other.

Chord The top or bottom member of a truss or girder (sometimes called boom).

Clam bridge A primitive single-span bridge consisting of immense irregular 'Cyclopean' stone slabs spanning between abutments made from carefully placed piles of rock. The multiple-span form is called a clapper bridge.

Clapper bridge A primitive bridge consisting of immense irregular 'Cyclopean' stone slabs spanning between piers and abutments made from carefully placed piles of rock. The single-span form is often called a clam bridge.

Coade stone An architectural stoneware product made by firing a specially formulated mixture of pre-fired clay and sand. It was ideal for producing weather-proof decorative external features such as swags (which inside a building would have been made of plaster) that were cheaper than stone carvings.

Cofferdam A temporary water-excluding structure constructed in water, the water then being pumped out to allow construction work to proceed in the dry. When the permanent structure is completed the cofferdam is removed (cf. Caisson).

Collared pier A masonry pier with encircling decorative moulding near its top.

Colonnade A row of columns supporting a horizontal beam, or a series of arches.

Column An independent vertical structural member that supports loads by its resistance to compression.

Concrete A mixture of carefully graded stone and sand (the aggregates), cement and water which sets into a rock-like mass.

Concrete cancer Degeneration of concrete structures by a process properly known as alkali aggregate reactivity (AAR) or alkali silica reactivity (ASR). This occurs when silicates in concrete aggregate react with alkalis, such as de-icing salt or salts in seawater, to produce an expanding gel that leads to severe concrete cracking and loss of strength. Montrose Bridge in Scotland had to be demolished in 2004 as a result of AAR.

Continuous beam A rigid beam spanning between three or more consecutive supports without any structural break, thus giving it greater strength than identically sized beams spanning the openings singly.

Console A supporting bracket, similar to a corbel in function, but usually greater in height and with a scrolled outward face.

Coping The protective capping to a wall designed to shed rainwater.

Corbel A projection from a wall that supports a beam, arch or statue.

Corbel table A continuous corbel course usually supporting a parapet.

Corinthian The classical order of architecture typified by fluted columns with highly decorated capitals.

Cornes-de-vache A winding splay or chamfer given to the outside of an arch barrel by constructing the intrados of the main part of the barrel to a smaller-radius curve than that used on the face of the arch, the two curves being tangential at their midspan crowns. The amount of splay is therefore greatest at the abutments and diminishes to nothing at the crown.

Cornice The horizontal moulded projection that crowns part of a structure.

Course A single continuous layer of masonry (stone or brick) of the same thickness.

Covered bridge A bridge of timber protected from the weather by a wooden barn-like enclosure.

Crib An alternative name for a grillage (q.v.).

Creep Long-term changes to a structure under continuous load.

Crescent arch A steel arch in which the top and bottom booms have different curves that meet at the arch's support points, the elevation thus looking like a crescent moon.

Crocket An ornamental carved projection on pinnacles and spires.

Crown The central, highest part of an arch.

Cutwater A pointed or rounded upstream or downstream extension to a bridge pier at water level to smooth the flow of water past the pier.

Cyclopean Masonry work consisting of immensely large irregular pieces of stone.

Dead load The self-weight of a structure and any loads permanently supported by it. See also Live Load.

Deck The upper structural part of a bridge, the top surface of which forms the roadway.

Deck arch A bridge in which the deck is supported on posts from an arch spanning along the length of the deck but beneath it.

Deck truss A bridge in which the deck is supported on the upper members (booms or chords) of a truss or girder bridge.

Dentil One of a series of small, usually square, projections from a wall that support a cornice.

Doric The classical order of architecture typified by fluted, sometimes baseless, columns with simple capitals.

Drawbridge A beam structure that is pinned at one end of its span so that the free end can be raised to prevent passage across the bridge or to enable large vessels to pass beneath.

Entablature The part of a classical structure standing immediately upon column capitals and including the architrave (the main lintel or beam), the frieze and the cornice.

Expansion joint A mechanism built into a bridge to allow deck sections to expand and contract as a result of temperature changes without inducing additional horizontal forces at the tops of piers.

Extrados The line (usually curved) followed by the outer edge of an arch ring. In a bridge with stepped extrados the outer edge of each voussoir is not cut to a radial curve but finished with vertical and horizontal faces to fit into the coursed masonry of the spandrels.

Extradosed bridge A type of post-tensioned box girder bridge in which the prestressing cables emerge from the deck and are anchored to low pylons located above the supporting piers. The bridge profile is thus superficially similar to that of a cable stay bridge, except that the prestressing tendons are at a shallower angle than cable stays would be.

Eyebar A flat metal bar, usually rectangular in cross section, enlarged at each end to accommodate holes allowing adjacent eyebars to be connected together by pins through these holes.

Fan An arrangement of several tension members in a cable stay bridge or struts under compression, for example in a timber bridge, radiating out from a common point in the manner of a lady's fan.

Fatigue The reduction in tensile strength of a structural material following repeated application and removal of heavy loading.

Finback bridge A bridge in which the main supporting structure takes the form of a rectangular beam built above the centreline of the bridge deck, the overall cross section thus appearing like an inverted T.

Finial A vertical ornamental feature that terminates a gable, pediment or similar.

Fink truss A structure in which a series of inverted king post trusses overlap, with the diagonals from one truss ending at the top of the vertical king post of the adjacent truss.

Fish belly beam A beam in which the top edge is straight but the depth increases from the ends to the middle to give the bottom edge a sagging curve (the opposite of hog-backed).

Flange The wide top or bottom horizontal sections of an I-shaped joist, truss or girder that are separated by the vertical web.

Fluting The decorative vertical grooving around the circumference of a classical-style column.

Folly A costly and unnecessary structure.

Four-centred arch A pointed arch, also known as a Tudor arch, in which each half span is formed by two arcs, an outer one of small radii and an inner larger arc, which meet together at a common tangent point.

Girder A made-up beam member either with a solid web (a plate girder) or a trussed web consisting of vertical and/or triangulated members. See also lattice girder, Vierendeel girder and Warren girder.

Glulam Structural timber sections made from planks that have been glued and laminated together (the joints being staggered), thus allowing the composite unit to be of larger section, longer and stronger than could be available naturally.

Gothic An architectural style, typified by pointed arches, common from the early thirteenth century.

Grillage A support system consisting of two or more layers of parallel beams (of timber, stone, concrete or steel), the layers at right angles to each other, that is placed on prepared ground to act as a simple foundation for heavily loaded columns or to support main bridge beams.

Grout A rich, binding slurry of cement and water that can be poured or pumped into a structure for filling small unwanted gaps between beams and the like.

Harp layout of cable stays Arrangement of stays in a cable stay bridge in which the stays are parallel to each other (rather than, as in the fan arrangement, radiating from a virtual common point).

Half-through bridge A bridge in which the deck is located about mid-way between the upper and lower members (booms or chords) of a truss or girder bridge.

Haunch The part of an arch between its springing and crown.

Helix A curve like a spiral but with a constant radius rather than the reducing radius of a true spiral.

Hog-backed girder A girder in which the bottom edge is straight but with a vertically curved upper member that does not meet the lower boom as in a bowstring girder (the opposite of a fish belly beam).

Hyperbolic paraboloid A three-dimensional, doubly curved surface looking like a saddle, with the special property that every point on the surface lies on two separate straight lines. A model can thus be constructed with string. It is mostly used in roofs.

Impost The top course of a pier or abutment, sometimes with moulded decoration, from which a semicircular or semi-elliptical arch springs. See also skewback.

Indulgences Remission of time spent in purgatory granted by the medieval church in return for funds for the building, repair or maintenance of bridges.

Intrados The curved line followed by the inner edge of an arch ring.

Invert The paved surface beneath an arch over which people, traffic or water will move.

Ionic Classical order of architecture typified by slender fluted columns with capitals decorated with scrolls (volutes).

Jack arch A small arch, usually of brick, spanning between the bottom flanges of adjacent steel joists to support a deck between them.

Jagger Early user of packhorse bridges, especially in mining areas.

Jettied A form of medieval building construction in which the first floor cantilevers out beyond the ground floor.

Jetting A technique for installing piles in sandy ground by using pressurised air or water to help the pile driving.

Keystone The central stone in a ring of voussoirs forming an arch. Structurally, the keystone is no more important than any other stone in an arch ring (as with the links in a chain) but, being at the crown of the arch and the last to be laid by being driven into place, is sometimes made visually more dominant.

King pier An enlarged pier built at intervals in a lengthy viaduct to increase stability during construction.

King post truss A triangular framework, mainly used for roofs, in which a vertical post connects the centre of the horizontal tie beam to the apex where the sloping rafters meet.

Knee brace A short diagonal strut between the underside of a beam and its support to increase the stiffness of the beam.

Ladder arch A bridge consisting of two arched ribs connected together by a series of horizontal beams, the whole structure looking something like a very large curved ladder.

Ladder beam bridge A bridge with a composite deck of concrete and steel where concrete slabs span longitudinally between steel crossbeams which themselves span between two longitudinal steel edge beams.

Lambswool finish Surface treatment of stonework involving the close scribing of parallel lines.

Lamination (of timber) The bolting or gluing together of several planks to make a single timber beam larger and stronger than could be formed from a felled tree.

Land bridge A structure, invisible from above, that carries parkland across it without a break.

Latex concrete Concrete to which polymers are added during mixing thus enabling it to become stronger and more waterproof.

Lattice truss or girder A fabricated structural beam in which the web space is made up of individual intersecting diagonal members in the form of overlapping Xs.

Lenticular bridge A bridge with curved top and bottom booms in the shape of an optical lens. The top boom is in compression and the lower one in tension, these forces cancelling each other out where the booms meet at the abutments, thus transferring vertical loads only into the abutments.

Lintel A horizontal beam spanning across a door, window or other rectangular opening and supporting the wall above.

Live load A load that can be temporarily applied to a structure and then removed again, such as from moving people and traffic, as well as occasional static loads such as snow. See also dead load.

Locked-coil cable A helically-twisted wire cable made up of concentric layers of individual wires, alternate layers being wound in opposite directions, and with the outer layers consisting of Z-shaped wires. When such a cable is put under tension the coils try to straighten out, causing the individual wires to tighten against each other thus sealing the inside of the cable. Locked-coil cables, manufactured to the exact lengths required, are usually used for the stays of cable stay bridges.

Lounge bridge A lounge bridge is the general name in countries such as China for a bridge to which some kind of house structure is attached that provides travellers with protection from the elements. Such bridges often include Buddhist or Taoist idols.

Machicolation A parapet supported on corbels and projecting forward from its support wall with openings in the floor between the corbels.

Masonry bridge construction The diagram on the following page shows the main parts of a stone bridge. All the terms shown have definitions included in this Glossary.

Merlon One of a series of higher parts in a battlemented wall of alternately higher and lower sections.

Network arch A bowstring bridge in which the hangers supporting the deck from the arch are not vertical but are inclined, crossing each other at least twice in a single plane, thus making the structure act somewhat like a lattice truss.

Niche A recess in a wall or pier, usually semicircular in plan, sometimes containing a statue.

Oculus A circular, decorative recess in a wall, usually with a solid back (blind), ringed with a moulded rim.

Ogee An S-shaped continuous double curve that is part convex and part concave.

Order In a medieval masonry bridge, the number of concentric rings of voussoirs that do not share the same plane of elevation. In architecture, the main divisions of classical style – Corinthian, Doric, Ionic and Tuscan – distinguished most

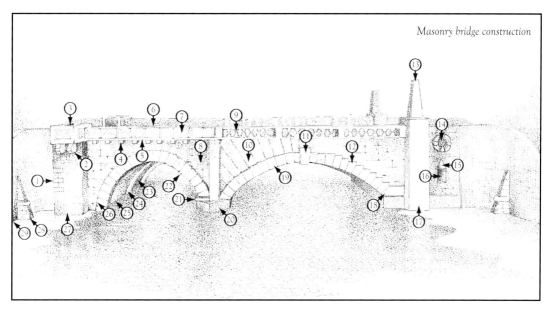

Masonry bridge construction

1. Quoin
2. Corbel
3. Refuge
4. Dentil
5. String course
6. Coping
7. Parapet
8. Spandrel
9. Balustrade
10. Spandrel with radiating masonry

11. Keystone
12. Steeped arch
13. Pylon
14. Oculus
15. Pediment
16. Niche
17. Pilaster
18. Skewback
19. Arch ring
20. Pier with pointed cutwater

21. Impost
22. Voussoir
23. Arch
24. Rib
25. Springing line
26. Arch ring
27. Abutment
28. Buttress
29. Wing wall

easily by their different types of column head decoration.

Orthotropic deck A steel plate deck stiffened and supported by two sets of steel joists or other sections, at right angles to each other, all welded together into an integral structural unit.

Overbridge A bridge across a road, railway or canal to carry another thoroughfare over it.

Palladian Describing a structure based on the drawings of Andreo Palladio (1508–1580), whose designs were often inspired by the buildings of ancient Rome, especially those of Vitruvius (c. 75–c. 15BCE).

Panelled girder A type of trussed girder in which the space between the top and bottom booms is divided by vertical members into a series of panels that, except for Vierendeel girders, are braced internally by additional diagonal members.

Parabola A curved line defined mathematically as that described by a point that is the same distance from a fixed straight line (the directrix) and a fixed point (the focus). This is also the shape assumed by a suspended cable of negligible self-weight when loaded with a weight distributed uniformly along the horizontal distance between the suspension points.

Parabolic arch An upright arch constructed as a mirror image of the sagging line of a parabola (q.v.) and thus the most economical arched shape to support the evenly distributed weight of a separate horizontal bridge deck.

Parapet The edge wall running along the length of a bridge deck.

Patera A decorative, low relief, circular ornament.

Pattress plate A circular, S- or X-shaped load-spreading device, sometimes seen on the spandrel wall of a masonry arch bridge, that is linked by a tie bar to a similar one on the opposite side of the bridge to prevent the walls bowing outwards under the pressure of the spandrel fill.

Pediment A triangular or segmental low-pitched gable above an archway or similar opening. In a broken pediment the central part at the top of the gable or part of the horizontal base is omitted.

Pier The intermediate support between two spans of a bridge (see also abutment). Also a seaside structure built mainly to provide landing stages and other leisure facilities for holidaymakers.

Pilaster A shallow decorative column attached to a wall.

Pile A timber, steel or concrete post inserted deep into the ground to carry loads that cannot be resisted by the ground either at its surface or at a shallow depth.

Pin joint A connecting joint between structural members that allows the members to rotate around the connection, thus making it impossible to transfer bending moments between them.

Plate girder A fabricated structural member made up of a solid vertical web with flanges attached to the top and bottom edges.

Plinth The square base of a column or the projecting base of a wall or balustrade.

Pointed arch An arch in which the curves on each side meet at the apex without this being a common tangent point.

Pontage Authority, by ancient custom or royal or parliamentary grant, to levy a toll for crossing a bridge in order to raise money for its maintenance or repair.

Pontoon bridge A bridge (often only temporary) constructed by anchoring boats across a river and laying a continuous roadway over and between them.

Pony truss A through truss but without any horizontal bracing between the trusses above deck level.

Portal frame A structural framework consisting of a beam rigidly connected to two upright posts.

Porte-cochère A covered entrance way to a grand house allowing coach passengers to get on and off in the dry.

Post-tensioned See prestressed concrete.

Precast concrete. Concrete structural elements such as beams that can be manufactured in controlled conditions off site.

Prestressed concrete Concrete in which tensile stresses induced by the self-weight of the structure, together with its expected load of vehicles and people, are reduced by compressing the concrete by means of high strength steel wire or cables. This prestressing is applied either by stretching the wires before the wet concrete is poured around them (pre-tensioned), or by passing wire cables through ducts formed in the concrete when it is poured, stretching these after the concrete has fully hardened (post-tensioned), then filling the duct with grout.

Pultrusion The process of manufacturing thermoplastic materials by pulling a mixture of reinforcement fibres impregnated with resin through a heated die.

Pumice A light stone cut from solidified volcanic lava.

Pylon In architecture, a plain tapering elongated column like an obelisk, used for decorative effect only. The word is also used for the above-deck extension of a pier in a cable stay bridge, to which the diagonal cable stays are anchored.

Quatrefoil A Gothic-style opening in stonework consisting of four small arcs meeting at pointed cusps.

Quoin A corner, or the line of bricks or stones, often decorative, that forms a vertical corner between two walls.

Raft A wide foundation platform, often made of timber, that sits on poor load-bearing ground and on which a structure such as a pier is then erected.

Refuge An alcove in the line of a bridge parapet, usually above a cutwater, to give pedestrians safety from passing traffic.

Reinforced concrete Concrete that contains steel bars or mesh that resist the tensile stresses in, for example, a beam supporting loads through its resistance to bending, thus preventing weakening cracks developing.

Reinforced earth The addition of cohesive strength to normal soil by burying within it tension-resistant materials such as mats of plastic netting.

Rib The part of a masonry arch that projects beneath the main barrel of the arch and was usually constructed first to economise on centring, the remaining part of the barrel being constructed by spanning stones between adjacent ribs; or an arched frame in metal or concrete.

Ribbon bridge A type of suspension bridge in which the roadway or footway is carried directly on top of the suspension cables rather than being hung beneath them.

Rigid frame bridge A bridge structure in which horizontal, vertical and diagonal members are fixed rigidly to each other in such a way that the resultant framework functions somewhat like an arch.

Rise (of an arch) The vertical distance between a horizontal line joining the springing points of an arch and the underside of its highest point.

Rod stayed bridge An early form of stayed bridge in which the stays were metal bars rather than cables.

Rolling lift bridge An opening bridge in which the main span is balanced by a shorter back span in the form of a quadrant which, as it opens, rolls backwards on tracks in the manner of a rocking chair.

Rubble Masonry work consisting of differently sized stones laid without any pattern (random rubble) or in courses (coursed rubble).

Rusticated masonry Coursed stonework laid with deeply recessed V-shaped joints.

Saddle A strengthening layer, these days usually of reinforced concrete, placed above the extrados of an arch and often constructed to increase the loading capacity of old stone arch bridges. Also, the curved member at the top of a suspension bridge tower over which the main cables pass.

Scour The erosive action of water currents on the bed or banks of a river.

Screed A thin levelled finish of sand:cement mortar laid over rough concrete.

Segmental arch An arch in which the shape is formed from a circular arc smaller than a semicircle.

Self-anchored suspension bridge A suspension bridge in which the cables are not fixed into gravity anchorages in the ground at each end, but instead are anchored into both ends of the deck, which then resists the tensile force in the cables by acting as a giant strut.

Semi-elliptical arch An arch in which the shape is a curve joining points located a constant distance from two fixed points, called foci. In a true semi-elliptical stone arch every voussoir in each half arch has to be cut individually and precisely. In order to simplify the stone-cutting and the setting out of the arch, many apparently semi-elliptical arches are in fact three-centred.

Sgraffito Decorative work produced by scraping white stucco off a dark ground.

Skew The obliqueness between the centreline along the length of a bridge and a line at right angles to the abutment face.

Skewback The angled stone or other structural element from which a segmental arch springs. See also impost.

Soffit The exposed underside of a beam or slab.

Span The distance between the supporting points of a structure. In stone and brick bridges the span is measured between the faces of the piers or abutments, whereas for steel and concrete structures the span is usually measured between the bearings that are inset behind the abutment face or on the centre of intermediate piers.

Spandrel The roughly triangular-shaped vertical area between the outside of an arch ring (the extrados) and the underside of the bridge deck structure.

Split bridge A bridge over a canal built as two separate half-cantilevers separated by a small gap. This allows a rope by which a barge is being towed to be passed through the central split without the horse having to be unhitched.

Springing The line along which the curved underside of an arch barrel meets the face of a pier or abutment.

Squinch arch An arch built across a corner to carry some superstructure, often one of a series of arches, each corbelled out from the one beneath.

Starling An artificial island contained by contiguous timber piling and formed of stones and gravel on which a pier will be constructed, or created after a pier has been built, to protect it from scour.

Stayed bridge A bridge in which the deck is supported by diagonal tension members (cables or rods) from a pylon above deck level.

Steel An alloy of iron and carbon with small amounts of special metals such as chromium, manganese, molybdenum and vanadium. Steel replaced cast and wrought iron as a structural material at the end of the nineteenth century.

Stepped arch An arch of stone voussoirs in which the outside edge of each voussoir is not cut to form part of an extrados curve but is shaped with a right-angled outside top corner, the vertical line abutting a horizontal course of masonry in the spandrel, the next course of spandrel masonry being continued above the horizontal top of the voussoir.

Stilted arch Arch in which the springing points are placed relatively high up on the supporting piers.

Strain The alteration in the shape of a structural member resulting from the application to that member of a compressive, tensile, torsional or shearing force.

Stress A force per unit area on a structural member, the result of which is strain.

String course A single course differentiated by colour, material, alignment or finish from the face of a vertical surface; typically a narrow projecting course, sometimes moulded, that serves to separate different parts of a wall.

Strut A structural member that contributes to a truss, girder or structural framework by resisting compression forces.

Stucco A type of plaster, normally painted, used as a hard-wearing external finish to lower quality brickwork.

Submersible bridge A type of vertical moving canal bridge in which the bridge structure is lowered to the bottom of the canal to allow vessels to travel over it rather than be raised high enough for vessels to pass beneath.

Suspension bridge A bridge in which the deck is supported by hangers from cables that are suspended between tall towers and anchored firmly into the ground at each end or, occasionally, into the ends of the deck structure itself (self-anchored). See also ribbon bridge.

Swag A carved decoration in the form of a length of drapery or foliage supported at the ends and sagging between.

Swing bridge A type of opening bridge in which the opening span rotates in a horizontal plane by pivoting on a vertical axis.

Table bridge A type of opening bridge across water in which part of the deck can be raised vertically on extending legs at each corner so that a vessel can pass beneath.

Three-centred arch An arch in which the shape is formed by three arcs, a central large radius arc and two outer ones of smaller radius, which meet at common tangent points. In order to simplify construction, many bridges that appear to be semi-elliptical in shape are in fact three-centred (with the smaller radius arcs being tangential to the vertical at the springings). Also, some segmental arches, instead of springing directly from a skewback, start with a very small radius arch off the abutment face.

Through arch A bridge in which the deck is suspended from an arch spanning along the length of the deck but above it.

Through truss or bridge A bridge in which the deck is supported on the lower members (booms or chords) of a truss or girder bridge.

Tie A structural member that contributes to a truss, girder or structural framework by resisting tensile forces.

Transporter bridge A fixed high-level structure that carries a suspended moving low-level platform at road level.

Transom A horizontal cross beam in a framework.

Trapezium A four-sided geometric shape with one pair of opposite lines being parallel.

Trefoil A Gothic-style opening in stonework consisting of three small arcs meeting at pointed cusps.

Trenail A cylindrical hardwood pin driven into preformed holes to connect together separate pieces of timber in structures exposed to the weather.

Trestle A lightweight frame of two or more legs supporting a further structure. A bent is a two-dimensional trestle.

Triglyph An architectural feature, derived from classical Greek temples and representing the ends of timber roof beams, now consisting of a stone block incised with three vertical grooves repeated in a series set into the frieze part of an entablature.

Trunnion The horizontal axis of rotation for a modern bascule bridge.

Truss A rigid fabricated structural framework, made up by connecting together smaller members carrying compressive or tensile forces only (struts or ties) and usually arranged in the form of triangles, that acts as a beam or girder.

Tudor arch A four-centred arch in which the two arcs that meet at the pointed apex are nearly flat (sometimes called a depressed arch).

Tufa A lightweight porous limestone mainly used for building exotic structures.

Turning bridge A particular type of drawbridge used in medieval castles that pivoted on a central horizontal axle.

Turnover bridge A bridge that carries a canal towpath from one side of a waterway to the other.

Tuscan The classical order of architecture typified by unfluted columns with little or no decoration to the capitals.

Two-centred arch A pointed arch in which the curves on each side meet at the apex without this being a common tangent point.

Underbridge A bridge under a road, railway or canal to carry that thoroughfare over an obstacle. A canal underbridge is called an aqueduct.

Under-deck cable-stayed bridge A cable-stayed bridge in which the stays, instead of passing over a pylon at each end of the deck and then descending to mid-span deck-level anchorages, pass between deck-level anchorages at the abutments and under one or more midspan posts beneath the deck.

Underspanned A suspension bridge in which the central lowest part of the cables is below the deck structure.

Vault An arched roof of stone or brick.

Vermiculated A masonry finish in which the stone surface is carved to give the appearance of worm tracks.

Versine The vertical distance a catenary cable sags at the midpoint between its supports.

Viaduct A generally lengthy road or railway bridge, usually flat or at a slight constant gradient, with a series of relatively short spans.

Vierendeel girder A made-up girder, the web of which has no diagonal members and consists solely of members at right angles to the top and bottom chords and connected rigidly to them.

Voussoir A wedge-shaped piece of accurately dressed masonry with opposite faces cut to slopes normal to the intrados curve and which, with its neighbours, makes up the structural part of an arch. For semicircular or segmental arches the voussoirs can be interchangeable; with true semi-elliptical arches, every voussoir in each half arch must be individually shaped.

Warren girder A made-up girder, the web of which consists solely of non-overlapping diagonal members, successively sloping in opposite directions, giving a repeated W-shaped elevation.

Web The central part of an I-shaped joist, truss or girder that separates the flanges.

Weir A low-level dam across a river with water flowing over its full length. Many old bridges are just upstream of weirs, built to reduce the speed of the water flowing under the bridge and thus to lessen the risk of scour damage to the pier and abutment foundations.

Wing wall A wall behind an abutment to support the sides of a bridge approach embankment or ramp.

Woven timber arch An arch structure, formed from a number of straight sides rather than a curved arc, in which each straight section, consisting of a plane of parallel 'warp' timber poles running along the line of the bridge axis, is locked against its adjacent section at the point where the planes cross by 'weft' timbers set at right angles to the bridge axis.

Wrought iron A type of iron that has been specially worked to reduce its carbon content to a minimum. It was first used as a structural material in the mid-nineteenth century, taking over from cast iron, before being replaced by steel. It is able to resist tensile forces (unlike cast iron) but suffers from fatigue and is more expensive.

The Bridges

For differently-named predecessor or adjacent bridges, not separately listed but included within another entry (where their names are shown in **bold**), see the relevant main entry as listed below:

Name of Predecessor or Adjacent Bridge	See Main Entry
15th July Martyrs Bridge, Istanbul	Bosphorus Bridge, Istanbul
A8 Motorway Bridge, Ventabren	TGV Bridges, Ventabren
Akbari Bridge, Jaunpur	Shahi Mughal Bridge, Jaunpur
Al Garhoud Bridge, Dubai	Dubai Creek Bridges, Dubai
Al Maktoum Bridge, Dubai	Dubai Creek Bridges, Dubai
Aleksotas Bridge, Kaunas	Vytautas the Great Bridge, Kaunas
Alfred Beit Road Bridge, Beitbridge	Beitbridge Bridges, Beitbridge
Alkaff Bridge, Singapore	Singapore River Bridges, Singapore
Allah Verdi Khan Bridge, Isfahan	Zayanderud Bridges, Isfahan
Anichkov Bridge, St Petersburg	City Bridges, St Petersburg
Anji Bridge, Zhaozhao	Zhaozhou Bridge, Zhaozhao
Annunciation Bridge, St Petersburg	City Bridges, St Petersburg
Arc River Bridge, Ventabren	TGV Bridges, Ventabren
Art Footbridge, Skopje	Skopje City Centre Footbridges, Skopje
Atlantic Bridge, Panama	Panama Canal Bridges, Panama
Baakenhafenbrücke, Hamburg	Speicherstadt Island Bridges, Hamburg
Bank Footbridge, St Petersburg	City Bridges, St Petersburg
Blue Bridge, St Petersburg	City Bridges, St Petersburg
Bolsheokhtinsky Bridge, St Petersburg	River Neva Bascule Bridges, St Petersburg
Bosch Bridge, Kiev	Metro Bridge, Kiev
Bouregreg Bridge, Rabat	Rabat Bridges, Rabat
BP Pedestrian Bridge, Chicago	Chicago Bridges, Chicago
Bridge of Augustus, Rimini	Tiberius Bridge, Rimini
'Bridge of Spies', Potsdam	Glienicke Bridge, Potsdam

Name of Predecessor or Adjacent Bridge	See Main Entry
Bridge of the Americas, Panama	Panama Canal Bridges, Panama
Britannia Bridge, Isle of Anglesey	Menai Straits Bridges, Isle of Anglesey
Broken Bridge, Hangzhou	West Lake Bridges, Hangzhou
Brooksbrücke, Hamburg	Speicherstadt Island Bridges, Hamburg
Brusio Spiral Viaduct, Switzerland	Rhaetian Railway Bridges, Switzerland
Business Bay Crossing, Dubai	Dubai Creek Bridges, Dubai
Caesar's Bridge, Shushtar	Shadorvan Bridge, Shushtar
Caiyuanba Yangtze River Bridge, Chongqing	Chongqing Bridges, Chongqing
Campo Volantin Bridge, Bilbao	Zubizuri Bridge, Bilbao
Canal Street Bridge, Chicago	Moving Bridges, Chicago
Carlisle Bridge, Dublin	O'Connell Bridge, Dublin
Carlisle Bridge, Londonderry	Craigavon Bridge, Londonderry
Cavenagh Bridge, Singapore	Singapore River Bridges, Singapore
Centennial Bridge, Panama	Panama Canal Bridges, Panama
Cermack Road (22nd Street Bridge), Chicago	Chicago Bridges, Chicago
Chaoyang Bridge, Chongqing	Chongqing Bridges, Chongqing
Chapel Bridge, Lucerne	Timber Bridges, Lucerne
Chicago & Northwestern Railway Bridge, Chicago	Chicago Bridges, Chicago
Chicago Sanitary & Ship Canal Swing Bridge, Chicago	Moving Bridges, Chicago
Chihai Bridge, Tianjin	Yongle Bridge, Tianjin
Chinese Bridge, Tsarskoye Selo	Tsarskoye Selo Park Bridges, Tsarskoye Selo
Cloak Bridge, Ceský Krumlov	State Castle Moat Bridge, Ceský Krumlov
Cortland Street Bridge, Chicago	Chicago Bridges, Chicago
Cuntan Yangtze River Bridge, Chongqing	Chongqing Bridges, Chonqing
Devil's Bridge, Tarragona	Tarragona Aqueduct, Tarragona
Doha Bay Crossing, Doha	Doha Sharq Crossing, Doha
Dongshuimen (Yangtze River) Bridge, Chongqing	Chongqing Bridges, Chongqing
Down Shore Line Viaduct, Antrim	Greenisland Railway Viaducts, Antrim
Dragon Bridge, Tsarskoye Selo	Tsarskoye Selo Park Bridges, Tsarskoye Selo
Dragon Bridge, Chi Kiaeng	Spean Praptos Bridge, Chi Kiaeng
Dubai Floating Bridge, Dubai	Dubai Creek Bridges, Dubai
Dusan Bridge, Skopje	Stone Bridge, Skopje
Egonyang Bridge, Chongqing	Chongqing Bridges, Chongqing

Name of Predecessor or Adjacent Bridge	See Main Entry
Egyptian Bridge, St Petersburg	City Bridges, St Petersburg
Elgin Bridge, Singapore	Singapore River Bridges, Singapore
Ellerntorsbrücke, Hamburg	Speicherstadt Island Bridges, Hamburg
Engetsu-kyo (Full Moon) Bridge, Tokyo	Koishikawa Korakuen Garden Bridges, Tokyo
Esplanade Bridge, Singapore	Singapore River Bridges, Singapore
Exchange Bridge, St Petersburg	City Bridges, St Petersburg
Eye Footbridge, Skopje	Skopje City Centre Footbridges, Skopje
Feenteichbrücke, Hamburg	Speicherstadt Island Bridges, Hamburg
Figurny Bridge, Moscow	Tsaritsyno Park Bridges, Moscow
Floating Bridge, Dubai	Dubai Creek Bridges, Dubai
Forth Bridge, South Queensferry	Firth of Forth Bridges, South Queensferry
Forth Road Bridge, South Queensferry	Firth of Forth Bridges, South Queensferry
Franz Josef Bridge, Budapest	Liberty Bridge, Budapest
Genoa San Giorgio Bridge, Genoa	Polcevera Viaduct, Genoa
Gongchen Bridge	Hangzhou Old Bridges, Hangzhou
Gothic Bridge	Tsaritsyno Park Bridges
Grotesque Bridge	Tsaritsyno Park Bridges
Grunwald Bridge, Wroclaw	City Centre Bridges, Wroclaw
Hayratiye Bridge, Istanbul	Galata Bridge, Istanbul
Helix Bridge, Singapore	Singapore River Bridges, Singapore
Homer M. Hadley Memorial Bridge, Seattle	Lacey V. Murrow Memorial Bridge, Seattle
Iron Bridge, Dessau-Wörlitz	Wörlitz Garden Bridges, Dessau-Wörlitz
Jialing River Bridge	Chongqing Bridges, Chongqing
Juan Pablo Duarte Bridge, Santa Domingo	Ozama River Bridges, Santa Domingo
Jubilee Bridge, Singapore	Singapore River Bridges, Singapore
Kalogeriko Bridge, Tymfi	Epirus Bridges, Epirus
Kampong Kdei Bridge, Chi Kiaeng	Spean Praptos Bridge, Chi Kiaeng
Kantutani Bridge, La Paz	Triplets Bridge, La Paz
King Hamad Causeway, Al Khobar	King Fahd Causeway Bridge, Al Khobar
Kirpa Bridge	Darnytskyi Bridge, Kiev
Köhlbrandbrücke, Hamburg	Speicherstadt Island Bridges, Hamburg
Krasnoluzhsky Railway Bridge, Moscow	Krasnoluzhsky Bridges, Moscow
Krasnoluzhsky Road Bridge, Moscow	Krasnoluzhsky Bridges, Moscow

Name of Predecessor or Adjacent Bridge	See Main Entry
Krestovy Bridge, Saint Petersburg	Tsarskoye Selo Park Bridges, Saint Petersburg
Lake Shore Drive Bridge, Chicago	Chicago Bridges, Chicago
Landwasser Viaduct, Switzerland	Rhaetian Railway Bridges, Switzerland
Langwies Viaduct, Switzerland	Rhaetian Railway Bridges, Switzerland
Lions' Footbridge	City Bridges, St Petersburg
Litevny (Foundry) Bridge	River Neva Bascule Bridges, Saint Petersburg
Lombardsbrücke, Hamburg	Speicherstadt Island Bridges, Hamburg
Lomonosov Bridge, St Petersburg	City Bridges, Saint Petersburg
Magi Bridge, Mtskheta	Pompey's Bridge, Mtskheta
Mainline Viaduct, Antrim	Greenisland Railway Viaducts, Antrim
Malleco Road Bridge, Collipolli	Malleco Railway Viaduct, Collipulli
Marble Bridge, Saint Petersburg	Tsarskoye Selo Park Bridges, Saint Petersburg
Menai Bridge, Isle of Anglesey	Menai Straits Bridges, Isle of Anglesey
Nalubaale Bridge, Njeru	Source of the Nile Bridge, Njeru
Nanko Bridge, Takamatsu	Ritsurin Garden Bridges, Takamatsu
Nepomuk Bridge, Heidelberg	Old Bridge, Heidelberg
New Darnytsia OR Kirpa Bridge, Kiev	Darnytskyi Bridge, Kiev
New Yalu River Bridge, Singapore	Singapore River Bridges, Singapore
Niagara Railway Arch Bridge	Whirlpool Rapids Bridges, Niagara Falls
Niagara Suspension Bridge	Whirlpool Rapids Bridges, Niagara Falls
Nicholas Chain Bridge, Kiev	Metro Bridge, Kiev
Nine-Turn Bridge, Yangzhou	Slender West Lake Bridges, Yangzhou
Niujiautuo Jialing River Bridge, Chongqing	Chongqing Bridges, Chongqing
North Viaduct, Savannah	South Viaduct, Savannah
Old Svinesund Bridge, Svinesund	Svinesund Bridge, Svinesund
Oued Sherrat Bridge, Rabat	Rabat Bridges, Rabat
Palace Bridge, Saint Petersburg	City Bridges, Saint Petersburg
Paul-Doumer Bridge, Hanoi	Long Biên Bridge, Hanoi
Pavilion Bridge, Hangzhou	West Lake Bridges, Hangzhou
Pennypack Creek Bridge, Philadelphia	Frankford Avenue Bridge, Philadelphia
Peter the Great Bridge, Saint Petersburg	City Bridges, Saint Petersburg
Poggenmühlen-Brücke, Hamburg	Speicherstadt Island Bridges, Hamburg
Pol-e-Khaju Bridge, Isfahan	Zayanderud Bridges, Isfahan

Name of Predecessor or Adjacent Bridge	See Main Entry
Pons Aelius, Rome	Ancient Roman Bridges, Rome
Pons Aemilius, Rome	Ancient Roman Bridges, Rome
Pons Fabricius, Rome	Ancient Roman Bridges, Rome
Pont de l'Iroise, Brest	Plougastel Bridge, Brest
Pont de les Peixateries Velles, Girona	Eiffel Bridge, Girona
Ponte Castelvecchio (Old Castle Bridge), Verona	Ponte Scaligero, Verona
Ponte Milvio, Rome	Ancient Roman Bridges, Rome
Ponte Rotto, Rome	Ancient Roman Bridges, Rome
Ponte San Angelo, Rome	Ancient Roman Bridges, Rome
Prachechny Bridge, St Petersburg	City Bridges, St Petersburg
President Juan Bosch Bridge, Santo Domingo	Ozama River Bridges, Santo Domingo
Puente Nicolas Avellaneda, Buenos Aires	Movable Bridges, Buenos Aires
Python Bridge, Amsterdam	High Bridge, Amsterdam
Qianximen Bridge (Jialing River) Bridge, Chongqing	Chongqing Bridges, Chongqing
Queensferry Crossing, South Queensferry	Firth of Forth Bridges, South Queensferry
Railway Suspension Bridge, Niagara Falls	Whirlpool Rapids Bridges, Niagara Falls
Rakots Bridge, Gablenz	Devil's Bridge, Gablenz
Ras Al Khor Bridge, Dubai	Dubai Creek Bridges, Dubai
Ras Darge's Bridge, Ethiopia	Portuguese Bridge, Ethiopia
Ripshorst Footbridge, Duisberg	Emscher Landscape Park Bridges, Duisberg
River Arachthos Bridge, Arta	Epirus Bridges, Epirus
River Arachthos Bridge, Plaka	Epirus Bridges, Epirus
Sand Bridge, Wroclaw	City Centre Bridges, Wroclaw
Royal Towers Hotel Bridge, Nassau	Paradise Island Bridges, Nassau
St Job's Bridge, Venice	Three-Arched Bridge, Venice
Sant Bartomeu Bridge, Martorell	Devil's Bridge, Martorell
Seven Nymphs Bridge, Seogwipo	Seonim Bridge, Seogwipo
Shahrestan Bridge, Isfahan	Zayanderud Bridges, Isfahan
Sheikh Rashid Bin Saeed Crossing, Dubai	Dubai Creek Bridges, Dubai
Shindagha Bridge, Dubai	Dubai Creek Bridges, Dubai
Shiqigong (Seventeen Arches) Bridge, Beijing	Yiheyuan Summer Palace Bridges, Beijing
Shoreline Viaduct, Antrim	Greenisland Railway Viaducts, Antrim
Siosepol Bridge, Isfahan	Zayanderud Bridges, Isfahan

Name of Predecessor or Adjacent Bridge	See Main Entry
Sir Otto Beit Bridge, Beitbridge	Beitbridge Bridges, Beitbridge
Sky Garden Bridge, Dubai	Dubai Creek Bridges, Dubai
Slinky Springs to Fame Footbridge, Oberhausen	Emscher Landscape Park Bridges, Duisberg
Sogakope Bridge, South Tongu	Lower Volta Bridge, South Tongu
Spreuer Bridge, Lucerne	Timber Bridges, Lucerne
Stone Bridge, Rimac	Rimac River Bridges, Rimac
Third Bosphorus Bridge, Istanbul	Yavuz Sultan Selim Bridge, Istanbul
Transbordador del Riachuelo Nicolas Avellaneda, Buenos Aires	Movable Bridges, Buenos Aires
Troitsky (Trinity) Bridge, St Petersburg	River Neva Bascule Bridges, St Petersburg
Tsuten-kyo Bridge, Tokyo	Koishikawa Korakuen Garden Bridges, Tokyo
Tuchkov Bridge, St Petersburg	City Bridges, St Petersburg
Tumski Bridge, Wroclaw	City Centre Bridges, Wroclaw
Twin River Bridges, Chongqing	Chongqing Bridges, Chongqing
Unkapani Bridge, Istanbul	Galata Bridge, Istanbul
Upper Steel Arch Bridge	Rainbow Bridge, Niagara Falls, Ontario, Canada
Upper Swan Bridge, St Petersburg	City Bridges, St Petersburg
Valerian's Bridge, Shushtar	Shadorvan Bridge, Shushtar
Webb Dock Rail Bridge, Melbourne	Webb Bridge, Melbourne
Wells Street Bridge, Chicago	Moving Bridges, Chicago
West Arch Footbridge, Moscow	Tsaritsyno Park Bridges, Moscow
White Bridge, Saint Petersburg	Tsarskoye Selo Park Bridges, Saint Petersburg
White Bridge, Dessau-Wörlitz	Wörlitz Garden Bridges, Dessau-Wörlitz
Wiesen Viaduct, Switzerland	Rhaetian Railway Bridges, Switzerland
Wilhemina Bridge, Maasatricht	St Servatius Bridge, Maastricht
Wuting (Five Pavilions) Bridge, Yangzhou	Slender West Lake Bridges, Yangzhou
Yalu River Broken Bridge, Singapore	Singapore River Bridges, Singapore
Yanpu Bridge, Shanghai	Nanpu Bridge, Shanghai
Yudai (Jade Belt) Bridge, Hangzhou	Hangzhou Old Bridges, Hangzhou
Yudai (Jade Belt) Bridge, Beijing	Yiheyuan Summer Palace Bridges, Beijing

The heading to each main entry below gives four names: that of the bridge itself, the name of the nearest location (village, town or city), the name of the relevant division or province of the country and, finally, the name of the country. Where these names are followed by an asterisk *, this indicates that there is no longer a bridge at that site.

A

4 de Abril Bridge, Benguela, Benguela Province, Angola

The main span of this cable-stayed road bridge over the Catumbela River connecting Benguela and Lobito is 160m long with 64m-long side spans and there are a total of five 30m-long approach spans. The prestressed concrete deck, which is 25m wide and carries four traffic lanes, is supported by a semi-fan arrangement of stays from 48m-tall towers that lean outwards gracefully. The bridge, which is named after the date of the peace agreement ending the Angolan Civil War, opened in 2009.

6th October Bridge, Cairo, Cairo Governorate, Egypt

Carrying road traffic across Gezira Island and both arms of the Nile, this 20,500m-long elevated highway and causeway took over from Lagos's Third Mainland Bridge (q.v.) as the longest bridge in Africa when it was finally completed in 1996 after twenty-seven years of construction. Its name commemorates the Egyptian-Israeli war in 1973 and the assassination of President Anwar Sadat in 1981. It consists of a number of different elements most of which are column and beam structures.

25th April Bridge, Lisbon, Estradura, Portugal

Originally called Salazar Bridge and opened in 1966 to cross the Tagus estuary, this structure was renamed after the Carnation Revolution on 25 April 1974. At first it had a single four-lane road deck and, at 1,013m, was the longest suspension span in Europe with a total length of 2,277m. In 1999 the towers were heightened and a second set of main cables was installed and the bridge now has six road lanes and, on a new lower level deck, twin railway tracks. Somewhat similar to the San Francisco-Oakland Bay Bridge (q.v.), it was also designed by the American

25th April Bridge, Lisbon

Bridge Company. In 1998 the Vasco da Gama Bridge (q.v.) was built nearby as a relief structure.

A13 Motorway Footbridge, Bologna, Emilia, Italy

Providing a crossing for cyclists and pedestrians over the A13 motorway between Bologna and Padova, this 100m-span bridge was opened in 2009. It consists of two straight-sided A-frames, which span 100m between abutments well set back from the roads, meeting at a central point from which two planes of cables radiate to support the edges of the deck.

Abdoun Bridge, Amman, Amman Governorate, Jordan

The road bridge across the Abdoun wadi was opened in 2006. It has a five-span cable stay structure with the stays in a harp format. The three Y-shaped piers are made from tubular steel sections, the two main spans are 132m long and the overall length is 417m. The deck, which is S-shaped in plan, is 40m above the bottom of the wadi.

Ada Bridge, Belgrade, Serbia

On its completion in 2011 this bridge over the Sava River became the world's longest single-pylon cable-stayed bridge. Its main span consists of a steel box-girder spine beam 376m long and the prestressed concrete back span is 250m long, with each of the spans being supported by eighty stay cables. The cables are anchored in a pylon that stands on the tip of Ada Ciganlija Island. This pylon is 207m tall and is split into two legs below the 98m level, one of which contains a lift shaft, the other stairs. There is an additional 388m-long reinforced concrete approach span at the New Belgrade end of the bridge and the 45m-wide deck carries six road lanes, two light railway tracks and two pedestrian/cycle ways.

Ada Bridge, Belgrade

Adana Roman Bridge, Adana, Adana Province, Turkey

The stone bridge at Adana across the Seyhan River was probably built by the Romans early in the second century CE and is one of the world's oldest bridges, even carrying motor vehicles until 2007. It is believed it originally had twenty-one or twenty-two arches, although some of the larger piers contained subsidiary arches. It was also probably only about 3m wide. Following construction of stone embankment walls to contain the river and widening on the bridge's downstream face during the early twentieth century, it is now 310m long and 11m wide with fifteen main arches and there are a further six narrow flood relief arches within some of the larger piers. The bridge is now restricted to pedestrians only.

Adana Roman Bridge

Adolphe Bridge, Luxembourg City, Luxembourg

This large masonry arch bridge, named after Grand Duke Adolphe of Luxembourg who opened it in 1903, was designed by Paul Séjourné and Albert Rodange to replace an earlier narrow structure. Its twin segmental main arches, which span 85m and support four subsidiary arches in the spandrels on each side, are flanked by smaller arches. Overall, the deck is 153m long. In 2018 the road bridge was widened and CBA Architects designed a 4m-wide lower deck, suspended between the original arches, for cyclists and pedestrians. The bridge is part of the Luxembourg Old Town World Heritage Site.

Adomi Bridge, Atimpoku, Asuogyaman, Ghana

The two-pin crescent-arched steel bridge spanning 245m over the Volta River was designed by Freeman, Fox & Partners and opened in 1957. With an overall length of 334m long, it is the longest bridge in Ghana. After cracks had developed it was refurbished and re-opened in 2015.

Adži-Paša's Bridge, Podgorica, Central Region, Montenegro

Originally built by the Romans, this hump-backed bridge over the Ribnica River was rebuilt during the eighteenth century and is now named after that builder. It has a single main segmental arch with a smaller arch to one side with, between them, a tall vertical opening.

Adzija's Bridge, Danilovgrad, Central Region, Montenegro

The beautiful stone bridge across the Susica River, a tributary of the Zeta, lies on an old caravan route between Shkodra and Niksic. Possibly dating from the sixteenth century, its semicircular arch has no parapets and the road over it is steeply ramped.

Afghan-Uzbek Friendship Bridge, Hairatan, Balkh, Afghanistan

The first permanent crossing on this site over the Amu Darya River was a timber structure built by the Russians in 1888 as part of the Trans-Caspian Railway. Problems with the foundations in the fast-running river soon led to closure of the bridge and work began in 1902 on a new twenty-seven-span railway bridge designed by the engineer S. Olshevsky. This, too, suffered damage over the years. In 1979, during their military intervention in Afghanistan, the Russians built a temporary pontoon bridge here and the current road:rail bridge was opened in 1982 to provide a link between the Afghan port of Hairatan and Termez in Uzbekistan. This is a 15m-wide nine-span girder structure with an overall length of 816m and carries a single railway line in the middle of its roadway. In 1997, after the Taliban had captured parts of Afghanistan, this bridge

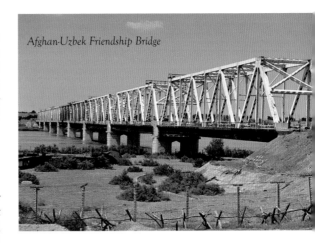

Afghan-Uzbek Friendship Bridge

was closed until being re-opened in 2001. (The area in which the bridge lies is perhaps best known from Robert Byron's justly famous travel book *The Road to Oxiana*, published in 1937.)

Aghakista Bridge, Castletownbere, County Cork, Ireland

The seventeenth-century packhorse bridge across the Kista stream formed part of the old road into Castletownbere from the east. It has two segmental stone arches with different spans, the northern one being slightly the larger at about 15ft, both being formed from noticeably thin voussoir slabs. The bridge has no parapets.

Águas Livres Aqueduct, Lisbon, Estremadura, Portugal

Although the water-carrying channel of this aqueduct is more than 36 miles long, the main above-ground arched structure crosses the Alcantara Valley and consists of thirty-five pointed arches over a length of 941m. The tallest of these is 65m high and spans 29m. Along this part of the structure there are also a number of intermediate decorative square Baroque towers with domed roofs topped by pinnacles that stand on semicircular arches above the water channel. In some of the other sections of the aqueduct the arches are round-headed. Commissioned by King Dom João V, the aqueduct was designed by the military engineer Custodio Vieira and construction lasted between 1731 and 1744.

Aguas Livres Aqueduct

It survived an earthquake in 1755 and remained in operation until 1967. Considered to be the last of the world's great classical aqueduct bridges to be built, it is now a National Monument in Portugal and a UNESCO World Heritage site.

Águila Aqueduct, Nerja, Andalusia, Spain

This striking structure (Eagle Aqueduct in English), commissioned by Francisco Cantarero between 1879 and 1880 to serve his sugar refinery, is distinctive because of its brightly coloured brickwork and its central decorative spire. Four tiers of tall, round-headed, red brick arches span between red brick piers, but with the spandrels above the arches in yellow brick. The top tier, which has eighteen arches and is 90m long, carries the water course 40m above the gorge.

Aguila Aqueduct

At its centre is a buttressed tower supporting a needle spire which in turn is crowned by the eponymous double-headed eagle. The aqueduct was damaged during a naval bombardment in the Spanish Civil War and restored in 2011.

Ain Diwar Bridge, Ain Diwar, Al Hasakah Governorate, Syria

The single-span ruined Roman bridge at Ain Diwar, built in the second century AD, originally crossed part of the Tigris River and was decorated with stone carvings. It was rebuilt around the end of the twelfth century although, later, the river changed its course leaving the bridge isolated in the desert.

Ain Diwar Bridge

Aioi Bridge, Hiroshima, Japan

The main 'head' part of this T-shaped bridge crosses between the east and west banks of the Honkawa River just to the north of an island in the river while the much narrower 'upright' part of the 'T' provides direct access to the Peace Memorial Park on the island. The steel girder bridge was originally built in 1932 and its distinctive shape led to it being the target point for the atomic bomb dropped on Hiroshima in 1945. The bridge, though badly damaged by the bomb, was repaired and eventually rebuilt in 1983. It now has steel girders standing on twin-arched concrete piers.

Ai-Petri Bridge, Yalta, Crimea, Ukraine

One of the peaks (called St Peter's in English) in the 4,000ft-high Crimean Mountains is topped by a stone cross and, instead of the cable car, can also be reached by two wire rope suspension bridges with slatted timber footways. These are each about 45m long and were built in 2015.

Airport Lighting Bridge, Bratislava, Slovakia
Bridges, of course, can be used to carry many different things but this one's function is particularly rare: to support an airport's runway approach lights over water. The structure crosses over the Little Danube River just outside Bratislava Airport without encroaching into the flight path space or hindering the water flow in any way. It consists of a 117m-long prestressed concrete ribbon with a sag of just 3.2m and is only 0.8m thick.

Aizhai Bridge, Jishou, West Hunan, China
Opened in 2012, this suspension bridge carries the motorway linking Baotou and Maoming between two tunnel mouths 300m over Dehang Canyon. Hangers from the suspension cables support a main girder structure beneath the bridge deck. This, which spans 1176m and is 336m above the river, has six traffic lanes with a pedestrian walkway beneath, the central section of which is glass-floored. Unusually, because of the mountainous locality, one of the two main towers is only 50m tall and stands on foundations that are at a higher level than the deck.

Ajuinta Bridge, Altiani, Corsica, France
The striking modern structure that carries the T50 road over the River Tavignano near to the seventeenth-century Ponte Novu (q.v.) was designed by Michel Virlogeux (of Millau Viaduct – q.v. – fame) to connect Aleria and Corte and was opened in 2011. The bridge, which is 115m long between abutments, has seven spans: a 56m-span central arch which is flanked on each side by three further spans supported by two intermediate Y-shaped columns. The thickness of the 12m-wide post-tensioned concrete deck slab varies between 60 and 80cm.

Akapnou Bridge, Akapnou, Limassol District, Cyprus
The Venetian bridge at Akapnou crosses a tributary of the Vasiliko River. It has a main

pointed arch over the stream with a second semicircular arch on one side. It is claimed that a crusader army crossed it in the twelfth century.

Akashi Kaikyo Bridge, Kobe, Hyogo Prefecture, Japan
In 1998 Japan's two largest islands were linked by this record-breaking double-decked suspension bridge across the Seto Inland Sea. The overall length of the structure is 3,800m with a main span of 1,991m, this significantly exceeding the world's previous 1,410m longest span of the Humber Bridge in England, opened in 1981 (see *AEBB*). The bridge's X-braced main towers are 283m high.

Akashi Kaikyo Bridge

Al Garhoud Bridge, Dubai, United Arab Emirates
The first Al Garhoud Bridge across Dubai Creek was a five-span beam structure. It was opened in 1976 but by 2007 was proving to be a major constriction to traffic as a total of twelve traffic lanes reduced to just six on the bridge. The new bridge, opened in 2008, has an overall length of 520m and carries fourteen lanes capable of taking 16,000 vehicles an hour.

Al Salam Peace Bridge, El Qantara, Ismaelia, Egypt
This road bridge across the Suez Canal, which links Africa and Asia, has an overall length of 3,900m and was opened in 2001. The main part is a cable-stayed concrete structure spanning 404m that is supported by pylons 154m high.

Alameda Bridge, Alameda, Valencia, Spain

Alameda's new railway station and the 163m-long bridge that crosses above it were designed by the architect:engineer Santiago Calatrava (who was born in the city) and completed in 1995. The bridge's main segmental steel arch, which consists of two differently-sized tubes, spans 130m across the old dry bed of the now-diverted River Turia and is inclined at an angle of 20° from the vertical. The weight of this arch, together with the pedestrian deck below it, counterbalances the road deck and its outer pavement on the other side of the arch. The connectors between the arch and the deck are hangers with an unusual slightly rhomboidal cross section.

Alamillo Bridge, Seville, Andalusia, Spain

Designed by Santiago Calatrava and opened in 1992, this bridge crosses the Canal de Alfonso XIII. It was built as part of a programme of infrastructural investment in advance of Expo 92 and linked the city centre to the Expo site. The 200m-long main span is supported by thirteen pairs of cable stays connected to a 142m-high back-leaning balancing anchor arm, which supports the deck through its own weight without the need for any ground anchorages. The bridge carries six vehicle lanes and a central footway.

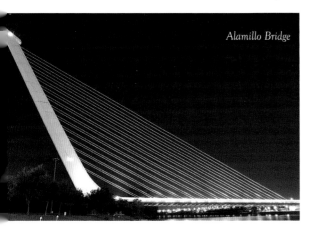

Alamillo Bridge

Alcántara Bridge, Alcántara, Cáceres, Spain

The 194m-long bridge over the Tagus River in the town of Alcántara (its name meaning simply 'the bridge' in Arabic), with its castellated triumphal 14m-high archway in the middle of the structure, is one of the most famous surviving Roman bridges. It was built between 104 and 106CE by the architect Caius Julius Lacer following an instruction by Emperor Trajan. There are six semicircular arched spans, which are roughly 14m, 22m and 28m long on each side, and the 8m-wide deck is about 48m above river level. The bridge has suffered war damage several times including the destruction of one span in 1809 during the Peninsular Wars. There is now a dam across the river just upstream of the bridge.

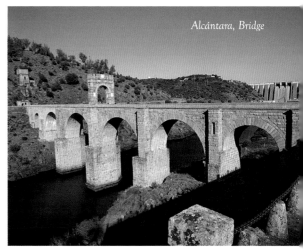

Alcántara, Bridge

Alcántara Bridge, Toledo, Castile-La Mancha, Spain

The present bridge across the Tagus River in Toledo was built in about 1258 on the Roman foundations from an earlier bridge. Its overall width is 5½m and it has two stone arches spanning 28m and 16m. These are separated by a massive pier with pointed cutwaters which extend upwards as half hexagons to form a large recess on each side of the roadway. The 127m-long bridge deck is approached from the east through a triumphal arch and, at the western end, entry into the old town is through a defensive gateway with two square towers. Entry to the UNESCO World Heritage town from the west is over the San Martin Bridge (q.v.).

Alexandra Suspension Bridge, Alexandra, Central Otago, New Zealand

The original single-lane underslung suspension bridge over the Clutha River, designed by Leslie Duncan MacGeorge, was completed in 1882. It had an overall length of 168m with a central span of 83m supported by eight 3in diameter cables spanning between its two main towers. These towers are in three main vertical sections with, below the main archway, two shallower stages each containing a low arch. The towers are capped with decorative low segmental pediments and the bridge was widely considered to be the most beautiful in New Zealand. A replacement steel truss arch structure was opened upstream in 1958 and only the two piers from the earlier bridge now remain.

Alexandre III Bridge, Paris, Ile-de-France, France

This triumphal bridge, which was built for the Paris Universal International Exhibition in 1900 and to commemorate the Franco-Russian alliance, is named after the then Russian Czar. Designed by the engineers Louis-Jean Résal and Amédée d'Alby, it is a three-pinned steel arch structure spanning 108m, with each of its fifteen ribs being made up from thirty-two sections. At 40m wide, it is the widest bridge in Paris. The ornate decoration includes tall Art Nouveau lampposts, bronze putti riding fish and four enormous corner pylons supporting gilded statuary of winged horses representing Arts, Science, Commerce and Industry.

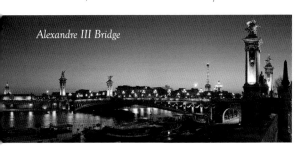

Alexandre III Bridge

Alf-Bullay Bridge, Bullay, Rhineland-Palatinate, Germany

The 314m-long bridge across the Moselle at Bullay was built between 1875 and 1878 as part of the Koblenz-Trier Railway. The six-span wrought iron trussed structure, with the railway above a road, was the first double-deck bridge in Germany and had to be rebuilt between 1928 and 1929 to carry heavier loading. It was destroyed by bombing in 1945 and re-opened in 1947. The bridge now also carries the Kanonenbahn, a hiking trail that passes a number of historical engineering structures.

Alf-Bullay Bridge

Allenby (King Hussein) Bridge, Jericho, Jericho Governorate, Jordan

The history of the crossings over the Jordan River that have been built at this site demonstrates the strategic infrastructural importance of bridges. The first modern-era road bridge was built in 1885 and in 1918, after General Edmund Allenby had been victorious in his military campaign here, he constructed a simple Warren truss crossing. However, the structure fell into the river in an earthquake in 1927 and was destroyed again during military actions in 1946, rebuilt and then destroyed again during the Six-Days War in 1967. Replaced in 1968 by a temporary wooden bridge, the crossing point gained its current bridge in 1994. This, which is also known as the King Hussein Bridge, is a three-span prestressed concrete bridge with extradosed tendons surrounded in concrete. It has an overall length of 110m and is 19m wide.

Alloz Aqueduct, Alloz, Navarre, Spain

The aqueduct over the River Salado in Alloz was built in 1939 to a design by Eduardo Torroja. Its overall length is 220m made up of 20m-long spans. Vertically-stretched X-shaped piers provide a saddle-type support to the parabolically-shaped water channel, which itself forms the beam structure between the piers.

Alsea Bay Bridge, Waldport, Oregon, USA

The first bridge here was a reinforced concrete arch bridge, designed by C. B. McCullough, that carried US Route 101 over the Alsea River estuary. Opened in 1936, it had an overall length of 3,011ft and consisted of three through arches, the central one with a span of 450ft, flanked on each side by three further deck-arch approach spans. Following major corrosion of the reinforcement steel, the bridge was demolished in 1991 after a slightly shorter replacement structure had been opened. This has only one main through arch but, to help protect against any future corrosion, all the piers are thicker and there is a latex concrete deck.

Alvord Lake Bridge, San Francisco, California, USA

The first reinforced concrete bridge in America was built in 1889 by Ernest L. Ransome and carries a public road over the pedestrian entrance into San Francisco's Golden Gate Park. The structure has a very flat segmental arch with rounded haunches, spans 20ft and is 64ft wide. The innovative reinforcement was steel bar that had been cold-twisted using a system patented by Ransome. The bridge is included in the ASCE list of historic bridges.

Amoreira Aqueduct, Elvas, Alentejo, Portugal

The first attempts to build an aqueduct to bring water to Elvas began in 1537, although the huge underground storage tanks were not built until 1650. The overall length of the water supply system is more than 7km with the arcaded section, which was designed by the architect Afonso Alvares in 1573, being 1683m long. Completed in 1622, there are 833 semicircular stone arches, many of them four levels high. The aqueduct is part of the Garrison Border Town of Elvas UNESCO World Heritage Site.

Amoreira Aqueduct

Ampera Bridge, Palembang, South Sumatra, Indonesia

The opening bridge across the Musi River in Palembang was completed in 1965 and its vertical lifting system was designed to increase the maximum headroom for vessels passing through the structure from 9m, when closed, to 44m with the movable span fully raised. This section, which is 61m long, was counter-balanced by a 500-ton counterweight in each of the two towers. However, it has not been possible to open the bridge since 1970, believed to be as a result of structural distortion following

Ampera Bridge

movement of the foundations. The bridge has an unusual elevation since the towers are braced, to prevent any lateral movement during raising or lowering operations, by two sets of suspension cables that stretch from shore to shore via the towers to which they were connected. One set of these cables is at tower-top level and the other at the level to which the opening section could be raised. The full length of the bridge with its approach spans is 224m.

Aného Bridge

Amstel Park Footbridge, Amsterdam, North Holland, Netherlands

This 27m-long pedestrian-only covered bridge, completed in 2006, is a simple girder structure with its cantilevered timber-covered roof being supported by G-shaped frames along one side of the footway. It was designed by Steven van Schijndel and Stefan Strauss.

Ancient Roman Bridges, Rome, Italy

There are believed to be more than 130 ancient Roman bridges still standing in Italy, of which four cross the River Tiber in Rome. The magnificent **Ponte Milvio**, which was completed by Emperor Hadrian in 134CE, now has four central arches spanning 18m with two 9m-long end spans, one of which is modern. The piers contain small flood channels. **Pons Aelius** (now known as Ponte San Angelo) was completed in 174CE and is the most decorative of the ancient bridges. It has five 18m semicircular spans supporting a level deck with ten seventeenth-century sculpted angels by Bernini and followers on the parapets. **Pons Fabricius**, dating from 62BCE, is the oldest and best preserved and has two main segmental spans with a wide central pier containing a raised flood channel. The brick facing on the spandrels replaced earlier marble in 1679. Only one arch, spanning about 25m, remains of **Pons Aemilius**'s original seven and the bridge is now called **Ponte Rotto** (Broken Bridge). Its two surviving piers date from 179BCE.

Aného Bridge, Aného, Lacs, Togo

The capacity of the N2 road crossing over the harbour entrance at Aného was increased in 2017 by the opening alongside the original bridge of a new structure. This has a single flat concrete arch 95m long with a cantilevered footway on the open side.

Anghel Saligny Bridge, Cernavoda, Constanta, Romania

The Romanian engineer Anghel Saligny designed this magnificent single-track bridge across the Danube, which was opened in 1895 to connect Cernavoda and Fetesti as part of the railway between Bucharest and Constanza. Its overall length is 2,632m but the principal structure has five main spans of through trusses which are distinctively higher at the piers: a central span of 190m – claimed to be the longest span in Continental Europe at the time – with four other spans of 140m. Clearance beneath the deck was 30m. The Romanian end is guarded

Anghel Saligny Bridge

by an imposing medieval-style gateway tower with mock machicolations above the entrance and giant statues of soldiers above the ends of the supporting pier. A smaller structure across the Borcea – a nearby subsidiary arm of the Danube – was blown up in 1916. The bridge, which had only a single track, was closed in 1987 when the new Cernavoda Bridge (q.v.) was opened and is listed in the Romanian National Register of Historic Monuments.

Angostura (Narrows) Bridge, Ciudad Bolivar, Bolivar State, Venezuela

This suspension bridge, built between 1962 and 1967, was the only fixed crossing over the main Orinoco River upstream from its mouth until 2006. It has a central span that is 712m long, with back spans of 280m, and the steel pylons are 119m high. The structure was designed by Venezuelan engineer Paul Lustgarten.

Anlan Suspension Bridge, Dujiangyan City, Sichuan, China

It is believed that the first structure across the Yuzui River on this site, was a multi-span bamboo, plank and rope suspension bridge built in the third century BCE, and it is considered to have been one of China's five great ancient bridges. The present structure, also a suspension bridge but with steel chains, was built in the 1970s and connects a mid-river artificial island to both banks. It has three main spans, the piers topped by pagoda roofs, and is 320m long and about 2m wide.

Anping (Peace) Bridge, Fujian, Hebei, China

The bridge over the Shiajing River at Anping, which was built in 1138–51, is 2,070m long and winds randomly along its length, presumably as suitable ground for the pier foundations was located. It has 332 spans, each consisting of several granite slabs 8-10m long spanning between heavy grillage stone piers. The edge parapets, also of granite, consist of vertical slabs with tenon joints into low supporting stone posts. There are five pavilions along its length where travellers can rest. The bridge is considered to be one of the five great ancient bridges of China.

Anping Bridge

Anshun (Peaceful and Favourable) Bridge, Chengdu, Sichuan, China

Earlier bridges across the Jinjiang River in Chengdu include one seen by Marco Polo (1254–1324) and another, built in 1746, that was destroyed by floods in 1947. The present structure, which dates from 2003, was not built for a road but specifically to carry a pagoda-style building that would act as a focus for the regeneration of the city centre. It has three segmental arches, the central one significantly higher than the flanking arches, and large tunnels above the piers complete with dragons on guard. It is all brightly floodlit in colour at night.

Anshun Bridge

Arenal Hanging Bridges, La Fortuna, Alajuela, Costa Rica

The 250-hectare Mistico National Park, on the site of an old volcano, is more than 600m above sea level and includes a 3km walk that allows eco-tourists to travel at high level through the rain forest. There are several wire suspension bridges, the longest spanning 265ft and 213ft, and the decks and sides are made of strong wire mesh panels. There are also zip wires.

Arenal Hanging Bridges

Arenales Suspension Bridge, Pantasma, Nueva Segovia, Nicaragua

This 130m-span suspension bridge was built in 2017 by Bridges to Prosperity (B2P) in association with Kiewit Corporation and KPFF Consulting Engineers. The structure won an Award of Merit in the Engineering News-Record's Global Best Project of 2017.

Arganzuela Footbridge, Madrid, Spain

This footway and cycleway is carried inside a cylindrical steel structure that crosses the Manzanares River and links Madrid's Arganzuela and Carabanchel areas. The bridge, which was designed by Dominique Perrault and completed in 2011, is in two separate

Arganzuela Footbridge

sections that are 150m and 128m long on slightly different axes and are separated by an open space. The hollow cylinder consists of longitudinal square-section hollow tubes connected by further similar tubes that form an enclosing helix, part of which is covered by a helical strip of stainless steel wire mesh.

Argenteuil Bridges, Argenteuil, Val-d'Oise, France

The footbridge, road and railway bridges in Argenteuil are best known for the Impressionist paintings of them by Caillebotte, Monet, Renoir, Sisley, Vlamink and others. The road bridge at the time had been newly built to replace one destroyed in the Franco-Prussian War. It had five spans, each of five cast iron arched ribs, that supported a level deck on spandrel posts, with additional diagonal bracing in the two end panels of each span. The river piers were of stone. The current road bridge, which was completed in 1947, is 19m wide and 232m long. It has three steel deck arches, each of six ribs, that have spans of 62m, 68m and 62m. The original railway bridge seen in the paintings had five wrought iron plate girder spans standing on circular cast iron columns braced together at mid-height. The present trussed structure stands on the original piers.

Arkadia Park Aqueduct, Lowicz, Lodz, Poland

The ruined Roman aqueduct folly in this stately park was built in about 1778 by the architect

Arkadia Park Aqueduct

Szymon Bogumil Zug for his patron Princess Helena Radziwill. It consists of two tiers of semicircular brick arches crossing over a stream, the lower tier with six complete arches and only two above.

Armando Emilio Guebuza Bridge, Chimuara, Zambézia Province, Mozambique

Named after a former president of Mozambique, this bridge across the Zambezi River is a box girder structure. It has an overall length of 2,376m, of which the main structure is 710m long with six spans and 16m wide. The bridge was designed by WSP Global and built between 2005 and 2009.

Arthur Kill Vertical Lift Bridge, Elizabeth, New Jersey, USA

The Baltimore & Ohio Railroad built this single-track bridge in 1959 to replace an earlier swing bridge dating from 1890. The main N-braced steel girder lifting section spans 558ft between two 215ft-tall steel towers, making it the world's longest vertical lift span. Normal clearance below the bridge is 31ft but, when the lifting span is raised, this increases to 135ft.

Arvida Bridge, Saguenay, Quebec, Canada

Built between 1948 and 1950 (shortly after the Hendon Gateway Bridge in Sunderland, the world's first aluminium bridge – see *AEBB*), this deck arch road bridge has an overall length of 154m with a span of 92m. The arch itself, together with the columns and beams of the supporting framework for the 10m-wide reinforced concrete deck, were all constructed by riveting together aluminium plates and angles into hollow box sections.

Ashford Castle Bridge, Clonbur, County Mayo, Ireland

Ashford Castle dates back nearly 800 years but the present mainly Victorian buildings now house a luxury hotel on the shore of Lough Corrib. This is approached by the medieval fantasy carriageway bridge, built in about 1865 by the Guinness family, which has six pointed stone arches that span between piers with pointed

cutwaters. The roadway is crossed at the inner castle end by the pointed arch of the gatehouse tower, which is decorated with a turret, and there are two smaller turrets flanking the road at the outer end.

Ashtabula Bridge, Ashtabula, Ohio, USA

The Lake Shore & Michigan Southern Railway's first bridge over the Ashtabula River was a timber structure, but this was replaced in 1865 by a wrought iron truss bridge 165ft long. On 29 December 1876 one of the worst-ever bridge disasters occurred here when the bridge collapsed as a train was crossing over. Apart from the leading locomotive, a second engine and all eleven coaches fell 76ft into the river and then caught fire. The exact death toll was never known but, of the 159 people on board, officially 92 people died and there were two later suicides by railway officials. The finding of the six-man coroner's jury was that the bridge had been improperly designed, poorly constructed and inadequately inspected. Later technical analysis has more specifically blamed fatigue in key materials and the cold weather causing cast iron members to become brittle.

Ashtarak Bridge, Ashtarak, Aragatsotn, Armenia

This three-span bridge, on the site of an earlier structure, crosses the Kasakh River where it flows against one side of the river valley and the pointed arches accordingly are smaller and

Ashtarak Bridge

lower as the roadway descends. The bridge was built in 1664 and the builder's name has been recorded as Motsakentz Makhtesi Khoja Grigor. It is 76m long and 6m wide and was extensively renovated in the mid-twentieth century and again in 2004.

Asparuhov Bridge, Varna, Bulgaria

The Asparuhov Bridge was opened in 1976 to provide a new crossing where the main canal linking Lake Varna to the Black Sea had been widened. It is a 2,050m-long continuous beam bridge with thierty-nine spans, but includes a 50m-high main arched span over the canal flanked by a half-arch at each end. There has been a 52m-high bunjee jumping site from the bridge deck for more than twenty-five years.

Atamyrat-Kerkichi Bridges, Atamyrat, Lebap Province, Turkmenistan

The first railway bridge across the River Amu Darya, a timber trestle structure designed by the Polish engineer M. Bielinsky, was built in the late 1880s and this was replaced by a steel truss bridge carrying the P-39 road and a railway line. The road bridge, which was built alongside between 2009 and 2011, is a 1,414m-long structure with twelve main spans of 110m. There is also a shorter approach span at each end and its deck is 12m wide.

Atamyrat-Kerkichi Bridges

Atatürk Bridge, Istanbul, Marmara Region, Turkey

Istanbul's modern Atatürk Bridge, which was completed in 1940, is named after Mustafa Kemal Atatürk, founder and first president of the modern Republic of Turkey. It crosses the Golden Horn between Cibali and Emekyemez, a short distance up the Golden Horn from Galata Bridge (q.v.). A simple column and beam structure, it is 477m long and its 25m width allows it to carry six lanes of road traffic.

Atatürk Viaduct, Gazientep, Izmir Province, Turkey

Like its namesake in Istanbul (q.v.), this curved structure is also named after Mustafa Kemal Atatürk, founder and first president of the Republic of Turkey. The viaduct opened in 1999 to carry part of the Trans European Motorway across the Olucak Valley. It consists of steel box-section beams that stand on tall reinforced concrete piers and support a composite reinforced concrete deck slab. It is 802m long and has eight spans, the longest being 110m, and at its maximum is 146m above ground level.

Athi River Super Bridge, Athi River, Nairobi, Kenya

This simple beam bridge, opened in 2017, carries the single-track, 609km-long Mombasa-Nairobi Standard Gauge Railway three times over the meandering Athi River on the railway's approach to Nairobi. With an overall length of 2,785m, the reinforced concrete structure is one of the longest bridges in Africa. It consists of twin beams spanning 18m between T-headed piers and allows free movement of animals below the bridge as the line passes through the Nairobi National Park. The railway replaces the earlier British-built line and is Kenya's biggest infrastructural project.

Attock Railway Bridge, Attock Khurd, Punjab Province, Pakistan

The first bridge across the River Indus on this site was a pontoon structure but, although work started in 1860 on a tunnel link under the river,

Attock Railway Bridge, Pakistan

this was abandoned in 1862. Next, a railway bridge designed by Sir Guildford Molesworth was opened in 1883. This consisted of 26ft-deep, double-intersection, wrought iron girders with the railway tracks above a separate lower deck that carried the Grand Trunk Road. The bridge had an overall length of 1,412ft with five spans: three that were 257ft long and there were two 312ft-long main spans. Because of its strategic importance, the bridge's approaches were heavily fortified with twin-towered gateways where the road joined the bridge. In 1929 a completely reconstructed bridge, designed by Sir Francis Callaghan, was opened. This reconstruction involved building intermediate piers in the middle of the shorter spans and strengthening the two longer spans by converting the original girders into deeper fish-bellied steel girders. In 1979 road traffic was transferred to a new bridge leaving the earlier bridge still to carry the railway.

Atyrau Bridge, Astana, Kazakhstan

A new 314m-long bridge for pedestrians and cyclists across the Ishim River in Astana was opened in 2018. The 11m-wide deck, which splays outward as it reaches one bank, is partly covered over for about half its length by a silvery triangular grid, these two characteristics making the structure look somewhat like a fish in plan, a fish being the region's symbol.

Auckland Harbour Bridge, Auckland, New Zealand

The famous bridge over Waitemata Harbour linking St Mary's Bay with Northcote was built

between 1955 and 1959 and, following widening in 1969, now carries Highway 1 motorway's eight lanes. It was designed by the British firm Freeman Fox & Partners and the 5,800 tons of steelwork was fabricated in Britain. Overall, the bridge is 3,348ft long with the main part of the structure consisting of an 800ft-long trussed central span over the navigation channel that is cantilevered out from a trussed anchor span on each side. There are six further approach spans. The clearance over the shipping channel is 142ft at high tide. Originally the 12.8m-wide structure carried only four traffic lanes but two additional traffic lanes were added to each side between 1968 and 1969 and the bridge now carries 200,000 vehicles a day. These additions consist of 'clip-on' orthotropic box girders which span between extensions to the original piers. Plans for the addition of a footway and cycleway SkyPath have been under consideration for many years but there is a bridge climb to the highest point of the bridge, 200ft above the water, as well as bunjee jumping. A new structure is now being considered.

Auckland Harbour Bridge

Augartenbrücke, Vienna, Austria

The first bridge here over the Danube Canal in Vienna was built in 1782 and was then rebuilt after being burnt down in 1809 during the Napoleonic Wars. This structure, in turn, was replaced in 1829 but, in 1873, a new bridge was opened as part of the World's Fair. This consisted of iron suspension chains and stays, but its most noteworthy features were the granite pillars at the ends of the trusses that were topped by large allegorical figures. The bridge was replaced in 1931 by a wider one carrying two tram tracks, four lanes for road traffic and two footways. This was destroyed in 1945 and replaced in 1946 by the present structure containing seven steel beams.

Augustus Bridge, Dresden, Saxony, Germany

The first recorded bridge over the River Elbe in Dresden was a timber structure built in 1070 and the first stone bridge was built in about 1222. By 1344 this had been replaced by a twenty-three-arch stone bridge and between 1727 and 1731 King August the Strong was responsible for building another structure that had twelve stone arches. In order to widen the arches for river traffic, a new nine-arch concrete bridge was built between 1907 and 1910 and named after King Frederick August III. This was 390m long and 18m wide and had nine spans, the biggest being 39m long. The Germans blew up this on the second last day of WWII but by 1949 it had been rebuilt to the original plans. The bridge was restored again between 2018 and 2019. The bridge has been a great favourite of painters, including Canaletto and Oskar Kokoscha.

Aveiro Circular Footbridge, Aveiro, Centro, Portugal

The unusual steel footbridge across the São Roque Canal, opened in 2006, was designed by Lius Viegas and Domingos Moreira. Its 2m-wide

Aveiro Circular Bridge

deck is circular and is supported by stays from an inclined mast shaped like a Greek letter alpha standing on its open end.

Aveyron Bridge, Belcastel, Aveyron, France

The lovely bridge over the Aveyron River at Belcastel was probably built in the early fifteenth century by Alzais Saunhac, Lord of Belcastel, to link the village with the church he had built on the opposite side of the river. It has five pointed arches, the two largest of which each span 10.5m, and has an overall length of 57m.

Aveyron Bridge

Avignon Bridge, Avignon, Vaucluse, France

St Benézét built a wooden bridge standing on stone piers over the River Rhône at Avignon between 1177 and 1185, but this was destroyed when the town was besieged in the early thirteenth century during the Albigensian Crusade. In the fourteenth century a new 5m-wide stone bridge was built, which crossed the river on a curve to take advantage of existing islands. This had twenty-two segmental arches, the largest spanning 36m, and the overall length was about

Avignon Bridge

920m. Because of its strategic importance, the crossing was protected by a gatehouse at each end. In around 1300, some of the arches were pulled down on the orders of Boniface IX during his troubled papacy and, although rebuilt, these or others collapsed during severe floods in 1603, 1605 and 1680. The bridge became unusable and was replaced by a ferry service. Only four arches now remain, one of which supports the two-storied St Nicholas chapel and another small chapel stands on one of the remaining piers near the Avignon bank. The bridge is within the Avignon World Heritage site.

Ayeyarwady Bridge, Mandalay, Myanmar

Also known as the New Ava Bridge, this bridge is just to the north of the existing Ava Bridge. Completed in 2008, it has three main segmental steel arches, each spanning 224m. Together with lattice girder approach spans its total length is 1711ft.

Ayjay (Khor Al Batah) Bridge, Sur, Ash Sharqiyah South, Oman

This suspension bridge, completed in 2009, has two main 115mm-diameter locked-coil suspension cables from which vertical hangers support steel cross-beams beneath the 10m-wide concrete deck. There are also three additional stay cables on each side of the deck at both ends. The overall length of the bridge is 204m and the main span is 138m long. Unusually, the normal portal-frame arrangement of the two pylons is augmented by a second slightly lower cross-beam.

Ayub Bridge, Sukkur, Sindh Province, Pakistan

The second railway bridge across the River Indus at this site is named after Field Marshall Mohammad Ayub Khan and was designed by David B. Steinman to take rail traffic from the existing Lansdowne Bridge (q.v.). It is an 806ft-span steel arch, with the deck hung from the arch on coiled wire rope suspenders. The bridge was completed in 1962 and its arch is 806ft long and 247ft high. Together with Lansdowne Bridge, only 100ft away, the two bridges now look from a distance like a single very complex structure.

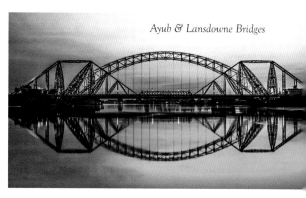

Ayub & Lansdowne Bridges

B

Bac de Roda-Felipe II Bridge, Barcelona, Catalonia, Spain

This 129m-long bridge in Barcelona was designed by Santiago Calatrava to provide a new road linking the city and the coast that passed across the railway lines into the city from the north. It was completed in 1987 for the 1992 Olympics. On each side of the deck are twin steel parabolic arches, one of which is vertical and stands between the roadway and the footway. The second arch of each twin is raked inwards at 30° from the vertical and converges with the first arch at their apices. Stays in the plane of each arch support the deck, the outer edge of which is curved in plan.

Bacunayagua Bridge, Santa Cruz del Norte, Mayabeque Province, Cuba

Opened in 1959 to carry the main road along the north coast of the island, this is the highest bridge in Cuba with its deck 110m above the valley floor. The 314m-long reinforced concrete structure has a polygonal arch spanning 114m that consists of four straight members.

Bad Kreutznach Bridge, Mainz, Rhineland-Palatinate, Germany

An earlier wooden bridge here was replaced in about 1300 by the count of Kreutznach. His eight-arched stone bridge, which crossed both the River Nahe and the adjacent Mühlenteich Canal, included defensive towers, one of which was used as a local prison. Houses were first built on the upstream end of the canal piers in about 1480 but the present four half-timbered buildings date from between 1582 and 1612 and are now used as shops and business premises. The river arm of the bridge was destroyed in 1945 and was replaced by a 50m-span concrete structure, later given a strengthening mid-stream prop.

Bad Säckingen Covered Bridge, Bad Säckingen, Baden-Württemberg, Germany

The first bridge here over the Rhine connecting the German city of Bad Säckingen and the Swiss village of Stein was built in 1272. The present structure, which was opened in 1700, is the longest covered bridge in Europe with a total length of 204m over nine spans between stone piers and is 5m wide. Since a new road bridge was opened in 1979 the bridge has been for pedestrians only.

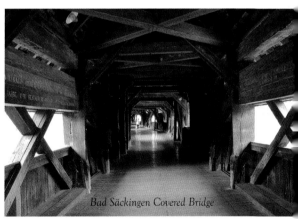

Bad Säckingen Covered Bridge

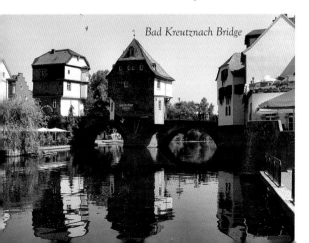

Bad Kreutznach Bridge

Bailey Island Bridge, Harpswell, Maine, USA

This 1,167ft-long bridge over Will's Strait links Bailey Island with Orr's Island and is believed to be the only granite cribstone bridge in the world. Designed by Llewelyn Edwards and built in 1928, it contains 10,000 tons of granite, each individual stone being generally one foot or more square in section and several feet long. These are laid in an open criss-cross framework fashion called a crib, and kept in place solely by their weight, thus allowing tidal flows to pass through the structure without damaging it. There is a short, traditional reinforced concrete beam section across the main channel. The crib carries a narrow, slightly humped two-lane roadway and a footway was added on the eastern side in 1951. The bridge is included in the ASCE list of historic bridges.

Baling Bridge, Weiyun, Gansu, China

This beautiful bridge, with its doubly-curved elevation, spans 30m over the Qingyuan River and was rebuilt in 1934. It is 4½m wide and, including the entry pavilion with upturned eaves at each end, the overall length is 45m. The structure is a timber woven arch, with each section consisting of ten parallel timbers.

Baling Bridge

Bamako Third Bridge, Bamako, Koulikoro, Mali

Bamako, capital city of Mali – one of the world's poorest countries, was founded in 1640. Its third bridge over the Niger River after the Bridge of Martyrs (1957) and King Fahd Bridge (1982 – q.v.) was completed in 2011 and represents China's largest investment to date in West Africa. The structure is S-shaped in plan and has a total of sixty-nine column-and-beam spans with an overall length of 1,627m. Its 24m-wide deck carries four traffic lanes, two motorcycle lanes and two footways. The bridge was built by China Gezhouba Group International.

Bamboo Bridge, Surakarta, Java, Indonesia

Designed by the Indonesian chapter of Architects Without Borders, the 18m-span bamboo bridge here was built in 2016 across the Kali Pepe River. The roofed structure consists of a main arch, which shares the same springing points as a lower deck arch, with both sets of curved bamboo being connected by radial stays.

Banab River Bridge, Madang, Papua New Guinea

Papua New Guinea is one of the world's poorest nations and, when a bridge over the Banab River collapsed in 2018, probably as a result of over-loading on a corroded structure, nearly 750,000 people were affected. Although a simple trestle timber footbridge was built fairly quickly, it was about a year before a temporary Bailey bridge 88m long had been erected to restore vehicle traffic. The country's Works Secretary has said that about 80 percent of the nation's bridges are ageing and posing a threat to the travelling public.

Band-e-Amir Bridge, Marvdasht, Fars, Iran

A weir was originally built across the Korr River in the tenth century to provide water for an area that was then mostly desert but was later able to support 300 villages. A replacement bridge, which was built in the nineteenth century along the crest of the weir, is about 25ft above the riverbed. It supports thirteen pointed arches which each span 15ft and carry a road bridge about 350ft long and 18ft wide.

Bandra-Worli Sea Link, Mumbai, Maharashtra State, India

The cable-stayed bridge across Mahim Bay, which was designed by engineer Seshadri Srinivasan, was built between 2000 and 2010 to link the southern and western parts of Mumbai. The overall length of the structure is 5,600m and there are two main spans that are each 250m long. Each side of the deck is supported by stays anchored into separate diamond-shaped 128m-high pylons. The deck itself, which is 40m-wide and carries eight traffic lanes, is constructed from 2,342 reinforced concrete precast deck sections. In the approach viaducts these are assembled into 300m-long box girder sections separated by expansion joints, each section of which has six 50m-long spans. The bridge includes special anti-seismic arresters to cope with earthquakes up to magnitude 7.0 on the Richter scale.

Banegas Bridge, Montero, Obispo Santistevan, Bolivia

Opened in 2017, this bridge over the Grande River is part of a 4,700km-long road link across Bolivia between the Atlantic and Pacific Oceans. Replacing earlier dangerous pontoon bridges, the new multi-span 1,440m-long structure is Bolivia's longest and was push-launched across the river from both ends.

Bangabandhu Bridge, Sirajganj, Rajshahi Division, Bangladesh

At 4,630m long, this prestressed concrete box girder bridge over the Jamuna River linking Bhuapur and Sirajganj is the longest bridge in Bangladesh. It has forty-seven main spans each 100m long and a shorter 65m-long span at each end, all consisting of 4m-long precast concrete segments craned into position on a two-span movable erection gantry and then prestressed together. The depth of these sections varies from 6.5m to 3.25m giving an arched soffit. The 18.5m-wide deck carries dual two-lane carriageways of National Route 405 and a mixed-gauge single rail track. In addition, there are power and telecommunication cableways as well as a gas pipeline. The bridge, which was opened in 1998 was designed by T. Y. Lin International and built by Hyundai Heavy Industries.

Banpo-Jaamsu Bridge, Seoul, Gyeonggi Province, South Korea

Jamsu Bridge, the first part of this double structure over the Han River, was completed

in 1976. It has a total length of 795m and is 18m wide. In order to keep the cost down, the bridge was deliberately built low above the water, meaning that the deck is submerged at times of heavy monsoon rainfall. It now carries only pedestrians and cyclists. In 1982 a steel box-girder upper deck for motor vehicles was built but, because its supporting columns are outside the deck of the existing bridge, the new deck is 25m wide. The main feature of the upper structure is the 570m-long Moonlight Rainbow Fountain, installed in 2009, which is claimed to be the world's largest bridge fountain. This consists of nearly 380 water jets, from which water is projected out nearly 43m on each side of the bridge at the rate of 190 tons a minute, with illumination from nearly 10,000 coloured LEDs.

Banpo-Jaamsu Bridge

Baodai (Precious Belt) Bridge, Suzhou, Jiangsu, China

A bridge here was first built in 816–819 but the current structure dates from 1446, when it was built to carry a towpath for the adjacent Grand Canal, now a World Heritage site. It is 317m long and contains fifty-three pointed stone arches. The normally flat deck, which is 4.1m wide and

Baodai (Precious Belt) Bridge

has a hump-backed profile over the three larger central arches, has no edge railing, only a slightly raised kerb. At one end, where there are no arches, the deck steps down to a low causeway.

Bar Aqueduct, Stari Bar, Bar Municipality, Montenegro

The stone aqueduct near the old fortress at Stari Bar was built by the Ottomans in the late sixteenth or early seventeenth century and was the largest aqueduct in the former Yugoslavia. It was cranked in plan and also had seventeen arches (both pointed and semicircular) of differing spans, presumably built in this way so the piers could be founded in the most suitable locations. It was destroyed by an earthquake in 1979 and rebuilt later.

Barqueta Bridge, Seville, Andalusia, Spain

The Barqueta Bridge was built over the Alfonso XII Channel of the Guadalquivir River between 1989 and 1992 to form the main road entrance onto Cartuja Island, site of the 1992 International Exposition. Designed by Juan J. Arenas and Marcos J. Pantaleón, the steel structure consists of a single tied arch above the bridge centreline. This splits into two straight struts at each end that also frame a triangular entrance onto the bridge. Overall, the bridge is 199m long with a span of 168m and the edges of the 21m-wide deck are supported on inclined hangers from the arch.

Barqueta Bridge

Barrington Bridge, Limerick City, Munster, Ireland

The 54ft-span road bridge over the Mulcair River was completed in 1818 to a design by J. Doyle in order to link two parts of Matthew Barrington's estate that were separated by the river. It is one of the oldest iron bridges in Ireland. The 14ft-wide slightly humpbacked structure has nine internal ribs consisting of 12in-diameter cast iron tubes in 6ft-long sections. These are connected longitudinally at flanged joints with, clamped between the flanges, iron plates that extend across the width of the bridge to stiffen the structure. Decorative external ribs support thin-spindled iron railings as well as additional S-braces to the railings.

Bastion Bridge, Lohmen, Saxony, Germany

A timber bridge was built nearly 200m above the River Elbe here in 1824 in order to help attract visitors to this mountainous area, but the present footbridge dates from 1851. It is 76m long and its seven semicircular stone arches span between tall rocky pinnacles across a 40m-deep ravine. The area is now a National Park.

Batman Bridge, Sidmouth, Tasmania, Australia

The deck of this cable-stayed bridge across the Tamar River spans between trussed edge girders. These, in turn, are supported on stays from a single 299ft-tall elongated A-shaped pylon at one end, which leans in towards the centre of the span. The structure was built between 1966 and 1968. Its overall length is 1,417ft, its longest span 676ft and its 34ft-wide deck carries two traffic lanes and a footway each side.

Batman Bridge

Batuan Bridge, Sanjiang, Guangxi, China

This structure, with its distinctive two levels of covered decks and its three pagoda-like pavilions all constructed in timber without nails, is known as a Dong 'wind and rain' bridge. It was built in 1910 on stone foundations, one on each bank of the Liujiang River and another in midstream. Overall, it is 50m long and 4½m wide and contains two separate passageways: the upper one for people and the lower for animals.

Bealaclugga Bridge, Spanish Point, County Clare, Ireland

The Gothic-style masonry bridge here was built by John Killaly in 1824 to carry a raised coastal road over the Annagh River. The 34ft-span pointed arch ring is separated from the spandrels by an archivolt and the spandrels themselves are decorated with blind oculi. The massive abutments have projecting towers on battered bases and include three levels of mock arrow-slits.

Beijian Bridge, Taishun, Zhejiang, China

This covered timber 'lounge bridge', constructed with timber estimated to have been over 600 years old, was built in 1674 and refurbished in 1849. The overall length of its covered passageway is 52m and, inside, it is 5.4m wide and 11m high. The woven arch spans 29m between its support points and is made up of five straight sections of parallel timber poles, although the protective cladding is in just three sections. The bridge's main feature is its double-eaved roof pavilion, the ridge line of which sags decoratively to the middle and is topped with a dragon at each end.

Beipan River Shuibai Bridge, Liupanshui, Guizhou Province, China

The Shuibai Railway crosses the Beipan River on one of the world's highest concrete arch bridges. Built in 2001, it has a main span of 236m in its overall length of 486m and is 275m above river level.

Beipan River Railway Bridge

Beitbridge Bridges, Beitbridge, Matabeleland South, Zimbabwe

The **Sir Otto Beit Bridge** across the Limpopo River linking South Africa to what is now Zimbabwe was built between 1937 and 1939. The first suspension bridge in South Africa, it was designed by Ralph Freeman, fabricated in Middlesbrough and built by Dorman Long. The bridge, which is 369m long with a main span of 328m, was also the first major suspension bridge outside the USA to have parallel wire rather than twisted wire cables. A campaign is under way for the structure to be given World Heritage status. Originally, the narrow deck carried a single-track railway within its single carriageway but road traffic now uses the **Alfred Beit Road Bridge**, which was completed just upstream in 1995. This is a 462m-long steel girder structure.

Belize City Swing Bridge, Belize City, Belize

The first permanent crossing over Haulover Creek off the Belize River was a timber swing bridge opened in 1818, and this was replaced by another one in 1859. In 1923 a new steel swing bridge was built that consisted of a slightly arched main plate girder either side of a central roadway. This was flanked by cantilevered footways edged by elegant iron parapet railings. The bridge is claimed to be the oldest opening

Belize City Swing Bridge

bridge in the world still to be operated by hand cranking. The bridge has been damaged several times by hurricanes, including in 1931, 1961 and 1998.

Bellon Viaduct, Coutansouze, Allier, France

There are masonry approach viaducts at each end of this 231m-long railway viaduct, built by Gustave Eiffel in 1869. The central section, however, consists of three over-lapping X-braced ironwork trussed spans, the middle one being 48m long and with a maximum height of 49m. To help increase the stability of the 4.5m-wide structure, its legs splay out at their bases in parabolic curves at right angles to the line of the viaduct, an interesting feature that Eiffel later used in his eponymous 1889 Tower in Paris.

Benito River Bridge, Mbini, Littoral Province, Equatorial Guinea

The bridge across the Benito River (also known as the Mbini River) links Bolondo and Mbini. Paid for by the country's oil wealth, it was built by the Chinese. The bridge is about 800m long and its cable-stayed deck is supported by a reinforced concrete inverted Y-shaped pylon.

Bercy Bridge, Paris, Ile-de-France, France

The first bridge over the River Seine at this site was a suspension bridge built in 1832 to replace a ferry service. This, in turn, was replaced by a stone road bridge, designed by the engineering firm Feline-Romany, that has five semi-elliptical arches and was opened in 1864. In 1904 the bridge was widened by 5½m to take a metro line on an elevated viaduct supported by forty-four semicircular arches. The overall length of the bridge is 175m. The bridge was widened again and re-opened in 1992.

Bercy Bridge

Besalu Bridge, Besalu, Girona, Spain

The first bridge over the Fluvia River at Besalu, which was probably built in the eleventh century, was swept away by floods in 1315. Its replacement is 105m long and has seven pointed arches. The bridge is in three parts, with each of the outer parts containing three arches and these are separated by a short central section and arch. The line of the bridge is cranked at the two junction points so suitable rock outcrops could be used for the piers. Approaching the old city, there is a 30m-tall hexagonal tower, complete with portcullis, one-third of the way across the bridge and a second gateway tower at the end. The bridge was blown up in 1939 during the Spanish Civil War and rebuilt in the 1950s.

Besalu Bridge

Betsiboka Bridge, Maevatanana, Betsiboka, Madagascar

An earlier crossing over the Betsiboka River was a suspension bridge with a 146m-long central span and a 4m-wide deck that was built in 1932. In September 1942 British commandos attacked Vichy-controlled forces on Madagascar who cut the cables, thus bringing down the bridge and leaving it lying in water only a metre deep, and Royal Engineers quickly had it usable again. The current structure, the longest bridge on the island of Madagascar, is a two-lane suspension bridge carrying the RN4 road. Its overall length is 350m and it spans between two steel towers.

Bévera Bridge, Sospel, Alpes-Maritimes, France

This railway bridge near Sospel, which crosses diagonally over the Bévera River, was built in 1962 to replace an earlier structure from 1927. It consists of a main steel plate girder 91m long

Bevera Railway Bridge

which is supported at its mid-point directly over the river by a concrete arch that itself crosses the river at right angles.

Bevera Viaduct, Varese, Italy

The reinforced concrete structure here, completed in 2014 to carry the new 8km-long Arcisate-Stabio Railway linking Italy and Switzerland, has seven 62m-long spans. These are supported by 12m-high piers that are elliptical in plan and flare out distinctively at the top.

Bhumibol Bridges, Bangkok, Krung Thep Maha Nakhon, Thailand

Bhumibol Bridge carries the city's seven-lane Industrial Ring Road across a loop of the Chao Phraya River. It actually consists of two very similar cable-stayed structures (Bhumibol 1 and Bhumibol 2) running north to south that are separated by a major elevated interchange junction with another road off to the west. The overall lengths of the bridges are 702m and 582m with main spans of 398m and 326m respectively, and the decks of both are 50m above water level. The diamond-shaped pylons, which are 173m and 164m high, are topped by golden domed caps. The reinforced concrete structures were designed by engineers Mott MacDonald and opened in 2006.

Bhumibol Bridges

Birchenough Bridge, Chipinge, Manicaland, Zimbabwe

The British firm Dorman Long built the bridge here across the Sabi River to a design

Birchenough Bridge

In 1909 it was dismantled for re-erection in the French port of Brest (but, following war damage in 1944, it was demolished in 1947). The present road bridge over the canal was built in 1980. It is a simple opening bascule bridge that consists of a 105m-long Warren truss girder which, when closed, is 15m above water level.

Black Dragon Pool Bridge, Lijiang, Yunnan, China

This little white marble bridge was built in 1737 and has five segmental arches under a gently hump-backed overall profile. At one end is a charming three-tiered Chinese pavilion and the whole scene is overlooked by the Jade Dragon Snow Mountain.

Black Dragon Pool Bridge

by Ralph Freeman and it was opened in 1935. It is a 1,080ft-span two-pinned steel arch and originally had an 18ft-wide road between two footways. Interestingly, because of the high cost of transporting the steelwork from the UK to the middle of Africa, much of the erecting plant and temporary works were built from pieces that later became part of the bridge itself. The bridge was widened and strengthened in 1984 but, since 2000, has been restricted to vehicles weighing less than 25 tonnes. The Zimbabwean Government is planning to build a replacement.

Bizerte Bridge, Tunis, Tunis Governorate, Tunisia

A transporter bridge with a main span of 109m was built over Tunis's Bizerte Canal in 1898.

Bizerte Bridge

Blenheim Covered Bridge, Blenheim, New York, USA

Built in 1855 over the Schoharie Creek with a 232ft-long span, this was claimed to be the world's longest single-span covered bridge. Consisting of three parallel Long trusses, the double-barrelled structure was built in 1855 by Nicholas M. Powers. It was 26ft wide and 30ft high and was designated a National Historic Civil Engineering Landmark in 1964. In 2011 the bridge was swept away by floods but a replica was built and opened in 2018.

Blenheim Covered Bridge

Bloukrans Bridge, Knysna, Western Cape, South Africa

This single-span reinforced concrete arch bridge on South Africa's famous Garden Route was opened in 1983 and carries the N2 highway 216m across Van Stadens Gorge above the Bloukransrivier. It has an overall length of 451m and the arch itself, which spans 272m, was built as a pair of stayed cantilevers until the final pour closed it at its crown. The deck is supported by cross walls standing on the arch and the space above its centre, which is accessed by a separate zipline from the side of the ravine, is used for one of the world's highest bunjee jumps.

Bloukrans Bridge

Blue Bridge, St Petersburg, Leningrad Oblast, Russia

The Blue Bridge over the Moika River in St Petersburg links St Isaac's Cathedral and the City Hall. The first section, a cast iron structure spanning about 33m, was built in 1818 to a design by the architect William Heste and was widened in 1844 to carry most of St Isaac's Square. Now 97m wide, it is the widest bridge in the city.

Blue Nile Bridges, Goha Tsion, Oromia, Ethiopia

The old Italian-built Haile Selassie road bridge here across the Blue Nile was completed in 1947. It is 205m long with a thick three-ribbed concrete arch spanning 125m that supports, in each spandrel, three sets of triple-column piers. This bridge is now only used by pedestrians and a new Bahir Dar Abay structure, built by the Japanese, was completed in 2008. This is 12m wide and, at 303m long over three spans, was until recently the longest bridge in the country. The 145m-long extradosed cable-stay main span is supported by two distinctive 'tuning fork' pylons, with each leg leaning out from the vertical. These reach to 14m above the deck level, 69m above river level. The bridge now carries National Highway 3 between Addis Ababa and Bahir Dar.

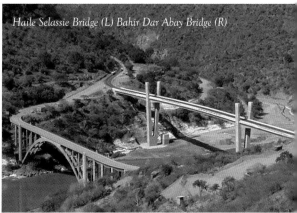

Haile Selassie Bridge (L) Bahir Dar Abay Bridge (R)

Bogdan Khmelnitsky Pedestrian Bridge, Moscow, Moscow Oblast, Russia

Initially called the Kiev Footbridge, this structure was built in 2001 and uses a steel arch previously incorporated in the railway part of the Krasnoluzhsky Bridges (q.v.) to span between massive masonry piers over the Moskva River. It is now named after a Ukrainian hero. The two-pinned crescent steel arch supports a deck at mid-height between the arch springings and the crown. Most of this deck is enclosed by glass, although there are open-air footways at each side. The glazing for the outer third of the span at each end is supported by stays from the tops of the piers. The enclosed space is accessed from embankment level by escalators and houses souvenir booths and the like.

Bogibeel Bridge, Dibrugarh, Assam, India

Opened in 2018, the Bogibeel Bridge carries a twin-lane road above a double-track broad-gauge railway across the Brahmaputra River. The Warren-trussed structure has an overall length of 4,940m and has thirty-nine 125m-long main spans. The bridge was built on previously prepared foundations during the four-month dry season over sixteen years between 2002 and 2018.

Bogibeel Bridge

Bogoda Wooden Bridge, Badulla, Uva, Sri Lanka

The 50ft-long, 6ft-wide Bogoda Bridge is a two-span structure with a central pier and crosses the Gallander Oya River. It was built entirely of timber, including wooden nails, in the sixteenth century, although some consider it to be the world's oldest surviving timber bridge. It has a

Bogoda Wooden Bridge

clay shingled roof that is supported on eleven pairs of timber posts and stabilised by wire guy ropes.

Bolzano Museum Footbridge, Bozen, South Tyrol, Italy

The Talvera River in Bozen is spanned by two separate steel girder structures, each 3m wide and spanning 52m, that provide pedestrian and bicycle access to the Bolzano Museum. Notable for their free-flowing shapes, the bridges were designed by KVS Berlin and opened in 2007.

Bombas Bridge, Bombas-Jijiga, Somali, Ethiopia

Completed by the Italians in 1940, this structure has a thick, reinforced concrete, parabolic arch spanning 30m that supports three tall masonry spandrel arches on each side. Behind the abutment there is a further masonry arch, making the bridge's overall length 59m. An

Bombas Bridge

unusual feature of the bridge is the series of upward projections to the parapet walls. In 2011 a new three-span bridge was built a short distance away. This is 11m-wide with a main span of 27m flanked by smaller 16m span and consists of four parallel reinforced concrete beams standing on masonry piers.

Bonabéri Bridge, Besseke, Littoral Province, Cameroon

The first crossing over the tidal estuary of the Wouri River, a road/rail bridge consisting of prestressed concrete beams, was built by the French between 1951 and 1954 to link the port of Douala with Bonabéri. At more than 1,200m long, it was claimed to be the longest bridge in central Africa and carried the N3 Highway. Despite being upgraded in 2004, it was proving increasingly inadequate and a second bridge was opened in 2018. This slightly curved new structure is 820m-long and 34m-wide and carries two rail tracks and five road lanes.

Boone Viaduct, Boone, Iowa, USA

The Chicago & North Western Railway's double-track railway viaduct at Boone, which was designed by George S. Morison, was built between 1899 and 1901. Although one of the longest and highest railway bridges in the USA, it is chiefly known because it commemorates the extraordinary heroism of a 15-year-old girl. In 1881, following a thunderstorm that caused a flash flood, an engine sent out to check the condition of the railway track brought down a local bridge in a great crash heard by Kate Shelley. Knowing that an express train was due later, she investigated and found two survivors of the four-man crew in the wreckage. She immediately set out to bring help, having to cross another bridge on hands and knees during her two-mile journey after her lantern went out. Her bravery was immediately acknowledged but, later, when the Boone Bridge was rebuilt, it became popularly known as the Kate Shelley Bridge. The structure, which has an overall length of 2,685ft and a maximum height of 185ft, was closed in 2009.

Borovnica Railway Viaduct, Ljubljana, Province of Ljubljana, Slovenia *

When it was completed in 1856 as part of the railway line between Vienna and Trieste, this was the largest masonry bridge in Europe. Founded on more than 4,000 oak piles, the two-tiered curved structure had twenty-two arches in the lower level with twenty-five above and was 561m long and 38m high. It was partly destroyed early in WWII by the Yugoslav army, following which the Italians replaced the missing sections with steelwork, and was finally rendered unusable by Allied bombing raids. Apart from a single remaining pier, the bridge had been completely removed by 1950.

Borovsko Bridge, Borovnice, Central Bohemian Region, Czech Republic *

The Borovsko Bridge across a tributary of the Zelivka River was originally built to form part of a Trans-European Highway south of Prague and was completed in 1950. However, the remainder of the project was cancelled and, in 1976, the river was dammed to form a reservoir. The 100m-high structure is now almost completely submerged and access to the area is prohibited.

Bosphorus Bridge, Istanbul, Marmara Region, Turkey

Opened in 1973, this 1,560m-long suspension bridge crosses the strait between the Sea of Marmora and the Black Sea and is famous for providing the first fixed link between the continents of Asia and Europe. It was designed by Freeman Fox & Partners, its main span is 1,074m long, the pylons are 165m high and the clearance between water level and the underside of the deck is 64m. In 2018, the bridge was officially renamed **15th July Martyrs Bridge** to mark the defeat of the 2016 attempted coup in which more than 250 people were killed and 2,000 wounded. A picture of the bridge has illustrated a Turkish 1,000-lira banknote.

Bosphorus Bridge

Boukei Bridge, Hokkaido, Japan

The concrete footbridge that crosses the Shimazaki River within parkland was built in 2011. The under-deck cable-stayed bridge has two spans, 23m and 33m long. A single external prestressing cable anchored in each abutment passes beneath the deck and supports it on V-shaped steel strutting units. Between the spans the cable passes above the intermediate pier within a solid concrete finback.

Boutiron Bridge, Vichy, Allier, France

The deck-arch road bridge over the River Allier at Vichy was designed by Eugène Freyssinet and opened in 1913. It has three slightly-pointed reinforced concrete arch spans, the central and biggest being 72½m long. Warren trusses, also in reinforced concrete, are integrated with these arches to support each side of the deck, the edges of which are protected by delicate concrete parapets.

Boyne Viaduct, Drogheda, County Louth, Ireland

The 1,760ft-long viaduct across the River Boyne, which carries the Dublin & Belfast Junction Railway, was completed in 1855. Designed by Sir John Benjamin MacNeill, it has three main wrought iron girder spans: a hog-backed central one spanning 267ft that is flanked on each side by a 141ft-long parallel boom span. These are approached by twelve stone approach arches from the south and three from the north. The structure underwent major reconstruction in 2015. The bridge is considered to incorporate the first large-scale use of wrought iron lattice girders and is on the ASCE List of Historic Engineering Milestones.

Boyne Viaduct

Brazzaville-Kinshasa Bridge, Brazzaville, Republic of the Congo

This planned road-rail bridge across the Congo River will provide a second connection between the country and the Democratic Republic of the Congo, the only other link being the 1983 Matadi Bridge (q.v.). The structure will have an overall length of 1,575m and include two main cable-stayed spans that will be 242m and 152m long. Construction, which will last about six years, was scheduled to start in about 2021.

Briare Aqueduct, Briare, Loiret, France

This 662m-long structure at the north end of the Canal Latéral à la Loire crosses 11m above the River Loire and the 1643 Canal de Loire and has been described as the 'most beautiful … metal aqueduct in Europe'. Its masonry piers support fifteen 40m-long spans of a 6m-wide steel channel. Curved brackets outside this channel carry paved towpaths producing an overall width between parapet railings of 11.5m. As well as elegant lamp posts along the line of the parapets, there are also ornamental columns at each corner of the aqueduct where it joins the overland waterway. The aqueduct was built between 1890 and 1896 to a design by Gustave Eiffel's company and was the world's biggest until Magdeburg Water Bridge (q.v.) was opened in 2003.

Briare Aqueduct

Bridge Between Two Continents, Midlina, Reykjanes, Iceland

This simple 15m-long steel joist footbridge spans the Alfagja rift valley, which is the gap between the tectonic plates beneath Europe and North

America. Previously, the bridge had been called Leif the Lucky Bridge, after the explorer Leif Ericson who was the first Icelander to reach North America.

Bridge of Apollodorus, Drobeta-Turnu, Mehedinti, Romania *

Although long since destroyed, this Roman structure remains perhaps one of the most famous bridges ever built. It was built between 103 and 105CE by Apollodorus of Damascus, a military engineer and architect, for the Emperor Trajan's crossing of the Lower Danube east of the Iron Gates into Dacia, and is commemorated on Trajan's Column in the Forum at Rome. The bridge is recorded as having had a main section of twenty-one segmental timber arches each spanning about 32m between masonry piers, but also as being over 1,100m long and 15m wide. The bridge was shown on coins of three different denominations that were struck in about 110CE.

Bridge of Augustus, Narni, Terni, Italy

The Roman Emperor Augustus directed that this bridge should be built to carry the Flaminian Way over the River Nera and it was completed in about 27BCE. Originally, it was 520ft long and nearly 100ft high with four semicircular arches and was one of the Romans' largest bridges. However, it had suffered earthquake damage over the years, including in 1855 which left just one 60ft-span arch standing, and another earthquake in 2000 caused further damage.

Bridge of Boyacà, Tunja, Boyacà, Colombia

The very normal-looking little bridge here was built to commemorate the Battle of Boyacà in 1819 which resulted in New Granada gaining its independence. The structure is a simple stone arch over the Teatinos River with white-painted spandrels and parapet walls.

Bridge of Elia, Kaminaria, Limassol District, Cyprus

The Venetian Bridge of Elia was probably built in the sixteenth century to allow camel trains to cross over the Diaryzos River on their journey between Troodos and Paphos. Its 2.4m-wide roadway is carried on a single pointed arch spanning 5.5m.

Bridge of Immortals, Tangkouzhen, Anhui, China

This bridge, built in 1987, leads to a rock cave high up in Mount Huangshan (Yellow Mountain). To reach it, it is necessary to walk along a pathway attached to the side of a cliff, from which many locals have fallen to their deaths – hence the name. The bridge itself is a short masonry arch bridge spanning perhaps 25ft between two cliff faces and, at an elevation of 1,320m (4,330ft) above sea level, is claimed to be the highest-located bridge in the world. Mount Huangshan is a UNESCO World Heritage Site.

Bridge of Immortals

Bridge of Boyacà

Bridge of Lies, Sibiu, Transylvania, Romania

An earlier bridge here was a creaky old timber structure, the movements of which gave rise to the belief that it would collapse if anyone told a lie while crossing over it and, when it was replaced by a stronger structure between 1859 and 1860, the old name stuck. The new bridge was designed by

Friedrich Hütt and carries a footway between the old and new towns over a lower road. It is 10m long and 6m wide and has four segmental cast iron arch ribs. There are also decorative cast iron lamp standards and wrought iron railings.

Bridge of No Return, Kaesong, North Hwanghae, North Korea

The footbridge that connects North and South Korea across the Military Demarcation Line between the two countries is more of political engineering interest than that of civil engineering. The bridge was given its name when it was used for exchanges of prisoners-of-war between the two Koreas from 1953 to 1976 following the Korean War of 1950–1953. Prisoners could opt either to remain in their country of imprisonment or return across the bridge to their homeland, but they could not later change their minds.

Bridge of Peace, Tbilisi, Georgia

The distinctive pedestrian Bridge of Peace across the River Mtkvari, which was designed by the Italian architect Michele de Lucchi and opened in 2010, links the old town of Tbilisi to a modern park. It consists of a 150m-long doubly-curved space frame, made of tubular members, that is arched both along its length and across its width and is supported at just two points on each side of the river. This framework is clad in glass panels and diagonal hangers from it support a narrower lightweight deck that also includes glass panels, and the side panels beneath the edge handrails are of glass too. The whole structure is illuminated with 30,000 LEDs embedded in the roof and side panels.

Bridge of Remembrance, Christchurch, Canterbury, New Zealand

The architectural firm Gummer & Prouse won a national competition for the design of this bridge, which is dedicated to the dead of World War I and also as a memorial to those who served in conflicts since. The bridge itself crosses the Avon River with a 15m-span semi-elliptical arch supporting the road and the structure's most significant feature is the tall masonry gateway arch standing on the abutment. The bridge, which is a Category I heritage structure, was badly damaged in the Christchurch earthquake of 2011 and was not re-opened until 2016.

Bridge of Shadman Malik, Samarkand, Samaquand, Uzbekistan

Built in about 1502 to replace an earlier structure destroyed by floods, this bridge across the Zeravshan River, the third longest river in Central Asia, now consists of three arches, roughly at 120° to each other in plan. Two of the arches are straight-sided and the other is a traditional pointed arch. Some authorities suggest that the bridge once had seven arches and was 200m long. Confusingly, other sources suggest that the structure was an aqueduct and that water flowed over one arch onto the structure and then into separate distribution channels over the other two arches. Nevertheless, it remains a most unusual historic structure.

Bridge of Shadman Malik

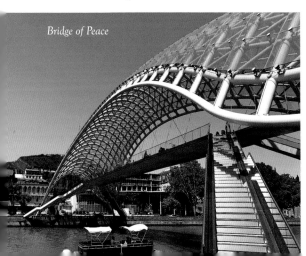
Bridge of Peace

Bridge of Sighs, Venice, Veneto, Italy

The world-famous Bridge of Sighs in Venice was built in 1603 to provide a secure crossing over the Rio de Palazzo Canal between the Doge's Palace and the city-state's prison, its name reflecting the despair of those forced to use it. Externally, though, it is very decorative, with a semi-elliptical arch spanning 11m between the buildings, the ring itself being decorated with carved stone faces. Above, there are two carved grillage openings to light the internal passageway and the segmentally arched parapet is decorated with scrolling ornamentation. The architect was Antonio Contino.

Bridge of Sighs

Bridge of the Americas, Balboa, Panama City, Panama

This steel through-arch road bridge, which has an overall length of 1,655m over fourteen spans, crosses the channel that leads from the Pacific to the port of Balboa and then on to the Panama Canal. Lattice girders in the approach spans deepen in the spans immediately adjacent to the main span and flow into a 344m-long arch. The road deck is 10.4m wide and the minimum clearance beneath it is 61m at high tide. The bridge was built between 1959 and 1962 and now carries the four-lane Pan-American Highway. Since then, two further bridges have been completed and work has started on another – see Panama Canal New Bridges.

Bridge of the West, Olaya, Western Antioquia, Colombia

This suspension bridge across the Cauca River was built between 1887 and 1895. It has four main suspension cables, the 1.2m-wide footway on each side of the bridge being hung from twin cables. The supporting cross beams beneath the footways continue across the full width of the bridge to support the single-lane 3.1m-wide road deck. The saddle for each pair of suspension cables is at the top of a 11m-tall pyramidal tower that itself has a pyramidal roof over the saddle. The underside of the deck near the suspension towers is also stayed back to the river banks. It is claimed that, when the bridge was completed, its span of 291m made it the longest suspension bridge in Latin America and third longest in the world. The structure has been submitted for consideration as a UNESCO World Heritage site.

Bridge Pavilion, Zaragoza, Aragon, Spain

Designed by Iraqi-British architect Zaha Hadid with engineers Ove Arup to form an entrance for the Expo 2008 in Zaragoza, this 270m-long enclosed and slightly-curved structure contains a footbridge and multi-level exhibition halls. These are supported by four main trusses that span 120m across the Ebro River via a mid-channel island. The building is clad in glass-reinforced concrete panels.

Bridge of the Americas

Bridge Pavilion

Bridgeport Bridge, Bridgeport, California, USA

This combination structure of Howe truss and timber arch was built in 1862 and, with its 208ft-long span, the bridge is claimed to be one of the world's longest single-span wooden covered bridges. Built to provide a road crossing over the South Yuba River, although now restricted to pedestrians only, it was designed by David Inglefield Wood and is a National Historic Civil Engineering Landmark.

Bridges of Eden, Port Vila, Shefa Province, Vanuatu

This tourist attraction offers a walk through tropical gardens and over three spindly suspension bridges across the Rentapau River before finishing at a mini zipline.

Britzer Garden Footbridges, Berlin, Brandenburg, Germany

The Britzer Gardens were originally established for a horticultural show in 1985. Crossing the water features are a number of timber bridges with simple beams spanning between piers, each of which consists of a pair of tall, narrow A-frames.

Bronislaw Malinowski Bridge, Grudziadz, Kuyavian-Pomerania, Poland

This bridge of eleven consecutive steel through trusses, each spanning 100m, carries both Highway 16 and a railway line across the Vistula River. It was built between 1947 and 1951.

Brooklyn Bridge, New York, New York, USA

The construction of this bridge is one of the most heroic sagas in the annals of bridge building. John Roebling had migrated to America from Germany and began making twisted wire ropes to replace hemp ropes before patenting the idea of making parallel wire cables in 1841. He used these on suspended aqueducts and then on the first bridge at Niagara (q.v.). In 1867 he was appointed chief engineer for the design and construction of the Brooklyn Bridge to provide a road connection between New York and Long Island across the East River. He died following an accident while he was setting out the position where the Brooklyn

Brooklyn Bridge

pier was to be built and was succeeded by his son Washington. Washington himself suffered caisson disease from working too long under high pressure during the construction of the New York pier foundations and, partially paralysed, supervised the remainder of the work from his house overlooking the site, relaying instructions through his wife Emily who also reported back to him on those site problems that could only be resolved by him. The structure itself, which was completed in 1883, is an 85ft-wide hybrid cable-stay / suspension bridge spanning 1,596ft between neo-Gothic stone towers and with 930ft-long side spans. Above the main deck there is also an elevated central promenade for cyclists and pedestrians. The bridge is included in the ASCE list of historic bridges.

Buchan (Stepping Toad) Bridge, Qingyuan, Zhejiang, China

This bridge was first built between 1403 and 1424, and rebuilt in 1917. The semicircular stone arch spans 18m but the full length of the enclosed timber corridor is 52m. In the centre of the main roof is a square raised pagoda-style roof.

Bunlahinch Clapper Footbridge, Louisburgh, County Mayo, Ireland

This unusual 164ft-long footbridge has thirty-eight spans and consists of flat stone slabs about 2ft wide, which span 3-4ft between small piers of boulders. There is a low protective wall on one side. The bridge was built in the 1850s by John

Bunlahinch Bridge

Alexander so people could cross a wide, shallow ford to reach a local church without getting wet and it is Ireland's longest clapper bridge.

Burgo Bridge, Pontevedra, Galicia, Spain

The first bridge over the River Lerez here was built by the Romans and it is considered that the eleven semicircular arches at the heart of the existing structure are reconstructions of the Roman originals. However, the bridge has been widened by the addition of later shallower segmental arches spanning between the pointed cutwaters at the ends of the piers. The bridge has been further widened by the construction of the parapet railings on corbelled extensions. The centre of each spandrel is enlivened by a carved scallop shell.

Burr Covered Bridge, Schenectady, New York, USA *

The timber covered bridges of North America have a special place in the story of bridges and bridge-building but, sadly, nearly all of them have long since disappeared. This entry (and that for Colossus Bridge – q.v.) represent those that have gone. The 997ft-long Burr Bridge over the River Mohawk connecting Scotia and Schenectady was completed in 1808. This was initially an uncovered truss structure developed by Burr that spanned between large masonry piers standing in the river. However, intermediate piers had to be added when the deflections under load became unacceptably large. Just like roof trusses in barns, these trusses included tension members

made of timber. In order to protect the trusses from the weather, and thus reduce the risk of rot, the superstructure was partially covered with protective sidings and roof making it look like a series of barns and these structures, too, were added to and altered during their working lives. In 1871 an iron bridge replacement was opened but this fell victim to age and was removed in 1936.

Butterfly Bridge, Copenhagen, Capital Region, Denmark

This bridge for cyclists and pedestrians in Copenhagen's Christiania docklands area is Y-shaped in plan with three arms that project out from a central fixed island. The shortest of these is a fixed span and the other two have independently-operated bascule arms hinged at their island ends which, when both are raised, look like the wings of a butterfly. Each of the two lifting arms is 23m long and the clear width available for boats when a bridge is open is 15m. The structure was designed by Dietmar Feichtinger Architects and completed in 2015.

Butterfly Bridge

Buzzards' Bay Bridge, Bourne, Massachusetts, USA

The single-track vertical lift railway bridge here over the Cape Cod Canal was completed in 1935 to replace a bascule bridge from 1910. The towers at each end of the 2200-ton 544ft-long lifting span are 271ft tall and the clearance below is 135ft when the span is fully raised. Each tower is topped by a pointed pinnacle supporting a globe. The bridge underwent a major upgrade between 2002 and 2003.

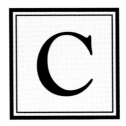

C

C.H. Mitchell Bridge, Port Edward, Kwa-Zulu-Natal, South Africa

This elegant crescent arch steel bridge was completed in 1966 to carry the R61 road over the Umtamvuna River. Its span of 206m makes it the second longest in the country (after Bloukrans Bridge – q.v.). An interesting feature is that it was constructed from special corrosion-resistant steel. In 2002 the bridge was damaged by two terrorist bombs but was re-opened after only three weeks.

Cabin John Aqueduct, Cabin John, Maryland, USA

The Cabin John Aqueduct, designed by Captain Montgomery Meigs to bring drinking water into Washington DC, was first known as the Union Arch and is sometimes called the Washington Aqueduct. The 450ft-long aqueduct's main segmental arch over Cabin John Creek has a span of 220ft with a rise of only 57ft and it was the longest masonry arch in the world between its completion in 1864 and 1903. Unusually, the spandrel areas include a second stepped arch ring immediately above the main ring of voussoirs. The MacArthur Boulevard is carried on a 17ft-wide deck above the waterway trough which, at peak flow, transported 40,000,000 gallons of water a day. The structure was designated a Historic National Civil Engineering Monument in 1972 and was placed on the National Register of Historic Places the following year.

Cadore Viaduct, Caralte, Belluno, Italy

Structurally, the main part of this crossing 184m high over the Piave River is a relatively simple three-span bridge consisting of a beam supported by A-shaped diagonal struts founded on high-level benches 255m apart on the ravine sides. However, the mechanics used in the erection of the bridge are particularly interesting. The 85m-long struts were fabricated horizontally behind the ravine faces, pushed forward to project over the ravine before being rotated into a vertical position and lowered onto their bottom hinged foundations. They were then jibbed outward to their final positions, restrained by temporary ties, and finally the box-girder deck, also pre-assembled behind the ravine, was pulled into place. Built between 1982 and 1985, the bridge carries the SS51 road, has an overall length of 535m with a longest span of 255m and its deck is 13m wide. It was designed by father and son Pietro and Giuseppe Matildi.

Cadore Viaduct

Campo Volantin Footbridge, Bilbao, Biscay, Spain

This steel pedestrian bridge over the Neruion River was designed by the engineer/architect Santiago Calatrava and completed in 1997. The main structural element is a parabolic arch, inclined to the vertical, that spans 75m with a rise of 15m. This arch has stays attached to it that support a longitudinal tube outside each edge of the 75m-long curved deck. These two tubes, in

Campo Volantin Footbridge

Cape Creek Bridge

turn, are each connected by welded brackets to a larger tube beneath the centre of the deck. The deck itself includes controversial glass bricks.

Canakkale 1915 Bridge, Lapseki, Cannakkale, Turkey

Under construction since 2017 and expected to open in the early 2020s, this bridge across the Dardanelles strait will be the longest suspension bridge in the world with a main span 2,023m long and a length with side spans of 3,563m. Together with its approach viaducts, the overall length of the crossing is 4,608m. The main towers are 318m high and the 45m-wide reinforced concrete deck, which is supported on steel box beams, will carry six motorway lanes. The bridge was designed by COWI. The bridge's name refers to the Gallipoli campaign of the First World War.

Cape Creek Bridge, Eugene, Oregon, USA

Designed by Conde Balcom McCullough and completed in 1932 to carry US Route 101, this is one of the more unusual reinforced concrete bridges of its time. Its main span is a parabolic arch 220ft long and this is flanked by two tiers of semicircular arches, those in the upper tier being half the span of those below in the manner of a Roman aqueduct. Alternate upper-tier bays and all those in the lower tier have

transverse X-braces. The overall length of the structure is 619ft and its 27ft-wide deck is on a gradient. In 1991 the bridge was one of the first in the world to have a special impressed-current cathodic system installed to protect the internal reinforcement from rust.

Capilano Suspension Bridges, Vancouver, British Columbia, Canada

The first Capilano Suspension Bridge, built by the Scottish civil engineer George Grant Mackay in 1889, had hemp ropes supporting a timber plank deck but this was replaced by a wire cable structure in 1903. The present 140m-long footbridge was built in 1956 and is now part of a park walk consisting of several suspension footbridges. The latest of these, opened in 2011, has a ½m-wide deck, hung off the cliff face by radiating tension cables, that forms the 210m-long nearly-semicircular and 90m-high cliff walk.

Capilano Suspension Bridge Cliff Walk

Captain William Moore Bridge, Skagway, Alaska, USA

This 110ft-span suspension bridge was built in 1976 to carry the Klondike Highway over the active earthquake fault-line that runs through the Moore Creek Gorge. The bridge has a single pylon on one side of the gorge that is inclined over the gorge and supports the full span of the bridge deck on three pairs of stays that are balanced by four pairs of backstays behind the pylon. This means that the bridge cannot be affected by different ground movements on each side of the gorge. However, because of difficulties with the pylon foundations, there are now plans to build a culvert for the Moore Creek water and backfill above it to ground level for the road. The bridge is named after a local riverboat captain and explorer.

Caracas-La Guaira Highway Viaducts, Caracas, Greater Caracas, Venezuela

At the time the Simon Bolivar airport at La Guaira was opened in 1953, the four-lane highway linking it to Venezuela's capital city of Caracas was considered to be the most important civil engineering work in South America since completion of the Panama Canal in 1914. The project included three major reinforced concrete arch viaducts across deep valleys, with the longest spans of each being 152m, 146m and 138m. Unfortunately, though, the Caracas earthquake of 1967 caused the foundations of Viaduct 1 to move, a difficult temporary detour was built and the bridge finally collapsed in 2006. A 900m-long replacement viaduct was opened in 2007.

Caracas-La Guaira Highway Viaduct under construction

Caramel Viaduct, Castillon, Alpes Maritimes, France

The tramway between Menton and Sospel was built in 1914 and, in order to cope with the difficult route through the mountains, had a steep ruling gradient of 6%. One of the main structures was the 120m-long Caramel Viaduct with eleven stone arches. A feature of it was that the line here curved back on itself like an omega in plan. The tramway closed in 1931.

Caramel Viaduct

Caravan Bridge, Izmir, Smyrna, Turkey

This bridge across the River Meles is a single segmental stone arch spanning 8½m. Listed by Guinness World Records as the oldest bridge in the world that is still in use, this structure was built in about 850BCE. (The 4,000-year-old baked-brick foundations at Girsu Bridge, Tello, West Azerbaijan, Iraq (q.v.) are considered to be the remains from the world's oldest bridge.)

Cardona Bridge, Cardona, Barcelona, Spain

This bridge over the Cardena River is included as a sop to all central and local government officials who, for whatever reason, have failed to complete a long-cherished infrastructural project. Work on the structure was started before 1350 and two 1m-thick arch rings were built, one spanning 25m, the other 21m. However, they have stood, unfinished, for more than six and a half centuries and nobody now seems to know why the work was stopped never to be restarted.

Carioca Aqueduct, Rio de Janeiro, Brazil

Work on building the first aqueduct here, to provide the capital city with fresh water from the Carioca River, lasted from 1706 to 1723 but major reconstruction was needed only twenty-one years later. The new structure, completed in 1750, has a double-decked arcade and its design is based on the Aguos Livres Aqueduct in Portugal (q.v.). Below the springing level of the arches the piers are wider and the piers for the lower level arches are also battered out on one side. Altogether, the aqueduct is 270m long, with forty-two arches, and up to 18m high. A distinctive feature is the large circular opening at the top of the piers. The aqueduct, which is now commonly called Arcos da Lapa (Lapa Arches), has been used for trams since 1900.

Carioca Aqueduct

Caroni Delta Bridge, Piarco, Saint George County, Trinidad & Tobago

An early road bridge across the Caroni River, built in 1873, was a single-lane iron structure and in 1997 a new temporary bridge was erected alongside to double capacity. An 81m-span Mabey Delta Bridge replacement for both structures was push-launched over the river in 2007.

Carpentras Aqueduct, Vaucluse, Vaucluse, France

The masonry aqueduct at Carpentras was built between 1720 and 1734 to replace a fourteenth-century aqueduct. It is 720m-long with forty-seven individual semicircular arches, which range in span from 12m to 25m, and has a maximum height of 24m. It is also noticeably narrow, being only 1.75m wide with a water channel just 0.25m wide, although the piers splay out below the springing level of the semicircular arches to give increased stability. At its south end, where the aqueduct crosses the River Auzon, a secondary structure built against the face of the aqueduct carries a footpath over the river.

Carpentras Aqueduct

Carrick-a-Rede Rope Bridge, Ballintoy, County Antrim, Northern Ireland

Fishermen have been using suspension bridges to reach the island of Carrick-a-Rede from the mainland for more than 350 years and the current wire rope one dates from 2008. Its span is 20m and the timber deck is about 30m above sea level. The bridge is owned and maintained by the National Trust.

Carrollton Viaduct, Baltimore, Maryland, USA

Considered to be the oldest railroad bridge in America, this bridge across the Gwynns Falls stream was built by the Baltimore & Ohio Railroad in 1829. The bridge is 312ft long and its single semicircular masonry arch, which spans 80ft, carries the railroad's twin tracks 65ft above water level. In 1971 the structure was listed as a National Historic Landmark and became a Historic Civil Engineering Landmark in 1982.

Castle Bridge, Kamianets Podilsky, Khelmnytskyi, Ukraine

An early bridge here 80ft above the Smotrych River connected the rocky citadel of the Old City to the rest of the mainland, and the fourteenth-century castle was subsequently built to protect it. This bridge consisted of seven timber trusses standing on stone piers. Following their siege in 1672 the Turks then rebuilt the bridge in 1687, replacing the timber trusses with stone arches, and gave it defensive fortifications. However, between 1841 and the end of the nineteenth century, it was reported to be in a poor condition. Later, a steel plate girder structure was built, with the steelwork standing on extensions above the original piers. The current bridge, which carries a two-lane carriageway and footways, suffered earthquake damage in 1986 and was included on the World Monuments Watch in 2000.

Caudal Bridge, Mieres del Camino, Asturios, Spain

Built in 1968 over the River Caudal, this 70m-span concrete arch was designed by Fernandez Casado. It is unusual because it includes short open back-arches that end on each bank at the vertical wall where the bridge deck meets the approach embankments. The whole spandrel area between the embankment end walls is in dazzling white concrete.

Cernavoda Bridge, Cernavoda, Constanta, Romania

Opened in 1987 alongside the existing Anghel Saligny Bridge (q.v.), which was then closed, this Warren truss bridge consists of a central section for the railway line between Bucharest and Constanza, on each side of which is one of the two-lane carriageways of the A2 motorway. The bridge has a total length of 2,622m, with a longest span of 190m, and is 33m wide. It was upgraded in 2003.

Cernvir Covered Bridge, Pernstejn, South Moravia, Czech Republic

The picturesque covered wooden bridge across the Svratka River was first built in 1718 and is considered to be the oldest wooden bridge in Moravia. The two-span structure is 32m long and 2.6m wide and has a massive stone pier in mid-river. The pitched roof is partly hipped over the entrances, is clad in shingles and the height to its ridge is 4m. A picture of the bridge was used to illustrate the Czech Kc11 stamp in 1999.

César Gaviria Trujillo Viaduct, Pereira, Risaralda Department, Colombia

This bridge over the Otun River is a 440m-long cable stay structure supported by rhombus-shaped piers. It is 26m wide and, when it was completed, its central span of 211m was the longest bridge of its type in South America. In the first eight years after the bridge was opened in 1997 there were about 88 suicides until protective barriers were built.

Pereira Viaduct

Chacao Channel Bridge, Chiloé, Chiloé Province, Chile

Work started in 2018 on a suspension bridge that will link Chiloé Island to mainland Chile, with completion expected in 2023. Clearance above water level will be 59m and the overall length of the structure, at 2,750m, will make it the longest bridge in South America.

Chain Bridge, Newburyport, Massachusetts, USA

The first bridge here over the Merrimac River was a timber arched-truss built by Timothy Palmer in 1792 that connected the river's right bank to Deer Island. This was replaced in 1810 by a ten-chain suspension bridge that spanned 244ft between stone towers. These were triangularly-shaped in elevation and had two separate carriageway arches. The structure was designed by James Finlay and was long considered to have been the world's oldest chain bridge that was still standing. In 1909 it was replaced by a new suspension bridge spanning 225ft between support pylons that have been clad in timber to act as gateway portals.

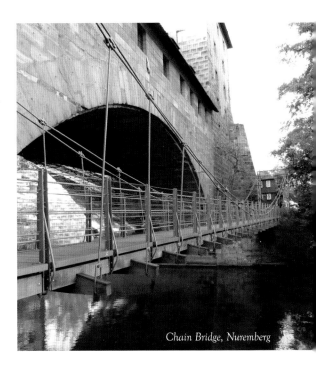

Chain Bridge, Nuremberg

Chain Bridge, Newburyport

Chain Bridge, Nuremberg, Bavaria, Germany

The footbridge here crosses the River Pegnitz just upstream from where the city walls also cross. An earlier two-span timber covered bridge had been painted in the early sixteenth century by Albrecht Dürer. The present chain suspension footbridge (called the Kettensteg in German), which was designed by Conrad Georg Kuppler and built in 1824, is claimed to be the oldest iron suspension bridge in continental Europe. It has two 33m-long spans and an overall length of 68m. In 2010 the old structure was completely restored, although with the addition of new stabilisation integrated into the timber deck.

Chains Bridge, Fornoli, Lucca, Italy

After an earlier stone bridge over the Lima River, dating from 1317, had been swept away by floods in 1836 the duke of Lucca commissioned the architect Lorenzo Nottolini (1787–1851) to design a replacement, although it was not completed until 1860 because of political difficulties nationally. It has two classical-style triumphal archways between which hang two double-chains made from long wrought iron links. The 50m-long deck consists of timber planks. The bridge was badly damaged in 1945 but rebuilt soon afterwards. However, after it was placed on the World Monuments Watch in 1999, it was fully restored in 2006.

Chamberlain Bridge, Bridgetown, Antigua and Barbuda

British settlers first arrived in the Bridgetown area in 1628. Here they found a primitive bridge across the Constitution River and the part of the harbour where the early indigenous people in the Caribbean had careened their boats, and called it Indian Bridge. The settlers built their first bridge here in 1654, but the present Chamberlain Bridge dates from 1872 and is named after the then British Colonial Secretary Joseph Chamberlain. It consists of two segmentally-arched masonry spans and a separate steel swinging span and is approached by the Independence Arch of 1987, built to commemorate the granting of the country's independence in 1966. A rebuild was completed in 2006.

Chamberlain Bridge

Chamborigaud Viaduct, Chamborigaud, Gard, France

The masonry railway viaduct at Chamborigaud, opened in 1867, carries the single-track Cévennes Railway across the Luech River in a horseshoe-shaped curve of 240m horizontal radius. The structure, which was designed by Charles Dombre, is 387m long and 45m high with piers that taper both laterally and longitudinally and, it has been calculated, contains 18,000 cubic meters of granite. The viaduct was opened in 1867 and is now a National Historic Monument.

Changhua-Kaohsiung Viaduct, Baguashan to Zuoying, Taiwan, China

Completed in 2007 and built to carry the Taiwan High Speed Railway, this bridge is 157,317m (97.8 miles) long and is the world's second longest bridge after the Danyang-Kunshan Grand Bridge in China (q.v.). It is 13m wide and the longest individual span is 35m. Because it crosses geological fault lines the viaduct has been designed to withstand potential earthquake damage.

Chaotianmen Bridge, Chongqing, Sichuan, China

When this two-pinned steel arch over the Yangtze River was opened in 2009, its central span of 552m was (by just 2m!) the world's longest through arch bridge, overtaking Lupu Bridge (q.v.). Together with its 190m-long half-arch side spans, the continuous steel truss structure is 932m long and the overall length, including the prestressed concrete approach spans, is 1,741m. It is 37m wide and 142m high. The lower deck has two road lanes outside a space reserved for a future metro line, while the upper deck has six road lanes and two footways. (Chongqing is known as the 'Bridge Capital of China' and has thirty-eight large river crossings, some of which are described under the heading Chongqing Bridges – q.v.)

Chaotianmen Bridge

Chapananga Bridge, Chikwawa, Malawi

Floods on the Mwanza River were regularly claiming many lives before the Chapananga Bridge was completed in 2018. The 180m-long bridge, Malawi's longest, has nine spans of steel beams supported on T-shaped columns.

Charles Bridge, Prague, Bohemia, Czech Republic

There have long been bridges across the Vltava River here to connect Prague Castle with the Old Town. In the tenth century there was a timber structure which was replaced by the stone Judith Bridge in about 1170. After this had been damaged by floods in 1342 the Czech king and Holy Roman Emperor Charles IV initiated work on the present structure in 1357 and, completed in 1402, it is now the oldest bridge in Prague. It is 515m long and 9.5m wide and has sixteen stone arches, the longest spanning 13.4m. It was defended by three towers and, although the Swedes besieged the bridge in 1648, they were unable to capture it. The bridge, which is decorated with replicas of thirty carved Baroque statues, is now restricted to pedestrians only. The historic centre of the city with its bridge is now a UNESCO World Heritage site.

Charles Bridge

Charles Kuonen Suspension Bridge, Zermatt, Valais, Switzerland

The Charles Kuonen Bridge, opened in 2017, has an overall length of 494m and is currently the world's longest suspension footbridge. Its greatest height above the valley is about 85m and the width of its footway is 65cm.

Château de Chenonceau, Chenonceaux, Indre-et-Loire, France

The original chateau at Chenonceaux was completed in 1521 and in 1559, on behalf of Catherine de Medici, Philibert de l'Orme completed a bridge of seven semicircular arches

Chateau de Chenonceau

across the River Cher to link the castle to a nearby mill. In 1570 the architect Jean Bullant was commissioned by Catherine de Medici to build a superstructure on the bridge consisting of a three-floor gallery. The result, completed in 1576, is considered to be one of the most beautiful buildings in the world.

Chateau Moat Bridge, Fère-en-Tardenois, Aisne, France

The architect Jean Bullant designed a Renaissance bridge here for King Francis I and it was built in 1555 to replace the early thirteenth-century castle's fortified drawbridge. A colonnade of five arches on tall columns crosses a 55m-wide moat and, originally, supported an enclosed passageway with, above, an observation gallery, although the gallery has long since been partially destroyed. The castle is now a luxury hotel.

Chateau Nové Mesto and Metuji, Krcin, Eastern Bohemia, Czech Republic

Between 1909 and 1915 the Slovak architect Dusan Jurkovic laid out the gardens at this sixteenth-century castle and designed the covered entrance bridge that links the upper and lower gardens. The bridge is a two-part timber structure about 100m long and 1m wide with a pitched roof. Separating the two parts is a decorative timber tower standing on a square stone pier and topped by a tall pointed roof with an elaborate 'sunburst' window in the

Chateau Nové Mesto Garden Bridge

gables centred on the bridge axis. The bridge itself has intermediate X-braced framework supports.

Chateau Veltrusy's Laudon Pavilion, Veltrusy, Bohemia, Czech Republic

This late-eighteenth-century enclosed bridge was built in memory of General Ernst Gideon von Laudon (1717–1790). Nearly square on plan, it crosses over a steeply-sloping water channel on a stilted semicircular arch and, above a short length of balustrade, is a handsome Venetian window. Each end of the passageway through the bridge is closed off by a glazed doorway fronted by a columned portal frame topped by a triangular pediment.

Chaumont Viaduct, Chaumont, Haute Marne, France

This three-tiered 600m-long stone viaduct carrying the railway line between Paris and Basle across the Suize River was opened in 1856. The maximum height above water level is 50m. All the viaduct's arches are semicircular, there being twenty-five in the bottom tier, forty-nine in the middle one and fifty along the top tier, and all of these have a 10m-long span except for the 19m-long span for the bottom tier arch over the river. The piers are tapered in cross-section and the two upper levels are pierced by transverse arches with a footpath running through the middle level. Four piers and the arches they supported were destroyed by German forces in 1944.

Chavanon Viaduct, Merlines, Corrèze, France

The 360m-long suspension road bridge across the Chavanon River valley, which was opened in 2000, has a main span of 300m between A-shaped piers. At the apex of each pier, where the raking legs of the 'A' meet, there is a large 'eye' through which the twin suspension cables pass. Vertical hangers from these cables support a central spine beam from which the deck cantilevers out on each side. The clearance below the deck is 100m. The bridge was designed by the French firm Jean Muller International.

Laudon Pavilon

Chavanon Viaduct

Châzelet Castle Moat Bridge, Châzelet, Centre-Val de Loire, France

A bridge, claimed to be the first reinforced concrete bridge in the world, was built in 1875 over the moat at this sixteenth-century castle. It was designed by the Frenchman Joseph Monier (1823–1906), who in 1873 had patented the concept of plant pots made of concrete reinforced by iron wire. His bridge, which spans 16.5m, consists of a number of shallowly-arched contiguous parallel beams of concrete reinforced by iron rods. Together, these beams make up a 4m-wide deck that is edged by concrete parapets.

Chenab Rail Bridge, Sala, Jammu and Kashmir, India

The bridge across the gorge of the Chenab River will carry the broad-gauge railway between Udhampur and Baramulla when all construction work on the line is completed in about 2022. The bridge deck will be 322m above water level, making Chenab Bridge the highest railway bridge in the world. The main steel arch will span 467m and altogether the structure will be 1,315m long and have seventeen spans.

Chengyang Yongji (Wind and Rain) Bridge, Ma'an, Sanjian, Guangxi, China

This covered bridge across the Linxi River has four 17m-long spans, the stone abutments and intermediate piers supporting five pagoda-like pavilions, all of which have double or treble roofs with upturned eaves. Inside the bridge, the main passageway broadens out under each pavilion with benches where travellers can rest and enjoy the views. Although the bridge was built between 1912 and 1924, it needed major restoration following severe flood damage in 1983. In 1996 it became part of a World Heritage Site.

Chicago Bridges, Chicago, Illinois, USA

Chicago is famous for its many moving bridges and was where the rolling lift bascule bridge was invented by William Scherzer. The first bridge across the Chicago River was a timber structure built in 1832 and this was followed by a drawbridge in 1834 and the first swing bridge in 1856. The city's first bascule structure, **Cortland Street Bridge**, was built in 1902 to replace an earlier swing bridge and in 1981 this structure was added to the ASCE list of historic bridges. The 39m-long structure has two lifting leaves but, with the greatly reduced water traffic these are now fixed together. The first appearance of Scherzer's new invention was the Van Buren Street Bridge, opened in 1895 and long since demolished. However, **Cermack Road (22nd Street) Bridge**, built in 1906, is a rare early surviving rolling lift bascule; it has a span of 216ft with a roadway width of 36ft. The **Chicago & Northwestern Railway Bridge**, one of the first steel railway bridges to be built in the city, was completed in 1908, but has been permanently closed to water traffic. At the mouth of the

Chengyang Yongji Bridge

Lake Shore Drive (Outer) Bridge

Chicago River the Outer Drive Bridge of 1937, designed by Joseph Strauss and now known as the **Lake Shore Drive Bridge**, was claimed to be world's longest (356ft) and widest (81ft) bascule bridge. One of the city's newest bridges is the **BP Pedestrian Bridge**, opened in 2004. Designed by the engineering firm Skidmore, Owings & Merrill with architect Frank Gehry, it has a curved 'snake-like' covering of stainless steel sculptural plates encasing a steel box girder structure and its exposed deck is covered with hardwood planking. It is 285m long and 6m wide. It has been calculated that, since the first bridge was built in Chicago in 1832, about 240 fixed and movable bridges have been constructed in the city.

Chiche Bridges, Quito, Pichincha, Ecuador

Opened in 2014 to relieve traffic on an earlier two-lane structure a short distance away, this 315m-long, six-lane bridge carries part of the Ruta Viva Motorway 137m above the Chiche River and is the third highest bridge in South America. The three-span structure is supported by raking struts from the ravine sides and has a 210m-long main span between the bases of these struts. The older bridge, 70m high, carries route 28C and is now also used for bungee jumping from platforms on its below-deck trusses.

Chillon Viaduct, Veytaux, Vaud, Switzerland

The Chillon Viaduct (named after the famous nearby Chillon Castle) is an unusual bridge structure in that it does not cross *over* an obstacle, such as a river, but runs *along* one – in this case the steep and irregular side of the mountains above Lac Léman. There are two separate 12m-wide prestressed concrete structures at different levels, one for each carriageway of the A9 autoroute.

Chillon Viaduct

These are each 2,150m long with twenty-three spans and the individual span lengths are 92m, 98m or 104m. The twin-column piers are between 3m and 45m high. The viaduct was built between 1966 and 1969 and has recently undergone remedial works to deal with water penetration into the structures and to upgrade its earthquake resistance.

Chinipas Bridge, Témoris, Durango, Mexico

When this bridge was built through Mexico's Copper Canyon in 1961 its 90m height made it one of the world's highest railway bridges. The three-span truss structure has a 90m-long central span and is supported on concrete piers. The valley was later dammed to form a reservoir and the lower part of each pier is now submerged.

Choate Bridge, Ipswich, Massachusetts, USA

This bridge was built in 1764 and is named after its builder, Colonel John Choate. The structure has an overall length of 72ft with two 30ft-span segmental stone arches. Originally, it was 20ft wide and was widened by a further 16ft in 1838. The bridge was entered on the National Register of Historic Places in 1972 and underwent substantial renovation in 1989.

Choluteca Bridges, Choluteca, Choluteca Province, Honduras

The first bridge here across the Choluteca River, built by the US Corps of Engineers between 1935 and 1937, was a suspension bridge designed by the American Conde B. McCullough. This 12m-wide and 300m-long structure had two main end-to-end suspension spans, each 100m long, with two 50m-long back spans, and its three main support towers had a distinctive Art Deco appearance. It was badly damaged by Category 5 Hurricane Mitch in 1998 but, fortunately, a replacement structure (Bridge of the Rising Sun) had been opened shortly before the hurricane struck and this suffered only minor damage. However, the flooding arising from the hurricane (more than two metres of rain fell in four days) caused the river to wash away the approach roads on each side and create a new channel for itself leaving the new bridge high and dry, becoming known as 'The Bridge to Nowhere'. New connecting roads were completed in 2003 and the new three-span bridge, a gift to the people of Honduras from Japan, now carries the Pan American Highway (route CA1). It consists of a prestressed concrete box girder with haunches at the pier supports. Meanwhile, the old suspension bridge was repaired and further rebuilt in 2002.

Choluteca Bridge (1937)

Chongqing Bridges, Chongqing, Sichuan, China

Chongqing's main bridge, the Chaotianmen Bridge, is described under a separate entry (q.v.) but this entry gives brief details about a few of the other bridges (more than 14,000 in total!) that have been built in the city, which since 2005 has been called 'The Bridge Capital of China'. The city's two rivers are the Yangtze River, the longest river in Asia, and the Jialing River, its major tributary in the Sichuan basin.

The city's first cross-river bridge, built between 1958 and 1966 to carry the No 3 metro line, was the **Niujiaotuo Jialing River Bridge**. With an overall length of 352m and a main span of 160m, it is claimed to be the world's longest continuous light rail girder bridge. The first **Chaoyang Bridge**, completed in 1969, was a most unusual road suspension bridge across the Jialing. Instead of the normal two cables, one above each side of the deck, that sagged symmetrically towards mid-span, there were four cables. The profile

of each pair was asymmetrical, with the lowest point of one pair being approximately one third of the way across the span and that for the other at the two-thirds point. The bridge had an overall length of 233m, with a main span of 186m, and pylons that were 64m high. The present Chaoyang Bridge, opened in 2016, is a double-decked cable stay bridge and is the first in the world to use corrugated steel webs.

The **Egongyan Bridge** suspension bridge was completed in 2000 and is the largest self-anchored suspension bridge in the world. It is 1,420m long with a 600m-long main span and was originally built with six traffic lanes and two footways. However, an expansion in 2013 converted the footways to traffic lanes. **Caiyuanba Yangtze River Bridge**, which was completed in 2007, has an overall length of 800m with a 420m-span arch and is 37m wide. The double-decked structure carries a six-lane highway above twin-tracked subway lines and has distinctive Y-shaped piers from which the arch springs.

The similar **Dongshuimen (Yangtze River)** and **Qianximen (Jialing River) Bridges,** located at the tip of China's Yuzhong Peninsula, are together known as the **Twin River Bridges** and were opened in 2014. Both are double-decked, with four traffic lanes above two rail tracks, and are steel girder structures supported by concrete towers. The larger Dongshuimen Bridge has a 445m-long main span. It is cable-stayed near the towers but the trussed box girder that supports the double decks is strong enough to allow a major gap between the cable-stayed sections and thus to keep the pylon height significantly under the 230m that would have been needed for a fully stayed span. Including approach spans, the Qianximen Bridge is 720m long with a fully cable-stayed main span of 312m and a 240m-long back span. The **Cuntan Yangtze River Bridge**, completed in 2016, is a suspension bridge with towers that are 195m high, the tallest in Chongqing.

Dongshuimen (Yangtze River) Bridge

Chords Bridge, Jerusalem, Israel

The 360m-long and 15m-wide cable-stayed Chords Bridge, which carries a glass-sided footway and the twin tracks of Jerusalem Light Rail's Red Line on a curve over the Shazar Boulevard, was designed by Santiago Calatrava and opened in 2008. The main span is 160m long and is supported by sicty-six stays from an angled steel pylon on one side of the main deck. This pylon, which is 118m high, is also cranked at mid-height. The bridge is shown on an Israeli postage stamp issued in 2016.

Chords Bridge

Chuha Gujar Mughal Bridge, Peshawar, Khyber Pakhtunkhwa, Pakistan

Like the Shahi Bridge at Jaunpur in India (q.v.), this bridge is in the Mughal tradition of having multiple pointed stone arches, usually twelve as here. Commissioned by Emperor Shah Jehan and completed in about 1629, the Chuba Gujar Bridge crosses the River Bara. Standing on every third pier are tall circular columns topped by onion domes. The intermediate pier ends are fronted by fluted half-columns topped by a half-dome set against the parapet wall. The bridge is about 300ft long and 20ft wide.

Cidi M'Cid Bridges, Constantine, Constantine Province, Algeria

A short distance downstream from the El-Kantara Bridge (q.v.), there are two interesting bridges named Cidi M'Cid that cross the 200m-deep

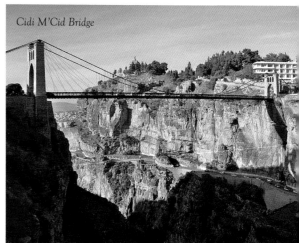

Cidi M'Cid Bridge

Rhumel River gorge. The higher one, designed by the French engineer Ferdinand Arnodin and opened in 1912, is a suspension and cable-stay hybrid that is 164m long and 5.7m wide. Both piers have a tall pointed archway for the traffic, above which is a tier of three similar but smaller arches. The bridge deck, at 175m above water level, was the highest bridge in the world until 1929. During restoration in 2000 the suspension cables were renewed. The lower road bridge, also called the Pont des Chutes, stands practically on the edge of the 80m-high waterfalls. Completed in 1925, this has five semicircular stone arches.

Citadel Bridge, Aleppo, Aleppo Governorate, Syria

The Citadel in Aleppo, which stands on a 50m-high mound, is considered to be one of the oldest and largest castles in the world and is now part of the UNESCO World Heritage Site of the Ancient City of Aleppo. Access to the Citadel is across an early thirteenth-century bridge between

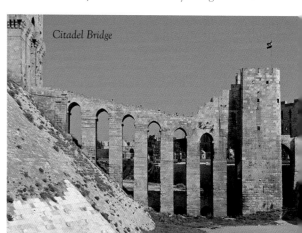

Citadel Bridge

an outer tower and the main gatehouse to the citadel. This bridge, which was built by Sultan Ghazi, has seven tall narrow arches which support a stepped walkway. Despite its strong defences, Aleppo was captured by the Mongols in 1260.

City Bridges, St Petersburg, Leningrad Oblast, Russia

St Petersburg, which was founded in 1703, has sixty-eight waterways dividing it into forty-two islands and is sometimes known as the City of Bridges. There are estimated to be about 350 within the city, the centre of which is designated as a World Heritage site. Of the bridges over the main river, three are included in the entry **River Neva Bridges** and there is another separate entry for the **Blue Bridge** over the Moika River, but brief notes on a few of St Petersburg's more interesting smaller bridges now follow. **Prachechny Bridge**, built in 1769, is a three-span stone structure with quite steep approaches on the outer spans and decorative circular tunnels above the pointed piers. **Upper Swan Bridge**, dating from 1768, replaced an earlier timber drawbridge and has a single semi-elliptical stone arch. **Lomonosov Bridge** over the Fontanka River is an elegant structure and was built in 1787, also to replace an earlier timber bridge. It has a semi-elliptical arched approach span on each side leading to a central section. Originally this consisted of two separate bascule leaves that were hoisted by chains supported from the four domed corner towers standing on two pointed piers, but the span is now a simple beam structure and the chains are purely decorative. Existing major bridges with twin bascules include **Annunciation Bridge**, **Exchange Bridge**, **Palace Bridge**, **Peter the Great Bridge** and **Tuchkov Bridge** (on which Dostoevsky had a character in one of his novels spend the last night of his life).

A feature of the city's bridges is that some have decorative animals. The **Egyptian Bridge** was built 1825–26 and the Bank Footbridge and the Lions Footbridge were both opened in 1826. It has sphinxes sitting on its abutments and the **Bank Footbridge** has winged griffins, the end rods of its wrought iron suspension system emerging

Bank Footbridge

from their mouths. On the **Lions' Footbridge** four cast iron lions undertake this function (these were perhaps the lions, mentioned in Pushkin's 1833 poem *The Bronze Horseman: A Petersburg Tale*, that survived the great flood of 1824). The mid-nineteenth-century **Anichkov Bridge**, also known as the Horse Tamers Bridge, has prancing bronze horses on its abutments at each corner and mermaids and sea horses in its iron railings.

City Centre Bridges, Wroclaw, Lower Silesia, Poland

Wroclaw is built on twelve islands separated by six rivers and, before 1945, was claimed to have had 303 bridges. One of the most famous of these was **Grunwald Bridge** over the River Oder. This suspension bridge, built between 1908 and 1910 to a design by the city council's engineer Richard Plüddemann, has a width of 18m and spans 127m between 20m-high stone piers. The first **Sand Bridge** was built in the twelfth century and by the fifteenth century the crossing included a drawbridge and timber gatehouse tower. The present iron structure, built in 1861, is the city's oldest such structure. **Tumski Bridge**, which was built in 1889 to replace an earlier timber bridge connecting Cathedral Island to the river bank, is a 52m-long two-span steel suspension bridge with an arched framework above its central pier. It is now restricted to pedestrians only and has become a 'love-lock' bridge with padlocks attached to the railings.

Cize-Bolozon Viaduct, Daranche, Tarn, France

This two-tiered stone viaduct crossing 73m above the Ain Gorge, built in 1875 to a design by Jean-François Blassel, was destroyed by the French Resistance in 1944 and then rebuilt in 1950 to the original design. The lower tier, carrying a local road, has five main arches while the upper tier of eleven arches now carries the Paris-Geneva main railway line. The overall length is 273m and the top deck is 73m above the river. Since 2018, the structure has been in the top ten on the World List of Tourist Attractions.

Cize-Bolozon Viaduct

Clairac Bridge, Clairac, Lot-et-Garonne, France

An earlier suspension toll bridge here over the Lot River was built in 1831. A design competition for a replacement structure was won by Nicolas Esquillian and the new road bridge was opened in 1939. This has three shallow reinforced concrete arches, each spanning 49m, with the deck supported on slim spandrel piers. With the water level of the river at the arch springing points, the bridge and its reflection produce a perfect lenticular image.

Clermont Ferrand Viaduct, Clermont Ferrand, St Flour, France

Carrying the A75 motorway across the Truyère River less than a mile upstream from Eiffel's famous railway bridge, the Garabit Viaduct (q.v.), the Clermont Ferrand Viaduct was opened in 1992. It consists of a concrete box girder structure with distinctive piers that are inclined outwards both in elevation and as seen in end view from underneath the bridge.

Cobblers' Bridge, Ljubljana, Province of Ljubljana, Slovenia

The Romans built a timber bridge here across the River Ljubljanica and a thirteenth-century covered bridge successor, named Cobblers Bridge after the workshops on it, was replaced by the cast iron Hradecky Bridge (q.v.) in 1867. In 1931 this was moved elsewhere in the city and replaced in 1932 by a wide new artificial stone footbridge designed by the architect Joze Plecnik to create an open public square over the water. The bridge has two spans with a mid-river pier consisting of two semicircular arches. Each end of this pier extends on either side beyond the bridge parapet line and supports an Ionic column crowned by a light. The decorative bridge parapets support Corinthian columns topped by stone balls.

College Footbridge, Kortrijk, West Flanders, Belgium

The 3m-wide footbridge across the Leie River in Kortrijk, built in 2007 to a design by Ney & Partners, is S-shaped in plan but its main feature is that it is a rare mono-cable suspension bridge. This cable spans between two raking pylons placed at the centre of each of the two horizontal curves of the 'S' and hangers from the cable support the sides of the deck's steel hollow-section spine beam. The length of the main span is 86m and the footbridge is 203m long.

College Footbridge, Lyon, Rhône, France

Built between 1843 and 1845, this bridge was named after the Great College of the Jesuits on the right bank of the Rhône and is the oldest surviving crossing in the centre of Lyon. An accident during construction resulted in eight workmen drowning. The 4m-wide structure has an overall length of 198m, with a 110m-long main span between arched masonry piers. The side spans are 42m and 46m long. The bridge was rebuilt after being partly destroyed in 1944 and, during a major restoration between 1986 and 1987, a non-slip aluminium deck replaced the earlier timber one.

Colonial Bridge of Tequixtepec, Huajuapan de Leon, Oaxaca, Mexico

This bridge is believed to have been built between 1570 and 1572 and is claimed to be Mexico's oldest bridge. Connecting Dominican settlements in the Oaxaca region of Mexico, it has a single segmental arch built of tufa that spans between the vertical cliff sides of the Tequixtepec River. Part of one cliff, together with the lower voussoirs and some of the spandrel on one side of the arch, have collapsed and the bridge was included in the 2012 World Monuments Watch.

Colossus Bridge, Philadelphia, Pennsylvania, USA *

Colossus Bridge is perhaps the most famous of America's very many traditional covered bridges. It was built in 1812 to a design by Louis Wernwag and, with its clear span of 340ft, was the longest single-span timber bridge in the country. The trusses included iron rod tension members. The bridge was burnt down in 1838 and replaced by a wire suspension bridge designed by Charles Ellet.

Columbia River Bridge, Castlegar, British Columbia, Canada

The 1965 road bridge here, which is supported by four intermediate V-shaped piers, has five 80m-long cantilever-and-suspended-span concrete box girder sections. Each of the inclined pier legs consists of three square columns linked together at their third points. The bridge was designed by Riccardo Morandi and carries Highway 3.

Confederation Bridge, Borden-Carleton, Prince Edward Island, Canada

The Confederation Bridge, which opened in 1997, carries the Trans-Canada Highway across the Northumberland Strait between Prince Edward Island and the mainland on a post-tensioned concrete box girder structure of sixty-three spans. Overall, the bridge is 12.9km long

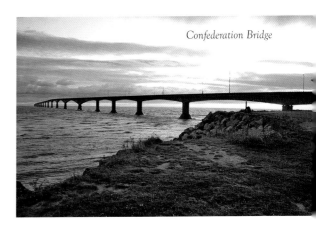

Confederation Bridge

and the longest span is 250m long. In order to protect the structure from ice damage, the piers have a conical ice shield which the ice rises up on and then breaks under its own weight.

Constitution Bridge, Venice, Veneto, Italy

Designed by Santiago Calatrava and completed in 2008, this is the first new bridge to be built in Venice for seventy years and is the fourth across the Grand Canal. Providing a pedestrian link between the city's car park, bus terminal and railway station, the structure has a 94m-long steel arched truss, its width tapering out from 3m at the ends to 9m in the centre. The footway includes glass steps, although these have proved to be slippery in wet weather and are being damaged through heavy use. There is a gondola cableway along one side of the bridge for people with walking difficulties.

Coppename Bridge, Jenny, Coronie District, Suriname

There are two timber-decked Bailey bridges over the Coppename River in Suriname that were completed in 1976 and 1999. The second of these, which carries the country's two-lane East-West Link across the mouth of the river between Jenny and Boskamp, has an overall length of 1570m.

Corinth Canal Submersible Bridges, Corinth, Peloponnese, Greece

An attempt to build a canal here in 602BCE was unsuccessful but the Hungarian civil engineer Béla Gerster (1850-1923) constructed the present 6.3km-long Corinth Canal between 1881 and 1893. This can be used by vessels up to 16.5m wide with a draught of up to 7.3m. A submersible bridge, consisting of steel beams 18m long, was built at each end of the canal in 1988, and these are lowered 8m below water level to allow vessels to pass.

Cornish-Windsor Covered Bridge, Cornish, New Hampshire, USA

Earlier bridges over the Connecticut River on this site were built in 1796, 1824 and 1828 but all were brought down by floods and the current two-span structure was built by James Tasker and Bela Fletcher in 1866. It is a 24ft-wide, 449ft-long Ithiel Town lattice truss with a longest span of 204ft and is claimed to be the USA's longest timber covered bridge and one of the longest in the world. It underwent major repairs in 1954, 1977 and again in 1986-8. In 1970 the ASCE listed the bridge as a National Civil Engineering Landmark.

Cornish-Windsor Covered Bridge

Coulouvrenière Bridge, Geneva, Switzerland

Some of the very early timber bridges over the River Rhone in Geneva were via an island which was also part of the city's defensive fortifications. The first modern structure on the site was a twin-span underslung chain suspension bridge that spanned between attractive little stone towers. It was designed by Guillaume Henri Dufour and built between 1856 and 1857 and was sometimes called the Bel-Air Bridge. When this proved too light to carry trams, it was replaced by a granite-clad concrete arch bridge, completed in 1900, that was 150m long and 19m wide. This consisted of two large segmental spans separated by a central smaller semicircular span and with a further small span at each end. It had a granite

balustrade and four 11m-tall granite columns stood above the piers on either side of this central arch, all giving the whole structure a great sense of civic distinction. In 1970 the bridge was widened to 27m and the decorative columns were removed.

Covered Bridge, Lovech, Lovech Province, Bulgaria

After an earlier bridge had been destroyed by floods in 1872, the first covered bridge over the Osam River was built by Kolyu Ficheto and opened in 1874. This was 84m long with six spans, carried both a roadway and a footway, and contained sixty-four shops. After it was destroyed by fire in 1925, a modern reinforced concrete replacement with a glass roof was opened in 1931. However, a reconstruction of Ficheto's original design was built between 1981 and 1982 and is now a great tourist attraction. This pedestrian-only bridge is 106m long and has fourteen shops.

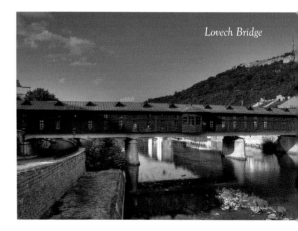

Lovech Bridge

Covered Bridge, San Martino, Pavia, Italy

The first permanent crossing over the Ticino River at Pavia was built by the Romans and was replaced in 1354 by a seven-arched stone structure with a covered arcade and a central chapel. This was badly damaged in the fighting of 1945 and part of it then collapsed in 1947. A new stone bridge, similar to its predecessor but with only five arches, was built between 1949 and 1951. This stands on round-ended piers above which, on the upstream face, are pointed cutwaters. Running along the full 216m length of the bridge is a tile-covered open walkway with, on the midstream pier, a simple chapel. At each end of the bridge there is also a small tower.

Craigavon Bridge, Londonderry, County Londonderry, Northern Ireland

The double-decked Craigavon bridge was built in 1933 to replace the earlier **Carlisle Bridge** of 1863, also double-decked, itself replacing an earlier timber structure from 1772. Originally, Craigavon Bridge carried both a standard and a narrow-gauge railway line on its lower deck with road traffic above, but the lower deck became a second roadway in 1968. These decks are supported by steel girders over the total bridge length of 1260ft and there are five main 130ft-long spans and shorter end spans. The Foyle Bridge (q.v.) was built in 1984 to provide additional capacity.

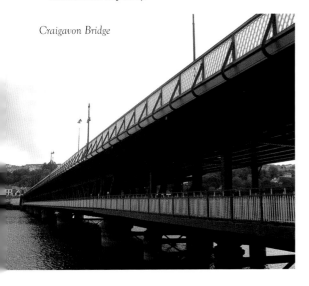

Craigavon Bridge

Craigmore Viaduct, Bessbrook, County Armagh, Northern Ireland

This stone viaduct, opened in 1852 to carry part of the Dublin & Belfast Junction Railway, is the highest railway bridge in Ireland at 137ft. It has eighteen semicircular arches each spanning 60ft.

Crni Kal Viaduct, Stepani, Koper, Slovenia

The Crni Kal Viaduct was built between 2001 and 2004 to carry the A1 motorway between Ljubljana and the coast across the Osp River valley. It is 1,065m long with twelve spans, the longest being 141m and, at the tallest pier, the structure is 95m high making it the country's highest bridge. The two separate carriageways are on haunched prestressed concrete box girders supported by Y-shaped piers.

Cross Bayou Bridge, Shreveport, Louisiana, USA

The rare A-frame iron railway bridge, which was designed by John Alexander Low Waddell, was originally built in 1896 across the Arkansas River in Oklahoma. It was moved here in 1926 to span Cross Bayou River for the Texas & Pacific Railway and is the central part of a 490ft-long three-span crossing, being flanked on each side by a deck girder span. The A-frame spans 100ft, has a deck width of 16ft and is 38ft high. In 1995 the bridge was placed on the National Register of Historic Places and there are now plans to restore it to carry the Fred Marquis Pinellas Trail for walkers and cyclists.

Cross Bayou Bridge

Cunene River Bridge, Xangongo, Cunene Province, Angola

This 880m-long bridge over the Cunene River linking the cities of Lubango and Ondjiva is Angola's longest bridge and was built to replace an earlier permanent bridge and various temporary replacements that were all destroyed during the twenty-seven years of the Angolan Civil War between 1975 and 2002. The structure consists of twin steel box girders, which were push-launched over a series of T-headed reinforced concrete intermediate piers spaced 50m apart. These girders in turn support the 11.6m-wide deck which is made of 1384 precast concrete slabs with site-poured concrete infilling over them. The bridge was opened in 2009.

Cuscatlan Bridge, San Miguel, San Miguel Department, El Salvador

A suspension bridge over the Lempa River, with a span of 820ft, was claimed to be the largest bridge in Central America when it was built by the Roebling's Sons Company in 1947 as part of the Pan-American Highway. Although guarded by government troops, it was blown up by guerrillas at the end of 1983 killing several defenders. A temporary Bailey bridge, called the San Marcos Bridge, was opened in 1986 and a new five-span concrete bridge, built a short distance downstream, was opened in 1999. This is 399m long and 18m wide.

D

Daedunsan Cloud Bridge, Daejeon, Gyeonggi Province, South Korea

The 50m-long Cloud Bridge is also known as the Gureumdari suspension footbridge. Its 1m-wide deck is 80m above ground level and connects to a steep (50°) 127-step stairway leading further up the mountain.

Danilo's Bridge, Rijeka Crnojevica, Crmnica Region, Montenegro

An earlier timber bridge here across the Crnojevica River was replaced by this very attractive 43m-long stone arched bridge in 1853, built by Prince Danilo. It has two wide segmental arches either side of a large central pier. This has pointed cutwaters and itself contains a high-level semicircular flood arch. The footway steps up to a level deck across the three river arches and then down onto a small subsidiary arch over a riverside footpath.

Danilo's Bridge

Danjiang Bridge, Taipei, Xinya District, Taiwan

Construction work on the Danjiang Bridge over the Tamsui River began in 2014 and is due for completion in 2023–24. The overall length of the bridge will be more than 12km and the main 450m-long cable-stayed span will be supported by a single 200m-high concrete pylon. The road bridge, which will also incorporate an extension of the Danhai Light Rail system, was designed by Sinotech Engineering Consultants and Zaha Hadid Architects.

Danube Bridge, Linz, Bundesland, Austria

The original railway bridge across the Danube at Linz, which was built in 1897, had three main segmentally-arched steel trusses to a design by Johan Schwedler. These spanned between stone piers with rounded cutwaters. The overall length of the bridge was 396m, the longest span being 113m. The bridge was later altered to take a road lane but this led to the structure being weakened by severe corrosion caused by de-icing salts and it was demolished in 2016. The new replacement bridge, completed in 2020, carries a single railway/metro line and a road.

Danyang-Kunshan Grand Bridge, Wuxi, Jiangsu, China

Considered to be the world's longest bridge, this rail viaduct on the Beijing-Shanghai High Speed Railway, built between 2006 and 2011, is 102 miles long and cost US$8.5 billion to build. The six-mile-long section that crosses Yangchen Lake is a viaduct supported by 2,000 concrete columns that includes a main cable-stayed span of 260ft over the water channel.

Darnytskyi Bridge, Kiev, Kievshchyna, Ukraine

The first railway bridge to cross the River Dnieper here was built in 1855. There were twelve spans, each 89m long, and its overall length of 1,068m made the bridge the longest in Europe. It was replaced in 1870 by a fourteen-span bridge constructed by Amand Struve. This was a huge box girder, through which the trains ran, that consisted of multiple-crossing double lattice trusses on each side. It was blown up in 1920, by Polish troops retreating after their joint attack with Ukraine against the new Soviet Russian state, and again in 1943, this time by the retreating Germans as had also happened to the nearby Nicholas Chain Bridge (see Metro Bridge, Kiev). Following the second demolition, Russian Army engineers built a temporary timber structure in the amazingly short time of just one month, and a permanent replacement was opened in 1949. The current 1100m-long crossing (sometimes called the **Kirpa Bridge**) carries a double-track railway and six lanes for road traffic. Work started in 2004 and was completed in 2012.

Dar-Tahar-ben-Abbou Bridge, Marrakesh, Marrakesh-Safi, Morocco

This stone road bridge over the Oued Tensift River in the centre of the country carries the N1 highway between Safi and Essaouira. It has seven semicircular arches with circular openings between the spandrels above each pier and there is a pronounced cornice below the parapet.

Deh Cho Bridge, Fort Providence, Northwest Territories, Canada

The Deh Cho Bridge carrying NW3 Highway across the Mackenzie River has been given the Canadian First Nations' name of the river. Designed by Infinity Engineering, the 1,045m-long structure consists of a main cable-stay span, with a back span each side, and three approach girder spans at either end. The cable-stayed section has a 150m-long central opening between A-shaped pylons. The bridge was opened in 2012, replacing a summer ferry and ice crossings that had to be made every winter, and has been awarded the Gustav Lindenthal Medal.

Deh Cho Bridge

Deir ez-Zor Suspension Bridge, Deir ez-Zor, Deir ez-Zor Province, Syria

Built in 1927 in an area that has been inhabited for more than ten thousand years, this four-span suspension footbridge across the Euphrates River linked Mesopotamia to the Levant. The elegant support towers had three tiers, the biggest span was 105m long and the bridge had an overall length of 500m. In 2013, during the Syrian Civil War when the city was besieged by ISIS, the bridge was destroyed by shellfire from Assad's forces. The Russians built a replacement structure in 2020.

Delal Bridge, Zakho, Kurdistan, Iraq

The old stone bridge over the Little Khabor River has a 16m-high, irregularly-shaped main arch flanked by three smaller arches on one side and a single arch on the other, giving a total length of 114m. It was probably built during the Abbasid era between about 750 and 1258AD. The World Monuments Fund worked with Google to help preserve the structure.

Delal Bridge

Delaware Aqueduct, Lackawaxen, Pennsylvania, USA

Completed in 1849, this is now the oldest suspension bridge in America. It was designed by John Roebling to carry the Delaware & Hudson Canal over the Delaware River and had an overall length of 535ft over four spans. After the canal was abandoned in 1898 the structure was converted to carry a road. It was then taken over by the National Park Service in 1980 and restored back

Delaware Aqueduct (pre-restoration)

to its original appearance. The aqueduct has been designated as both a National Historic Landmark and a National Civil Engineering Landmark.

Delaware Memorial Bridge, Wilmington, Delaware, USA

This bridge over the Delaware River, the lowest along the river's course, consists of two separate suspension structures that were built to identical dimensions, although differing slightly in structural details, with each one carrying four traffic lanes. The first structure, now carrying eastbound traffic, was opened in 1951 and the second, built 250ft to the north for westbound traffic, in 1968. The overall length of each structure is a little over 10,750ft and the main suspension span is 2150ft long. The decks are 59ft wide and the clearance between the undersides of the deck and water level is 174ft. The top of the steel towers is 440ft above river level. The bridge is dedicated to the 15,000 servicemen and women from Delaware and New Jersey who lost their lives in action in World War II and wars thereafter.

Demerara Harbour Bridge, Georgetown, Demerara-Mahaica Region, Guyana

The 6,074ft-long floating bridge at Demarara, which crosses the Demerara River a short distance upstream from its mouth, links the city with Demerara's West Bank and was completed in

1978. This structure, the world's longest floating bridge, was designed and built by Thos. Storey Group (famous as the original manufacturers of the Bailey Bridge). The bridge has a short, raised section providing 26ft clearance above a channel used by fishing boats, but a 240ft-long section can be retracted to allow free passage for large vessels. Altogether, the single-lane bridge has sixty-one spans. In 2019 a tug and barge hit the bridge, severely damaging the structure.

Denham Suspension Bridge, Mahdia, Potaro-Siparuni Region, Guyana

This rare railway suspension bridge was built over the Potaro River in 1933 by John Aldi, a Scottish civil engineer, and named after the then Governor of British Guiana. The single-track railway line connected the towns of Mahdia and Bartica. The bridge is now restricted to light traffic and is part of the National Trust and Heritage of Guyana.

Dessau-Wörlitz Garden Kingdom Bridges, Dessau-Wörlitz, Saxony-Anhalt, Germany

Following a study visit to England, Prince Leopold III created the first landscape garden in continental Europe here between 1769 and 1773 and there are several bridges. The most distinctive of these are a simple arched iron footbridge, the railings of which are in the style of successive sunbursts, and an eleven-panelled arched timber footbridge. The garden kingdom now covers fifty-five square miles and was designated a UNESCO World Heritage Site in 2000.

Detroit-Superior Bridge, Cleveland, Ohio, USA

This major 2,880ft-long bridge over the Cuyahoga River (also known as the Veterans Memorial Bridge since 1989) was built between 1912 and 1917. Its main span is a through steel arch 591ft long that has an overall height of 196ft and a clearance between the underside of the deck and water level of 96ft. The approach viaducts have concrete arch spans: three on the west and nine on the east, with spans ranging from 58ft to 181ft. Originally the lower deck had tracks for a streetcar service but this ended

Detroit-Superior Bridge

in the 1950s and the space is now open. When built, the upper deck was 75ft wide, but in 1969 work was completed to add two extra road lanes by narrowing footways and cantilevering out a widened deck beyond the main arch.

Devil's Bridge, Ardino, Smolyan, Bulgaria

Devil's Bridge was constructed between 1515 and 1518 on the site of an earlier Roman bridge over the Arda River by a builder called Dimitar. It has one central semicircular arch spanning 13m which is flanked by a smaller semicircular arch on each side. The intermediate piers have tall pointed upstream cutwaters and, above the piers, are further flood relief arches. The 3½m-wide bridge is 56m long and has been a national cultural heritage monument since 1984.

Devil's Bridge Ardino

Devil's Bridge, Céret, Pyrénées-Orientales, France

When this bridge over the Tech River was completed in 1341 after twenty years of construction, its 45m-long span was thought to make it the largest stone arch bridge in the world. This main span is flanked by side arches and there is also an unusual canted arch in each spandrel.

Devil's Bridge, Gablenz, Saxony, Germany

Completed in 1860, this semicircular stone bridge, also called **Rakots Bridge**, is located in the town's Kromlau Park and was built purely to provide a focal point as, with its reflection, it forms a near-perfect circle in the park lake. In fact, climbing onto the bridge is prohibited. At each end of the bridge, and also in the distance seen through it, are completely unnatural-looking faux-rocky pinnacles.

Devil's Bridge Gablenz

Devil's Bridge, Martorell, Barcelona, Spain

The first bridge here across the River Llobregat was built by the Romans for their Via Augusta and given a grand triumphal archway. The bridge itself was replaced in 1283 by a new three-span structure (also known as **Sant Bartomeu Bridge**), that was built on the original foundations with a large central pointed arch spanning 37m. This was destroyed in 1937 during the Spanish Civil War but was accurately rebuilt in 1965. There was a small chapel above the crown of the bridge but this no longer exists, and the Roman triumphal

Devil's Bridge Martorell

arch stands just behind the eastern abutment. (An elegant modern bridge now crosses the river a short distance away.)

Devonshire Bridge, Dungarvan, County Waterford, Ireland

The duke of Devonshire financed this bridge over the River Colligan, which was built in 1813–16 to a design by William Atkinson. The segmental stone arch spans 23m between panelled stone abutments and, unusually, the rusticated stonework is laid radially across the spandrels.

Dhangali Bridge, Rawalpindi, Punjab, Pakistan

Dhangali Bridge, which crosses the Jhelum River in the upper reaches of Mangla Reservoir, was completed in 2011 as part of compensation work arising from the raising of Mangla Dam. The three-span prestressed post-tensioned balanced cantilever box girder is 340m long and 8.5m wide.

Djurdjevica Tara Bridge, Zabljak, Northern Region, Montenegro

This bridge near Zabljak carries a road 172m over the Tara River. Work on the structure started in 1937 and, on its completion in 1940, it became the biggest concrete bridge in Europe. Slightly curved in plan, it has an overall length of 365m and contains five arched spans. After the Italians had occupied the country in 1942 one of the

engineers – Lazar Jaukovic – who had previously been working on constructing the bridge then helped partisans blow up the central arch. He was captured and executed.

Dodhara Chandani Bridge, Bhimdatta, Kanchanpur, Nepal

This steel footbridge across the Mahakali River, which was opened in 2005, is 1,453m long and 1.6m wide. It has four main 225m-long suspension spans, with the deck section arched in each, and straight backstays over the 70m-long spans between the 327m-tall pylons and the anchorages. There is an anchorage at each river bank and three larger common anchorages in the river and the metal grille deck over all these anchorages is also arched in a series of 70m-long spans. The sides of the deck are protected by rope handrails and wire mesh and the bridge is stiffened against vibration and deflection by further underslung cables.

Dodhara Chandani Bridge

Doha Sharq Crossing, Doha, Qatar

A planned bridge scheme next to Doha International Airport, formerly known as the **Doha Bay Crossing**, was originally developed by Santiago Calatrava in 2013 and was to have a bridge at each of the three ends of a tunnel system, all at an estimated cost of US$3 billion. A recent project redesign encompasses two bridges, two islands and two tunnels, with completion expected in around 2024 at a cost of about US$12 billion. The main bridge section will consist of a tubular structure with eight cable-stayed spans supported by seven tall pylons. A separate bridge over the West Bay will include an attached cable car system.

Dom Luis I Bridge, Porto, Norte, Portugal

Opened in 1886, this bridge across the River Douro was designed by the Belgian engineer Théophile Seyrig, a pupil of Gustave Eiffel. Apart from its additional low-level deck, it is somewhat similar to Eiffel's railway crossing, the Maria Pia Bridge (q.v.) of 1877, which is about half a mile further upstream. Seyrig's bridge has a parabolic arch that spans 564ft and supports at its crown two 1,264ft-long high-level decks – the upper for main road traffic with metro tracks below. Hanging from the arch there is also a further deck at arch springing level that is for local road traffic and pedestrians.

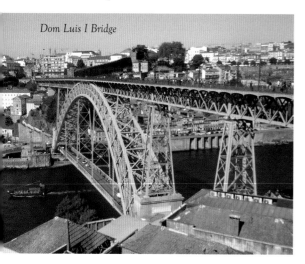
Dom Luis I Bridge

Dongguan City Aqueduct Bridge, Guandong, Guangzhou, China

The world's longest steel box-beam suspension aqueduct bridge was completed in 2019 and carries eight traffic lanes. It has a main span of 1,688m and contains 176 separate steel box units.

Donghai Bridge, Shanghai, China

This bridge was opened in 2005 to link the new Yangshan deep water port to metropolitan Shanghai. Its overall length is 32km, most of it being a low-level column-and-beam viaduct, and the longest individual opening is a cable-stayed structure spanning 420m.

Double Lift Bridge, Portland, Oregon, USA

Believed to be the only bridge of its type in the world, the Union Pacific Railroad bridge over the River Willamette in Portland has two separate lifting decks, one above the other. The lower deck carries twin railway tracks and the 71ft-wide upper deck a roadway. The bridge has three spans – a fixed 287ft-long approach span at each end and the 220ft-long double central lifting spans. There is 26ft headroom between the water and the lower lifting span when the river level is low. When this is insufficient for relatively small ships, the lower deck can be lifted on its own so that it telescopes inside the structure supporting the upper deck, thus giving 72ft of headroom above low water level. Alternatively, to provide the maximum space for shipping, both decks can then be raised further together to give a maximum clearance of 165ft. The bridge, which replaced an earlier structure from 1888, was opened in 1912.

Double-Arched Bridge, Gelsenkirchen, North Rhine-Westphalia, Germany

The 112m-long and 5.5m-wide footbridge, which crosses diagonally over the Rhine-Herne Canal in Gelsenkirchen, was built for the 1997 Federal German Garden Show in Nordstern Park. The deck is supported by bar hangers from two separate and parallel 79m-span tubular steel arches. These arches are 32m apart and cross at right angles to the canal, therefore each passing over one end of the deck.

Dragon Bridge, Da Nang, Hai Chau, Vietnam

The 666m-long road bridge over the River Han in the port city of Da Nang is the longest bridge in Vietnam. It is 38m wide and carries six traffic lanes and two footways. There are three arches, the longest spanning 200m, which look in elevation like the back of a three-humped symbolic dragon from local folklore, and there are projections like the head and tail of the

Dragon Bridge, Da Nang

an early reinforced concrete structure, with the reinforcement including an internal trussed iron framework. The bridge has a single 33m-long span across the Ljubljanica River, the world's third longest concrete span at the time. Its architectural treatment is in the Vienna Secession style and it is decorated with four, winged copper dragons, one at each corner.

dragon above each shorter end span, which are simple beam structures. The arches are formed from five tubes covered by decorative dragon 'scales' and there are short inverted arches at the piers to produce the flowing dragon shape. The head 'breathes' fire and water every Saturday and Sunday night and the rest of the 568m-long dragon is illuminated with 2,500 colourful LED lights. Construction of the bridge began in 2009 and it was opened in 2013 to mark the 38th anniversary of the city's capture during the Vietnam War. The bridge, which was designed by Louis Berger and Ammann & Whitney, received the 2016 ASCE Outstanding Civil Engineering Achievement Award.

Dragon Bridge, Ljubljana, Province of Ljubljana, Slovenia

Built in 1901 to replace an earlier timber crossing, the bridge was originally named after the Austro-Hungarian emperor Franz Joseph, but was renamed Dragon Bridge in 1919. It is

Dragon Bridge, Ljubljana

Dragon Bridge, Ubud, Bali, Indonesia

A tributary stream of the Petanu River in a small ravine is crossed by the 3ft-wide stepped footway of the segmentally-arched stone Dragon Bridge. Standing on a plinth at each corner of the bridge is a large stone dragon and the undulating parapets, which represent the backs of the dragons, are carved to look like scales.

Drava Footbridge, Osijek, Osijek-Barania, Croatia

Built in 1981, this elegant suspension bridge across the Drava River was damaged in the Battle of Osijek during the 1991–95 war in which Croatia fought for independence from Yugoslavia. It was repaired in 1993 and then fully renovated in 2007. The deck of the 5m-wide footbridge arches upwards in its 210m-long span between 30m-tall pylons on each bank. The pylons themselves consist of four gently curved legs which join together at the saddle where the suspension cables are supported.

Drava Footbridge

Dromana Bridge, Villierstown, Waterford, Ireland

The most interesting feature about the bridge at Dromana Castle is not the structure itself but the Hindu-Gothic gateway that is integral with it and forms the entrance into the castle's park. The first version of the gateway was a temporary timber and canvas building, erected

Dromana Bridge and Gateway

in 1826. This later masonry version of 1849, which was designed by the architect Martin Day in the Hindu-Gothic style of the Royal Pavilion at Brighton, consists of a pointed archway, topped by an oriental copper-clad dome with minarets, that is flanked by flat-roofed lodges with pointed windows. The original bridge across the River Finisk was replaced in the 1970s by a three-span prestressed concrete structure, but its railings are decorated by panels with pointed-arch openings in the style of the gateway.

Drum Bridge, San Francisco, California, USA

The Drum Bridge was designed and built in Japan by Shinshichi Nakatani before being re-erected in the Japanese Tea Garden for San Francisco's Midwinter Fair of 1894. The semicircular timber footbridge spans 20ft and is notable for its exceptionally low handrails.

Dubai Creek Bridges, Dubai, United Arab Emirates

There are several important road bridges over Dubai Creek. The first of these, opened in 1963, was the **Al Maktoum Bridge**, which consists of two parallel bascules each spanning 37m. The second, the original **Al Garhoud Bridge** was opened in 1976 and demolished after a steel girder replacement structure was completed in 2008 to carry the increasing amount of traffic

between Dubai and Sharjah. This bridge is 520m long and its 64m width enables it to carry fourteen traffic lanes.

The 22m-wide **Dubai Floating Bridge**, completed in 2007, carries six traffic lanes and two footways. It consists of two separate concrete pontoon structures, each 115m long, that are connected together end-to-end by a central steel linking unit which contains a gateway section that can be floated out of the way so water traffic can pass. Special link sections, which compensate for differential movements, connect the outer ends of the pontoons to the ramped approaches on each side of the creek.

The **Business Bay Crossing** (also known as the **Ras Al Khor Bridge**) was also opened in 2007. This has separate parallel bridges for each direction, with tall, steel, decorative A-frames linking the two structures along each pier line. The main span for river traffic is 60m long.

Another crossing, now called the **Sheikh Rashid Bin Saeed Crossing**, consists of two arches separated by an artificial island. The eastern arch spans 560m and the western 667m. Plans put forward for other crossings include **Shindagha Bridge**, which will be 380m long, 295m long and 56m wide, and a **Sky Garden Bridge** which, with its reflection, will be in the shape of the lemniscate symbol ∞ introduced by John Wallis in 1655 to represent the concept of infinity.

Dubai Floating Bridge (bottom) and Al Maktoum Bridge (top)

Duck Creek Aqueduct, Metamora, Indiana, USA

The aqueduct at Metamora carries the Whitewater Canal over Duck Creek, the difference in water levels being 16ft. The structure, which is America's last surviving covered timber aqueduct, is a Burr-type through arch and truss. An aqueduct was originally built here between 1839 and 1843 but the present structure dates from 1846 and was fully restored between 1946 and 1949. It spans 71ft and has an overall length of 82ft and width of 25ft. The aqueduct was designated a National Historic Civil Engineering Landmark in 1992.

Duck Creek Aqueduct

Duge Beipanjiang Bridge, Liupanshui, Guizhou Province, China

When it opened in 2016, this cable-stayed bridge carrying the G56 Expressway between Qujing and Liupanshui 565m above the Beipan River, became the highest in the world. The main pylons are H-shaped above the deck and its eastern tower, at 269m tall, is also a near record holder. The bridge has an overall length of 1,340m and its central span, at 720m long, was the second longest in the world when the bridge opened. It has since been awarded the Gustav Lindenthal Medal.

Duluth Lift Bridge, Duluth, Minnesota, USA

The first structure here across the 1871 Duluth Ship Canal was the Duluth Aerial Ferry Bridge, built in 1905 to a design by Thomas McGilvray,

that was the only transporter bridge to be built in America. However, in 1929, the structure was converted into a vertical lifting bridge by replacing the original travelling platform with two separate full-length trussed girder spans 390ft long that can be raised 135ft vertically up the existing but modified bridge towers.

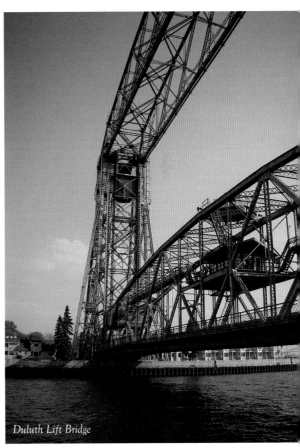

Duluth Lift Bridge

Dunajec Footbridge, Stromowce Nizne, Nowy Ttarg, Poland

When this 112m-long footbridge over the River Dunajec was built in 2006, it had the longest glued-laminate timber deck in the world. The cable-stayed structure has a main span of 90m and is supported by a 27m-tall pylon made of steel tubes that is inclined 15° from the vertical. This pylon has a very elongated H-shaped end elevation although with five cross bars, not one, between the two main legs and to which the stays are attached.

Dungeness River Bridge, Sequim, Washington, USA

The first railway bridge across the Dungeness River was a timber structure built in 1916 by the Seattle, Port Angeles & Western Railway. It was replaced in 1930 by a timber Howe truss 156ft long and 22ft high and is flanked by timber approach trestles on each side. Since 1995 the bridge has been part of the Railroad Bridge Park, carrying a bicycle way that is linked to the Olympic Discovery Trail, and is now listed on the National Register of Historic Places.

Duy Xuyen Bamboo Bridge, Hoi An, Qang Nam, Vietnam

This bamboo bridge across the Truong Giang River consists of X-shaped piers, each one made from two pieces of bamboo driven into the muddy river bed and tied together immediately below deck level. Long lengths of bamboo are then tied to these X-frames to provide supports for the simple 1.8m-wide deck and the handrails. The bridge is about 300m long and, typically, it takes 150 days to build each year and the structure is then washed away during the October to December rainy season.

E

Eads Bridge, St Louis, Missouri, USA

The construction of this bridge is one of the epic stories of bridge building. Designed by James Buchanan Eads and built by him between 1867 and 1874, this 6,442ft-long structure across the Mississippi River has three main spans, a central one 520ft in length flanked by a 502ft-long span on each side. The bridge foundations were the deepest yet built, being more than 100ft below water level, with the workers suffering from the then-unknown problem of compression sickness – the 'bends'. The bridge is included in the ASCE list of historic bridges.

East Taiheng Glasswalk

Eads Bridge

East Taiheng Glasswalk, Zhangjiajie, Hebei Province, China

One of the star attractions in the Tianmen Mountain National Park is the 60cm-thick glass-floored walkway which seemingly cracks when stepped on. It is, of course, an audio and visual special effect set off by sensors to frighten visitors. Although the walkway itself, which is 266m long and 2m wide, is sometimes called a bridge, it does not span across a gorge but is cantilevered off the side of a cliff 1,200m above sea level. The structure was designed by architect Haim Dotan and was opened in 2016.

Eastern Scheldt Bridge, Westenschouwen, Zeeland, Netherlands

The 5022m-long bridge across the Scheldt Estuary carries the N57 road about 5 miles west of the Zeeland Bridge (q.v.) and was the longest in Europe when completed in 1965. The reinforced concrete structure has forty-eight main spans 95m long, which consist of heavy arched beams spanning between massive piers that are in the shape of inverted Vs.

Ebro Bridge, Osera, Zaragoza, Spain

This unique structure was designed by Javier Manterola and completed in 2000 to carry the high-speed railway line across the River Ebro between Zaragoza and the French border. Although structurally it is a trapezium-shaped box beam, within which the railway runs, it is actually more of an open U-shape, with struts across the top and with circular openings along the slightly outward-sloping sides. The depth of the box is 9.2m and its maximum width is 16m. The viaduct is 546m long and the main section over the river has six spans, the biggest being 120m long.

Ebro Bridge

Edgar Cardoso Bridge, Figueira da Foz, Coimbra, Portugal

Now named after its Portuguese civil engineer designer, Professor Edgar Cardoso (1913–2000), this cable-stayed bridge was built between 1978 and 1982. Carrying the N109 road across the Mondego River, it has an overall length of 1421m with a 225m-long main span and back spans that are 90m long. The 83m-high pylon is A-shaped in elevation.

Eiffel Bridge, Girona, Catalonia, Spain

The footbridge here, which is also called **Pont de les Peixateries Velles**, replaced earlier timber bridges that had been swept away by floods. Completed in 1877 to a design by Gustave

Eiffel's company in Paris, it spans 42m and is 2.6m wide. Following a major restoration in 2008, the bridge now has an open box section with light, lattice girder sides and, above the footway, X-braced panels between arched portal beams.

Eiffel Bridge, Ungheni, Moldova

An earlier railway bridge over the River Prut, crossing the border between Romania and Moldova, was opened in 1874 but this was badly damaged by floods in 1876 and a replacement structure designed by Gustave Eiffel was completed in 1877. This is a tall girder structure with single- and double-height X-braced bays and carries a single-track railway line. There is a border checkpoint near the end of the bridge where there is also a change in track gauge.

El Ferdan Railway Bridge, Ismailia, El Qantara, Egypt

Several previous swing bridges carrying a railway across the Suez Canal here have suffered damage or destruction, including from a ship strike in 1947 and the Six-Day War in 1967. The present single-track steel structure, which was completed in 2001, has eight spans including an 1,100ft-long span consisting of twin balanced cantilever swinging sections – the biggest such structure in the world. The completion of the New Suez Canal in 2015 has left this bridge isolated but there are now plans for it to be widened and for a new swing bridge to be built across the new cut.

El Ferdan Railway Bridge

Elhova Footbridge, Elhovo, Yambol Province, Bulgaria

The simple chain-link suspension footbridge over the Tundzha River collapsed in 1996 when it became overloaded by people watching a religious ceremony in which a cross is thrown into the river. Nine people were killed.

Elisabeth Bridge, Budapest, Central Hungary, Hungary

The first bridge across the Danube on this site was the largest chain suspension bridge in the world when it was opened in 1903. Work had started in 1897 to a design by Aurél Czekelus and Antal Kherndl. The bridge had an overall length of 379m with a main span that was 290m long and the road on its 18m-wide deck was paved with timber blocks. The bridge was blown up by the Germans in January 1945. A replacement 27m-wide bridge, built between 1960 and 1964 to a design by Pál Sávoly, is a wire suspension structure with the cables spanning between the original piers (although now without their original decorative pinnacles).

Elisabeth Bridge

El-Kantara Bridge, Constantine, Constantine Province, Algeria

The El-Kantara Bridge is the latest and highest of a series of crossings, one above another, over the Rhumel River Gorge in Constantine, a city founded about 2,500 years ago. The lowest of these bridges is a natural rock arch left when the river originally formed the gorge, but immediately above this is the oldest man-made structure – a multi-arch stone bridge built by the Roman soldiers of the Third Augustan Legion.

Later, this structure was then used to support the three piers of a new bridge with two segmental stone arches on which was a level footway. On each side of these arches are tall half-arches that support steps leading up the gorge sides.

The next bridge at the site provided a high-level route 410ft above water level between the cliff tops on either side of the gorge. This had two tiers of tall, semicircular stone arches that made the bridge look rather like a Roman aqueduct. This partially collapsed in 1857 and, in 1862 under the auspices of Napoleon III, the top tier of arches was removed and replaced by a single three-ribbed cast iron segmental arch designed by Georges Martin. This spanned 187ft between massive stone piers and had distinctive iron latticework in the spandrels. It was flanked on each side by smaller semicircular stone arches. This was replaced in 1951 by the current El-Kantara Bridge, which has a single segmental concrete arch spanning between the stone piers from the previous bridge and supports the roadway on spandrel columns.

A short distance along the gorge are the Sidi M'Cid Bridges (q.v.).

Emir Bayindir Bridge, Ahlat, Harabesehir District, Turkey

The tall, pointed, stone-arched footbridge over the Keş or Harabesehir stream, which is named after its builder, was probably constructed in the fifteenth century and is considered to have once been part of the Silk Road network. The overall length of the bridge is 45m, it is 4m wide and the 7m-high arch spans 8.7m. Both the footway and the parapets are stepped, the bridge approach on each side is cranked and there is a small opening in one abutment. The bridge was repaired in 1954.

Emscher Landscape Park Bridges, Duisberg, North Rhine-Westphalia, Germany

The Emscher Landscape Park, which stretches from the source of the River Emscher in Holzwickede to its junction with the Rhine at Dinslaken, has been created by the redevelopment of a once heavily industrialised area of the Ruhr and covers an area of 175 square miles. The park

Slinky Springs to Fame Footbridge

and carries both road and rail traffic on its two separate 20m-wide decks. There are three parallel 75m-long arches over each of the two main spans and these are connected by an inverted arch over the shorter middle span to produce a wave effect. The arches on the centre-line of the structure between the two separate decks are a little deeper than the outer two, which are inclined inward slightly. The bridge was designed by engineers W.S. Atkins & Partners with architects Nicholas Grimshaw & Partners and was opened in 2001.

Enneus Heerma Bridge

contains several interesting new bridges. One of these is the 406m-long S-shaped **Slinky Springs to Fame Footbridge**, Oberhausen, which crosses the 50m-wide Rhine-Herne Canal. Although it looks like the child's Slinky spring toy invented in the 1940s, it is actually a stressed ribbon structure and consists of two high-strength steel bands beneath the 12cm-thick concrete deck which are stretched between slim V-shaped piers. With its conceptual design by artist Tobias Rehberger, the Slinky-like appearance is created by a 5m-diameter aluminium helix with 496 turns that encircles the 2.7m-wide deck along its length. Engineering design was by Mike Schlaich.

Also crossing the Rhine-Herne Canal is the 130m-long **Ripshorst Footbridge** designed by Schlaich Bergermann & Partner. The 3m-wide deck is supported by tubular props from a 78m-span main arch made from a steel tube that varies in diameter from 370mm to 550mm. As well as its vertical curve, this arch is doubly-curved in plan as well. Other footbridges in the park include one hung beneath two asymmetrical arches and an S-shaped cable-stay bridge with two single-post pylons.

Enneüs Heerma Bridge, Amsterdam, North Holland, Netherlands

This 230m-long tied-arch structure was built to link seven new artificial islands to the mainland

Enz Viaduct, Bietigheimer, Baden-Württemberg, Germany

The Württemberg Western Railway, one of the oldest railway lines in Germany, was ceremonially opened in 1853 and one of its major structures was the stone viaduct over the River Enz at Bietigheimer. This distinctive structure, designed by the architect Karl Etzel, is considered to be among the most elaborate structures on German railways. The viaduct, which is 287m long, is 26m high and has two rows each of twnty-nine arches. Those in the upper row are semicircular while those in the lower row are segmental and narrower from front to back. Six arches were destroyed at the end of World War II and, as part of the reconstruction work, one of the double arches was completely filled with concrete.

Enz Viaduct

Epirus Bridges, Epirus, Greece

There are several interesting bridges in Epirus, a mountainous region of north-west Greece. After an earlier bridge collapsed in 1863, the new **River Arachthos Bridge, Plaka** was completed three years later. Its 40m-long nearly semicircular central span is claimed to be the country's biggest stone arch structure and its 3m-wide deck follows a very hump-backed profile. The bridge was rebuilt in 2020 after the central section collapsed in 2015 during flash floods. There has also been a bridge over the River Arachthos at **Arta** for many centuries (possibly since the third century BCE) but the present structure probably dates from about 1615. It has four main arches, the largest spanning 24m at one end, with four subsidiary arches lightening the structure above the main piers. A legend tells how the work completed during the day collapsed each night until the master craftsman sacrificed his beautiful wife to propitiate the gods. The **Kalogeriko Bridge, Tymfi**, which was built in 1814 to replace an earlier timber structure, has three semicircular stone arches supporting a noticeably sinusoidal-shaped deck. This is about 55m long and 3m wide.

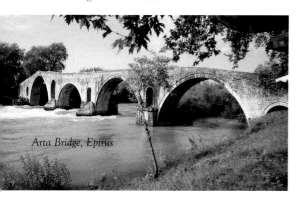

Arta Bridge, Epirus

Erasmus Bridge, Rotterdam, South Holland, Netherlands

The Erasmus Bridge over the Nieuwe Maas, opened in 1996, is nicknamed locally 'The Swan' because of the dominant profile created by its cranked and inverted Y-shaped pylon above the back-span deck and this shape is now the official logo for the city. Apart from an approach viaduct from the north, the 34m-wide and 802m-long bridge contains two main elements: the cable-stayed section with a 280m-long main span, and a southern section containing an opening bascule. The pylon supporting the cable stays has a lower part that leans away from the main span to help balance its weight. Above this the rest of the 139m-high pylon is vertical and anchors the thirty-two semi-fan system stays that support the main span. There are only two large-diameter backstays which link the top of the pylon to anchorages in the south foundation. The bascule section has an opening leaf in the shape of a parallelogram, with side lengths of 52m and 36m, that is skewed out sideways by 22° from a rectangle and, most unusually, the horizontal axis for this is not at right angles to the centreline of the bridge. The bridge was designed by Ben van Berkel.

Erasmus Bridge

Eshima Grand Bridge, Matsue, Shimane, Japan

The bridge across Nakaumi Lake was opened in 2004 and replaced a previous drawbridge by taking the road 45m above water level so road traffic no longer has to stop when ships are passing. The main part of the 1,700m-long reinforced concrete structure consists of a 250m-long central arched span flanked on each side by a half-arch. There are further approach spans on each side with road gradients of 6.1% and 5.1%, these looking frighteningly steep in end-on photographs taken with a telephoto lens.

Essinger Footbridge, Essing, Bavaria, Germany

The stressed ribbon timber footbridge across the Main-Danube Canal in Essing was completed in 1986. It has an overall length of 190m and consists of two 73m-long spans with a V-shaped pier at each end and the central pier having a double vee. The deck is 3m wide and 0.65m deep. The footbridge is considered to be the longest timber bridge in Europe.

Essinger Footbridge

Europa Bridge, Innsbruck, Tyrol, Austria

When the Europa Bridge across the Sill River was opened in 1963, after four years of construction, it became Europe's highest bridge, retaining this title for ten years (a title held since 2004 by the Millau Viaduct – q.v.). It carries the six traffic lanes of the A13 Brenner Autobahn and European route E45 at a height of 192m above the Sill River and is 777m long with a longest span of 198m. The structure consists of steel box girders on concrete piers and there has been bunjee jumping from a platform just beneath the bridge deck since 1998.

Eurymedon Bridge, Aspendos, Antalya Province, Turkey

The Roman stone bridge over the River Eurymedon at Aspendos is believed to have had originally nine semicircular arches and been about 260m long and 9m wide, with the biggest arch spanning about 24m. The bridge having collapsed at an unknown date, it was rebuilt in the early thirteenth century but cranked and with a shorter length and narrower width. This rebuild also had just seven stone arches, these now being slightly pointed. It was restored in the late twentieth century.

Eurymedon Bridge

F

Fades Viaduct, Les Fades, Puy-de-Dôme, France

The 470m-long and 12m-deep triple lattice truss steel viaduct at Les Fades, which carries a single-track railway line 133m above the Sioule River, was the tallest bridge in the world when it was completed in 1909 and, at 92m-high, its biggest pier is still the largest such stone construction ever built. There are three main spans, a central one that is 144m long with 116m-long side spans, and these are approached by a masonry arch over a cliff-top road at one end and a shorter truss span at the other.

Faidherbe Bridge, Saint-Louis, Saint-Louis Region, Senegal

The French General Louis Faidherbe, who was governor of Senegal, had initiated an early ferry crossing to access the island of Saint Louis but in 1865 a pontoon bridge was built to provide a fixed link. This, which was 350m long and 4m wide, consisted of forty pontoons connected by a wooden deck and was called Faidherbe Bridge after the General. Three of these pontoons could be disconnected and floated to one side to leave a 60ft gap for shipping to pass. A new road bridge over the Senegal River, which linked what had become the city of Saint-Louis to the mainland, was opened in 1897. This is 507m long and has eight segmentally-arched steel girder spans, one being a short swing span. Between 2008 and 2011 all the original spans were replaced.

Faidherbe Bridge

Faisal Bridge, Ksar el Kebir, Larache, Morocco

The world's longest single railway bridge, built between 2011 and 2018, is one of twelve on the new high-speed twin-track line between the cities of Tangier and Kenitra in Morocco. The structure, which consists of a series of long steel beams across the Allokos basin, is 2,300m long – five times longer than strictly necessary – in order that it would not restrict the existing water systems.

Fanambana Bridge, Fanambana, Sava, Madagascar

The five-span road bridge over the Fanambana River was completed in 1964 and now carries road 5A. The cable-stayed structure, its stays in a fan arrangement, has an overall length of 195m with a 120m-long main span.

Fanga'uta Lagoon Bridge, Nuku'alofa, Tongatapu, Tonga

Work is in progress on a new bridge across the narrowest part of Fanga'uta Lagoon that, by 2024, will link Nuku'alofa, Tonga's capital city, to Folaha. The bridge will be 720m long and is expected to cost US$55 million.

Fangsheng (Liberate Living Things) Bridge, Zhujiajiao, Shanghai, China

This 72m-long bridge over the Caogang River, built in 1571, has five semicircular arches carrying a lengthy stepped footway up either side to a short flat section at the crown. The parapet for this central section is decorated with four stone lions.

Father Bernatek Footbridge, Cracow, Lesser Poland, Poland

The new footbridge across the Vistula River, opened in 2010, is located where the earlier

Podgorski road bridge had been. It consists of a single tubular steel arch spanning 148m and with a rise of only 15m, from which an X-shaped arrangement of cable hangers supports twin decks that are held apart by horizontal props. Father Laetus Bernatek was a monk who constructed Cracow's Bonifrater Hospital.

Father Mathew Bridge, Dublin, County Dublin, Ireland

The first bridge on this site is thought to have been ordered by King John in 1214 but this was pulled down during Edward Bruce's invasion of Ireland between 1315 and 1318. After rebuilding, it was destroyed by floods in 1385 and rebuilt again in 1394. It was rebuilt once more, this time by Dominican Friars, in 1428 and this structure was replaced by the present bridge in about 1818. This latest structure is a three-span masonry bridge with semi-elliptical arches and has an overall length of 140ft.

Favazzina Bridges, Scilia, Calabria, Italy

The first motorway viaduct here, which was opened in 1974, was designed by Riccardo Morandi. It consisted of two curved ten-span prestressed concrete beam structures standing on tall inverted-V columns, that were sometimes described as 'clothes-peg' columns. Each was 147m high with 50m-long spans. However, the carriageway decks were narrow and tightly curved. They were therefore replaced in 2013 by two three-span cable-stay structures. These were 440m long and 147m high with 197m-long main spans. Following their completion, the older structures were demolished in 2015.

Favela da Rocinha Footbridge, Gavea, Rio de Janeiro, Brazil

Designed by the architect Oscar Niemeyer (1907–2012), this unusual footbridge serves to link Rocinha, Rio's largest favela (a labyrinthine slum area ruled by drug gangs), and a more open area containing a sports complex, that are divided by a dual carriageway road. A narrow concrete arch, which spans diagonally over the road between abutments set well outside its edges, has a thin

central hanging concrete tie supporting the midpoint of a U-shaped concrete footway. After crossing over the two carriageways in a straight line, the footway returns to ground level at each end by winding through a series of sharp curves. The footbridge was built as part of infrastructural improvement for the 2014 World Cup.

Fibreglass Bridge, Friedberg, Hesse, Germany

This innovative bridge at Friedberg, which carries a minor local road over the B3 highway, includes a lightweight glass-reinforced polymer structure. Two steel beams, each 21.5m long, span between cantilevered concrete abutments and support a pultruded multi-cellular deck which is adhesively bonded to the steel. This arrangement allowed rapid installation and it is expected that future maintenance costs will be greatly reduced. The bridge, which was completed in 2008, was designed by Knippers Helbig Consulting Engineers.

Fink Deck Truss Bridge, Lynchburg, Virginia, USA

The Fink truss was patented by Albert Fink in 1854 and this rare cast iron version was built by him in 1870 to carry the Norfolk & Western Railroad tracks over a road. The structure was taken down and rebuilt in 1893 to carry a road over another set of tracks and was moved again in 1985 to become an historical exhibit in Lynchburg's Riverside Park. The bridge's span is 53ft and its deck is 13ft wide. The bridge is now an ASCE National Historic Civil Engineering Landmark.

Firth of Forth Bridges, South Queensferry, City of Edinburgh, Scotland

The magnificent **Forth Bridge**, now a World Heritage site, is known throughout the world as a masterpiece of Victorian engineering and represents the first large-scale British use of steel as a construction material. When it was built, it had the world's longest span, an honour it kept until the completion of Canada's 1,800ft-long Quebec bridge (q.v.) in 1917. Tentative proposals for earlier fixed crossings of the Firth of Forth included, in 1806, a tunnel and, in 1818, a chain bridge. In 1873 the Forth Bridge Company, jointly owned by four railways, was set up to promote a rail bridge, and work on a suspension bridge designed by Thomas Bouch began in 1873. However, this was quickly abandoned when Bouch's Tay Bridge collapsed in 1879. Work eventually started again in 1882 on the bridge designed by John Fowler and Benjamin Baker, and it was opened in 1890. The structure consists of three towers 361ft tall with double cantilevers, each arm being 680ft long. Between the ends of these cantilever arms suspended sections 350ft long create two equal main spans between the towers of 1,710ft, with headroom over the shipping channels beneath of 150ft. This part of the bridge contains 51,000 tons of steelwork and is 5,350ft long. Internal double Warren girders propped off the lower booms of the cantilever arms support the twin rail tracks within the main structure. The approach viaduct on the south consists of ten 168ft-long lattice girder spans and four 66ft-span masonry arches, that on the north five similar lattice girder spans and three small arches. Including these approach viaducts the overall length of the bridge is 8,400ft. In 2011 a decade-long project to restore and repaint the bridge ended. The new paintwork, a two-coat epoxy and a top coat of acrylic urethane, should last for at least twenty years, thus finally making history of the 'painting the Forth Bridge' metaphor for a never-ending job. Network Rail is planning to open a visitor platform on the top of the Forth Bridge's Queensferry cantilever, with access from under the south approach span and then round the east face of the southern masonry tower. The bridge is included in the ASCE list of historic bridges.

The name Queensferry commemorates Queen (later Saint) Margaret (c. 1045–1093), wife of King Malcolm III of Scotland (and niece of England's Edward the Confessor), who encouraged pilgrims to St Andrews in Fife to use a ferry that crossed the Firth of Forth at the narrowest point between Kincardine, 15 miles upstream, and the sea. The ferry service had continued for about 900 years when the **Forth Road Bridge** finally replaced it in 1964. The Grade A structure, which was designed by Mott, Hay & Anderson and Freeman, Fox & Partners, has a main span of 3,300ft and side spans of 1,340ft, and the tops of the triple box steel towers are 512ft above mean water level. The cables, 2ft in diameter and consisting of 11,618 parallel high tensile wires 5mm in diameter, sag about 300ft below their support saddles to give a sag:span ratio of 1:11. The deck structure, which is suspended on hangers from the cables, consists of two Warren truss girders 78ft apart and 28ft deep, giving a span:depth ratio of 120:1. The approach viaduct on the south side has eleven spans, that on the north six. When the toll bridge was completed it was the largest suspension bridge in Europe and the fourth largest in the world. The discovery of corrosion in some of the wires as a result of moisture penetration required the installation of a dehumidification system for the cables. Other work has included the replacement of the expansion joints and the repair of a broken connection between the north-east tower and the adjacent main span truss. With these problems in mind and also to increase capacity, the decision was taken to build a second road bridge.

Designed by the Forth Crossing Design Joint Venture and built by the consortium Forth Crossing Bridge Constructors, the **Queensferry Crossing** bridge is located 700m upstream of the Forth Road Bridge and relieves that bridge of much of its heavy commercial traffic. It is a 40m-wide, 2.7km-long multi-span bridge with a three-tower cable stay structure at its heart, this having the unusual feature of a 146m overlap of the stays in both 650m-long main spans. During construction, the structure became the largest free-standing balanced cantilever in the world, with the arms extending 322m either side of the central tower. This, at 210m-high (690ft), is the tallest bridge structure in the country. The north and south approach viaducts were incrementally launched. Work began on site in mid-2011 and the toll bridge, which now carries the M90, opened in 2017. The total cost of the project, including 22km of new motorway link roads and upgrades, was £1.35bn. (See *AEBB*).

Five Circles Footbridge, Copenhagen, Capital Region, Denmark

This 40m-long footbridge crossing the southern end of the Christianshavn Canal consists of a series of five slightly differently-sized interlinked circular decks. The free outer edges of these decks are supported by a total of 110 inclined stays from the tops of tall masts located at the centres of the decks. Two of the decks swing open together to allow water traffic to pass. The structure was designed by Olafur Eliasson and opened in 2015.

Five Circles Footbridge

Flat Bridge, St Catherine, Middlesex, Jamaica

The 45m-long Flat Bridge over the Rio Cobre is one of Jamaica's oldest bridges and the Jamaica National Heritage Trust has declared it to be a National Monument. Built by slaves, it was completed around 1770 and in 1881 the bridge deck was washed away. The two stone piers have large triangular cutwaters at each end. Although it had metal and, later, timber handrails, these have also been washed away and it is now called Flat Bridge because there is only a low kerb at each edge of the deck. It carries a single one-way traffic lane, controlled by lights, of the A1 road.

Flisa Bridge, Flisa, Innlandet County, Norway

Completed in 2003, the three-span 196m-long Flisa Bridge over the River Glomma is one of the biggest glulam bridges in the world. Its 70m-long central span consists of a trussed section that is deeper over the mid-river piers and cantilevers out beyond them at each end by a further 17m. The free ends of these cantilevers then each support

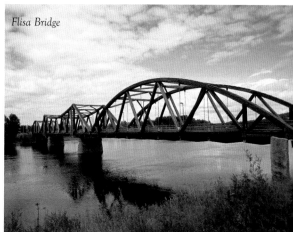

Flisa Bridge

one end of an arched truss, the other end of which stands on an abutment on the river bank.

Floating Bridge, Enshi City, Hubei Province, China

Although not as well-known as the Governor Albert D. Rosellini Bridge in Seattle (q.v.), this 500m-long floating bridge along a winding river in China is more beautiful. Opened in 2016, it is 4.5m wide and the passage of cars along it creates undulating waves. It is also claimed that, when closed to cars, the structure can accommodate 10,000 people walking on it together.

Fontevecchia Bridge, Valdragone, Borgo Maggiore, San Marino

A 32km-long narrow-gauge electric railway, which connected the independent sovereign state of San Marino to the Italian railway network at Rimini in Italy, started operating in 1932 but was never re-opened after the war. The bridge at Valdragone is an attractive structure with four semicircular stone arches carrying the railway over a small valley. In 2018 postage stamps showing a painting of a train crossing over the Fontevecchia Bridge were issued.

Foreshore Freeway Bridge, Capetown, Western Cape, South Africa

Construction began in the early 1970s on an eastern boulevard (part of the M62) for Capetown that included this flyover bridge, but work stopped in 1977 and has never been completed. The structure is now used as an elevated car park and the ends of the two separate carriageways project out into empty space like a huge modern sculpture.

Foreshore Freeway Bridge

Fourth Thai-Lao Friendship Bridge, Ban Houayxay, Bokeo Province, Laos

Opened in 2013 and making the final link in Asian Highway 3, this bridge across the Mekong River connects Laos and Thailand. The 630m-long concrete box girder structure has a main span of 480m and carries a 15m-wide deck.

Foyle Bridge, Derry, County Londonderry, Northern Ireland

Opened in 1984 to relieve the pressure of road traffic on Craigavon Bridge (q.v.), Foyle Bridge has a main span of 234m, the longest of any Irish bridge, flanked by side spans of 144m. The deck is carried on twin steel box girders and there are reinforced concrete approach spans.

Francis Scott Key Bridge, Baltimore, Maryland, USA

This 8,636ft-long bridge, which carries the I-695 road across the Patapsco River as it enters Baltimore Harbor, has a main 2,644ft-long three-span continuous through-truss structure. This provides a central 1,200ft-long opening across the shipping channel, with 185ft clearance above

water level. The bridge has a 52ft-wide roadway and was built between 1972 and 1977. Its name commemorates the man who, during the war of 1812, wrote the words to the *Star-Spangled Banner*, that has been the national anthem of the USA since 1931.

Franjo Tudjman Bridge, Kantafig, Dubrovnik-Neretva, Croatia

Opened in 2002, the shortened D8 route between Dubrovnik and Split now crosses the River Ombla on the 481m-long Franjo Tudjman Bridge at the point where the river meets the sea inlet to Dubrovnik. The cable-stay structure has a single A-shaped pylon 143m tall and supports a 304m-long main span. Clearance between water level and the underside of the deck is 52m.

Franjo Tudjman Bridge

Frankford Avenue Bridge, Philadelphia, Pennsylvania, USA

The Pennsylvania General Assembly passed a law in 1683 enacting that bridges were to be built wherever the King's Highway crossed rivers and creeks in the state. Frankford Avenue Bridge across Pennypack Creek was built in 1697 and is generally considered to be the oldest surviving bridge in the whole country. The 10ft-wide and 73ft-long structure consisting of three nearly-semicircular stone arches was widened in 1893. It is now 37ft wide, although a cantilevered concrete footway has been added more recently. The bridge is also known as **Pennypack Creek Bridge**. In 1970 it became the seventh bridge to be listed by the American Society of Civil Engineers as an Historic Engineering Landmark and in 1988 it was entered on the National Register of Historic Places. It has recently been thoroughly rehabilitated.

Fredrikstad Bridge, Fredrikstad, Viken, Norway

The two-hinged crescent arch steel bridge carrying Road 110 over the River Glomma was opened in 1957. Including the twenty-two approach spans the overall length of the bridge is 824m long, the arch itself spans 196m and the deck is 13m wide. The top of the arch is 104m above water level and the clearance beneath the deck is 40m. Since 2008 the bridge has been one of the cultural heritage sites of Norway.

Friendship Bridge, Ciudad del Este, Alto Paraná Department, Paraguay

This bridge spanning 290m across the Parana River to link Paraguay with Brazil was opened in 1965. The thick reinforced concrete deck arch supports four pairs of distinctively slim columns in each spandrel and there are further similar columns in the approach spans. The overall length of the bridge is 552m and its 14m-wide deck is 78m above water level.

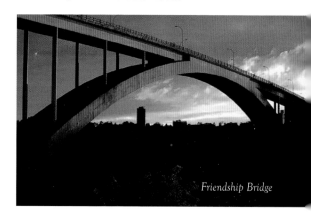

Friendship Bridge

Gabriel Tucker Bridge, Monrovia, Montserrado, Liberia

The story of this bridge across the Mesurado River in Liberia's capital city shows that bridges can suffer from civil unrest in small countries just like those in big countries during world wars. The reinforced concrete bridge, which was first built between 1972 and 1976 and named after the minister of public works, links the town centre to one of the suburbs via Providence Island. (This island is famous for being one of the first places where freed slaves landed in 1822, leading to the creation of Liberia.) During the series of Liberian Civil Wars the bridge was the front line in much fighting in 2003 between forces of former president Charles Taylor (the National Patriotic Front of Liberia) and those of Liberians United for Reconciliation and Democracy (LURD), leading to its collapse in 2006. It was rebuilt in 2018 when Turkey provided help in reconstructing Liberia's infrastructure.

Galata Bridge, Istanbul, Marmara Region, Turkey

An early bridge across the mouth of the Golden Horn in Istanbul was probably built in the sixth century. The first modern bridge there, called **Hayratiye Bridge**, was completed in 1836. It was a timber structure spanning between floating pontoons, was about 400m long and 10m wide and contained an opening bascule span. A replacement timber bridge, built in 1863 and known as **Unkapani Bridge**, was itself replaced in 1875 by an iron bridge 480m long and 18m wide which was also supported by floating pontoons. This version of the bridge, which introduced the concept of a lower deck lined by cafes and shops, was in turn replaced in 1912 by another floating crossing, which was 466m long and 25m wide. After this had been damaged by fire in 1992 it was towed away to another site about one mile further up the Golden Horn where diners and shoppers can still visit it. Its function as a bridge at the original site was continued by a replacement double-decked Galata Bridge, which was opened in 1994. This is a 465m-long road bridge with a main bascule span of 80m and is 42m wide. Above the lower deck, which houses cafes and restaurants, it carries six road lanes and two tram lines. The bridge continues to be a long-standing symbol of the city.

Galata Bridge

Galindo Bridge, Bilbao, Biscay, Spain

The bridge across the Galindo River in Bilbao is a complicated structure in geometric terms: it is curved in plan with a radius of 250m, has a cross-fall from side to side of 5% and a longitudinal slope of 3%. Its deck is 27m wide and carries dual two-lane carriageways and two footways of different widths. The deck is supported by a 110m-span tubular steel tied arch, the plane of which leans outward and is also tied back by stays to cantilever arms projecting outward and upward from the inside edge of the deck. On the outside edge of the deck the 6m-wide footway is covered by a canopy supported on curved cantilever frames. The bridge was designed by Javier Manterola and was opened in 2007.

Ganter Bridge, Brig, Valais, Switzerland

The reinforced concrete, extradosed Ganter Bridge carries the Simplon Pass road high over the Ganter River. Designed by Christian Menn and opened in 1982, the principal structure has a straight 174m-long main span flanked on each side by a single curved side span 127m long. The cable stays that radiate out each side from the tops of the main piers are enclosed in concrete in a flat triangular shape. In 1997 Menn was awarded the John A. Roebling Medal.

Ganter Bridge

Garabit Viaduct, St Flour, Auvergne, France

The wrought iron bridge across the Truyère River was designed by Gustave Eiffel and completed in 1884. Broadly similar in appearance to his earlier Maria Pia Bridge in Portugal (q.v.), it consists of a two-pinned parabolic arch with a main span of 165m. The shape of this arch is distinctive as it is wider from side to side at the base than at its apex but deepens towards mid-span. It supports a 448m-long trussed beam that carries the single-track railway line 120m above the water, and there are masonry arch approach viaducts at each end. The viaduct was the tallest bridge in the world for twenty-five years.

Garden Bridge, Grenoble, Isère, France

What is considered to be the world's first mass concrete bridge using Portland cement was built in 1855 by brothers Joseph and Louis Vicat

in the botanical gardens at Grenoble. It has a single span segmental arch, the formwork for which had been fabricated to include moulded panels so that the resulting finished concrete was marked in imitation of nine stone voussoirs.

Gazela Bridge, Belgrade, Serbia

Just before the Sava River in Belgrade flows into the Danube it is crossed by this 28m-wide six-lane bridge carrying the E-75 road. The steel structure, in the form of a slightly-arched continuous beam, was built in 1975 and is 332m long, the 210m-long central span being supported at each end by two inclined struts founded on the river banks. Because of the shallow angle of these supports the bridge was said to be leaping the river like a jumping gazelle, and this became the name of the crossing. It soon became apparent that there were problems of excessive deflection of the bridge under loading and, although the deck was immediately lightened, major reconstruction was undertaken between 2010 and 2011.

Gazela Bridge

General Belgrano Bridge, Corrientes, Corrientes, Argentina

The prestressed concrete bridge here was built between 1968 and 1973 to carry National Route 16 at a height of 35m over the Paraná River. The main part of the bridge is 1700m long with a central 245m-long cable-stayed span and back spans of 164m. The two A-shaped towers are 84m high. The year after it was opened, Argentina issued a 4.50 pesos postage stamp showing the bridge. Two people were killed on the bridge in 1999 during protests against the government.

General Rafael Urdaneta Bridge, Maracaibo, Zulia, Venezuela

Previously called the **Maracaibo Lake Bridge**, this cable-stayed reinforced and prestressed concrete structure carries the country's Trunk Road 3 across the Tablazo Strait between Lake Maracaibo and the Caribbean Sea. It was designed by Riccardo Morandi and was built between 1958 and 1962. Overall, the bridge is 8.7km long but it has five 235m-long main spans between 92m-tall piers. These have an A-frame outside the deck edges, between which is a double X-shaped frame. There are 130 approach spans, the main piers of these consisting of four parallel frames. In 1964 a tanker collided with the bridge leading to its partial collapse and the deaths of seven people. The bridge is conceptually similar to Morandi's Polcevera Viaduct in Genoa (q.v.) which collapsed in 2018.

General Rafael Urdaneta Bridge

George P. Coleman Memorial Bridge, Yorktown, Virginia, USA

The 3,750ft-long steel bridge here over the York River is famous for having two swing spans in tandem. It was built as a two-lane structure in 1952 before being reconstructed and widened in 1995, when the two 450ft-long swinging sections were floated in by barge. These are each 500ft-long cantilevered trusses. The bridge now carries the four-lane US Route 17 and, in the centre of the bridge, the road is nearly 90ft above water level. Parson, Brinkerhoff, Hall & Macdonald was the design firm. The bridge is named after a head of the Virginia Highway Commission.

Geumcheongyo Bridge, Seoul, Gyeonggi Province, South Korea

This little stone footbridge across the Myeongdangsu (also known as Geumcheon) stream at Changdeokgung Palace was completed in 1411 and is considered to be the oldest bridge in South Korea. In ancient Asian civilisations crossing a bridge was seen as a purifying act and bridges were therefore often built at the entrance to temples and palaces. Classified as a National Treasure and a World Heritage Site, the structure is 42ft long and has two semicircular masonry arches. The parapet is in three sections separated by tall stone pillars with lotus caps and there are also carvings of imaginary animals.

Gignac Bridge, Gignac, Hérault, France

The bridge carrying the N109 road across the Hérault River is widely considered to be the finest eighteenth-century bridge in France and has been a National Monument since 1950. It was built between 1776 and 1810, construction being delayed first by difficulties with the foundations and then by the French Revolution. It has an overall length of 175m and its 10m-wide deck is 20m above river level. There are three stone arches: a central one spanning 48m flanked on each side by a 26m-span side arch that is set forward slightly and has unusual splayed arch rings.

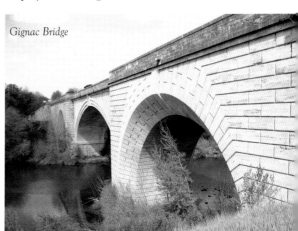

Gignac Bridge

Girsu Bridge, Tello, West Azerbaijan, Iraq

The baked-brick remains of a 4,000-year-old structure in the ancient Sumerian city of Girsu were rediscovered in 1929. These include walls up to 11ft high and it is now believed that they served to reduce the width of an ancient canal so that it could be spanned by timber planks. The remains are therefore considered to be parts of the oldest bridge in the world. New conservation work, by a joint team of British and Iraqi archaeologists, began in 2018. The Iraq State Board of Antiquities & Heritage is working with the British Museum to conserve the site. (The world's oldest bridge still in use is Caravan Bridge, Izmir, Smyrna, Turkey – q.v.)

Giuliana Bridge, Benghazi, Benghazi Province, Libya

Giuliana Bridge consists of two parallel prestressed concrete haunched box girders and carries Benhhazi's Algeria Street across the 23rd July Lake. It has a central span of 120m flanked by 80m-long side spans. It was built in the 1970s and refurbished in 2005. In 2011, Libyan forces loyal to Colonel Muammar Gaddafi fired on a crowd of protesters on the bridge and many were killed when they attempted to escape by jumping off it.

Gladesville Bridge, Sydney, New South Wales, Australia

The first bridge here across the Parramatta River, a short distance upstream from Sydney Harbour Bridge (q.v.), was a 274m-long trussed girder bridge with a swing span that was built in 1881. This was replaced by the present reinforced concrete deck arch road bridge, which was opened in 1964. The structure has a 305m-long main span and its overall length is 580m. The 27m-wide deck originally had six traffic lanes but this was later increased to seven by reducing the width of the footways. The bridge was registered as a state heritage structure in 2014.

Glanworth Bridge, Ballyquane, County Cork, Ireland

The bridge across the River Funshion at Ballyquane is considered to be narrowest public bridge still in everyday use in Europe. It has a broadly symmetrical slightly hump-backed appearance with three central semicircular arches each spanning about 16ft flanked on each side by five slightly smaller arches that are also semicircular. The arch at one end of the bridge is hidden by a later mill race and there are cutwaters on the upstream pier ends. Altogether the bridge is about 240ft long and is only 9ft wide. Beneath the bridge a weir runs diagonally across the river. The bridge was probably built in the late sixteenth or early seventeenth century and is overlooked by the thirteenth-century Glanworth Castle.

Glicnicke Bridge, Potsdam, Brandenberg, Germany

The first bridge here across the Havel River was a timber structure built in about 1660. Following a design competition, the present bascule structure, which does not impede river traffic like its predecessor did, was opened in 1907, but it needed extensive reconstruction following war damage in 1945. It later became famous as the 'Bridge of Spies' when it was used for the exchange of agents captured during the Cold

Gladesville Bridge

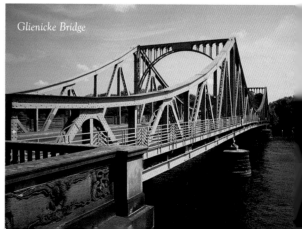

Glienicke Bridge

War. It is a three-span trussed steel structure with a central span of 74m flanked by side spans each of 37m, and its profile is somewhat akin to a suspension bridge. The bridge was shown on a German stamp issued in 1998.

Goat Canyon Trestle, San Diego, California, USA

The curved redwood trestle bridge across the Goat Canyon was built between 1932 and 1933 by the San Diego & Arizona Railroad to carry its single-track line and is considered to be the world's largest all-timber trestle structure. Timber was deliberately chosen because of the risk of high temperature variations causing metal fatigue in steelwork. It was designed by the railroad's chief engineer Carl Eichenlaub to have a length of 633ft at deck level and a maximum height of 186ft in ten tiers. The trestle was closed for repairs between 1976 and 1981 and the line itself was closed due to damage elsewhere between 1983 and 2004 and again since 2017.

Golden Bridge, Da Nang, Hai Chau, Vietnam

This unique 5m-wide and 150m-long curved bridge is part of a footway linking two mountain-top cable car stations. The underside of the gold-painted bridge deck, which is semi-tubular in section, is supported by inverted tetrapods standing on tall circular columns. However, visually these are overwhelmed by what appear to be a series of stone hands holding up the deck (but, in fact, are made out of fibreglass and wire mesh on an internal structure) to create the image of 'giant hands of gods pulling a strip of gold out of the land'. The bridge was designed by TA Landscape Architecture and was opened in 2018. In 2020 it was the subject of the Agora Prize for best architecture photo.

Golden Bridge

Golden Gate Bridge, San Francisco, California, USA

The suspension bridge across the narrow strait connecting San Francisco Bay and the Pacific Ocean was for many years the world's longest suspension bridge and it remains one of the most famous bridges in the world. Designed by Joseph Strauss in 1917, but not opened until 1937, the bridge's 90ft-wide deck carries six lanes of traffic on Route 101. It has an overall length of 8,980ft and its 4,200ft-long main span provides 220ft of clearance above the shipping channel at high tide. The bridge is included in the ASCE list of historic bridges.

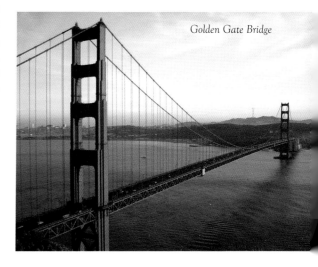
Golden Gate Bridge

Göltzschtal Viaduct, Mylau, Saxony, Germany

This double-track railway viaduct across the Göltzsch River, built between 1846 and 1851 to a design by Johann Andreas Schubert, is the world's largest brick bridge and contains more than 26 million bricks, although the arch rings themselves are of stone. It is 1,833ft long, 30ft wide at deck level and its greatest height is 256ft above river level. There are four tiers of arches and, because of the sloping valley sides, the top tier is the longest with twenty-nine arches. For all except the top tier, where the arches are full width to support the main deck, each arch consists of two separate ribs. Over the river itself there are only two arch tiers, each with an enlarged span of 101ft. All the masonry was fully restored between 1955 and 1958. The Elster Viaduct, 6 miles to the south, is similar but smaller.

Gongchen Bridge, Hangzhou, Zhejiang Province, China

The Gongchen Bridge, a stone structure with three semicircular arches, was built across the Grand Canal in 1631 and renovated in the 1880s. It is 92m long, 6m wide and 16m high and in 1986 it was registered as having great local heritage value.

Goodwill Bridge, Brisbane, Queensland, Australia

Built for Brisbane's Goodwill Games in 2001, the Goodwill Bridge for cyclists and pedestrians crosses the Brisbane River to link South Bank with the city centre. Together with approach spans, the overall length of the 21ft-wide structure is 1,475ft and the main span of its steel supporting arch is 328ft. The bridge was designed by Ove Arup & Partners, built off-site and floated into position.

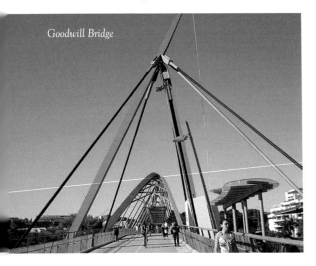

Goodwill Bridge

Gorbaty Bridge, Minsk, Belarus

This bridge was built in 1964 to provide a connection across the Svislach River to the Island of Tears, called after the monument there dedicated to the nation's 800 soldiers who had died in Afghanistan. The 51m-long structure consists of a simple arched concrete deck with an additional sloping strut from water level at the abutment to about the fifth-points of the span's soffit. The bridge is 6m wide.

Gorica Bridge, Berat, Berat County, Albania

The first bridge here over the Osumi River was a timber structure built in 1780 by Ahmet Kurt Pasha of Berat. This was replaced in 1927 by the existing 129m-long stone bridge which has seven segmental stone arches. The broad piers have a tall internal arch, with a further smaller arch within the spandrel on each side. The level roadway slopes down at both ends and each panel of the parapet wall is lightened by 'starburst' openings. The bridge was damaged by explosions in 1918 and was fully renovated again in 2015.

Gothic Bridge, Vilomara, Barcelona, Spain

An early timber bridge over the Llobregat River at Vilomara was followed by a Roman bridge but the present 130m-long structure has been dated to the twelfth century. It has a high central semicircular arch with the rounded pier ends on either side rising through the spandrels to finish at parapet level. However, although the parapet line slopes down each side from this arch, the structure is not symmetrical. One of the seven side arches is pointed and there is also a small flood arch high up above one of the piers. The bridge is now restricted to cyclists and pedestrians.

Gothic Bridge

Gouritz River Bridges, Albertinia, Western Cape, South Africa

The first bridge across the Gouritz Gorge, built between 1891 and 1892, was a steel cantilever-and-suspended-span lattice girder bridge that carried a single-lane road and in 1906 the New Cape Central Railway laid a railway track within the roadway. As traffic congestion built up a second bridge was built for the railway between 1930 and 1931 and the original crossing reverted to road-only use. This new structure is a five-span N-braced girder with three main five-tier towers based in the valley bottom and two further piers on the valley sides. The third bridge over the Gorge, an elegant rigid frame concrete structure carrying highway N2, was built between 1973 and 1976. The 270m-long level deck has four spans, the main one being 105m long between the tops of its two inward-leaning struts, the bases of which are 170m apart. This is flanked on each side by a 65m-long side span and an additional 45m span at the eastern end. The 14m-wide deck, which is supported on an 8m-wide full-length prestressed concrete box girder that varies from 3m to 6m in depth with gently arched soffits, is a maximum height of 65m above the river. The original steel bridge was the site of the first commercial bungy jump in Africa, opened in 2002. A more recent attraction is a high-speed pendulum swing between the two parallel steel lattice girder bridges.

Gouritz River Bridge 1976

Governor Albert D. Rosellini Bridge, Seattle, Washington State, USA

The first floating bridge carrying road traffic across Lake Washington, which was in use between 1963 and 2016, was the longest floating bridge in world. A new structure, also known as **Evergreen Point Floating Bridge**, opened in 2016 and is now the world's longest (7,710ft) and widest (116ft) floating bridge. It consists of seventy-seven concrete pontoons, the largest being 75ft wide and 360ft long, which are held in place by fifty-eight anchors and support the bridge deck that is made up of 776 precast concrete units.

Grand Canyon Skywalk, Peach Springs, Arizona, USA

Although hardly a bridge, in the sense of a connecting passageway, the horseshoe-shaped skywalk does carry people over a great void and is included here as effectively a modern version of a nineteenth-century seaside pier. Completed in 2007, it projects 70ft out from the west rim of the Grand Canyon more than 3,500ft above the Colorado River. The experience the skywalk offers is enhanced even more because both the floor and the side parapets are made from special glass panels. The deck is supported on side beams that are 32in wide and 72in deep and are fixed to posts, the foundations for which are held in position by rock anchors grouted 46ft deep into the canyon wall.

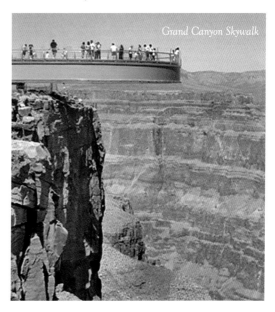

Grand Canyon Skywalk

Grand Maître Aqueduct, Sens, Yonne, France

Built between 1850 and 1865, this structure was the first to use unreinforced mass concrete in bridge structures. The purpose of the aqueduct was to convey water from the River Vanne north-westwards to Paris. The main part had a large segmental arch, with three smaller arches in each spandrel, and was approached by long approach viaducts of semicircular arches.

Grandfey Viaduct, Fribourg, Fribourg Canton, Switzerland

The first railway bridge here, built in 1862, consisted of four parallel wrought iron lattice girders. These were 343m long and crossed the Sarine River valley 82m above river level. The two railway tracks were laid above the trusses and a lower passageway for pedestrians and small vehicles was included within the structure. In 1892, in order to carry heavier trains, the bridge was strengthened with concrete encasing and reduced to a single track. In 1925 a completely new reinforced concrete structure was built consisting of seven wide parabolic main arches spanning 42m. Arcades of smaller arches then support the main deck on which the railway tracks are laid. A lower deck below carries cycling and walking trails called the Route de Grandfey and there is also a modern sculpture. The whole structure is one of Switzerland's bigger bridges.

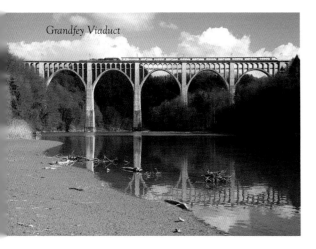

Grandfey Viaduct

Great Belt East Bridge, Korsør, Zealand Region, Denmark

Built between 1991 and 1998, this suspension bridge is part of the eleven mile-long Great Belt Fixed Link which also includes a railway tunnel and a separate box-girder bridge. The overall length of the East Bridge between Hallskov and Sprogø is 6,790m which includes seventeen spans of approach viaduct. The suspension bridge itself has a main span of 1,624m between 254m-tall pylons and has a deck width of 31m.

Great Belt East Bridge

Greenisland Railway Viaducts, Antrim, County Antrim, Northern Ireland

The **Mainline Viaduct** has three 89ft-span arches, each supporting spandrel columns, and there is a series of smaller approach arches at each end. Opened in 1934, this is claimed to be the largest reinforced concrete railway viaduct in the British Isles. Its arches pass 70ft above the stream in Valentine's Glen. A similar, but lower structure, the **Down Shore Line Viaduct**, passes under one end of the Mainline Viaduct. Both viaducts feature in Norman Wilkinson's famous railway poster of about 1935. The viaducts are now part of the Newtownabbey Way cycle path and walk, which itself is part of the Sustrans Cycle Route 93.

Groot Aub Bridge, Groot Aub, Hardap Region, Namibia

The bridge here over the Uiseb River is a proprietary Compact 200 structure, developed from the original Bailey Bridge of the Second World War, that was supplied and erected by the Mabey Bridge company. It is 173m long with four spans and was completed in 2013.

Guadiana International Bridge, Ayamonte, Huelva, Spain

This cable-stayed bridge across the Guadiana River was opened in 1991 to carry the A-49 road connecting Spain and Portugal. The overall length of the structure is 666m and the central span between the 95m-tall, inverted Y-shaped pylons is 324m. In 2019 a major upgrade of the bridge included inserting shock absorbers on the longest stays and installing decorative floodlighting.

Guadiana International Bridge

Guangji / Xiangzi (Great Charity) Bridge, Chaozhou, Guangdong, China

This bridge over the Han River, one of the five great ancient bridges of China, dates back to 1170, when it was a floating pontoon bridge supported by eighty-six boats connected together. After flood water had broken it in 1174, work started from both banks to build permanent piers with a central floating section of boats that could be disconnected to allow passage for river traffic. In 1435 it was reconstructed again, finishing with twelve fixed spans at the east end, nine at the west end and twenty-four boats between them, all providing a total bridge length of 518m. Pavilions were built on the piers and, at one stage, these contained 126 rooms.

Guangji Bridge

Inevitably, further repairs and changes followed until, in 1989, a new bridge was opened, now with an opening section of eighteen boats. The old bridge, complete with thirty bridge pavilions, was then fully re-furbished, re-opening in 2009.

Guyue Bridge, Yiwu, Zhejiang Province, China

The single-span traditional woven timber arch bridge across Dragon Creek was built in 1213 and it is claimed that it has not been repaired or rebuilt since then. The bridge, which has an overall length of 31m, spans 15m with a clear mid-span height above water of 4m.

Guyue Bridge

H

Ha'penny Bridge, Dublin, County Dublin, Ireland

Ireland's first cast iron bridge was built in 1816 by the Coalbrookdale Company of England to provide a pedestrian link across the River Liffey near the centre of Dublin. The segmental arch has three ribs that span 138ft between angled masonry abutments, with each rib consisting of six sections. A feature of the bridge is the trio of lamps mounted on distinctive doubly-curved ornamental supports that span between the tops of the railings and also serve to brace them. The structure was fully renovated in 2001.

Ha'penny Bridge

Halil Rifat Pasha Bridge, Yeslikale, Alacahan District, Turkey

The Turkish statesman Halil Rifat (1820–1901) was a grand vizier who was responsible for much road building in the Ottoman Empire during the late nineteenth century. His bridge here, built 1886–89, crosses the river at a skew, which was achieved by constructing a series of fifteen separate horseshoe-shaped arch rings spanning 3.8m, each successive one of which steps forward from its predecessor. The height inside the culvert is 4.2m and its length is 10.5m.

Hanging Bridge of Ghasa, Lete, Mustang District, Nepal

The modern wire rope suspension bridge here across the Kali Gandaki River was built to help move cattle and reduce congestion both in the village and on narrow mountain tracks. It is 137m long and is 70m above water level.

Hangzhou Bay Bridge, Hangzhou, Zhejiang Province, China

The Hangzhou Bay Bridge, built between 2003 and 2007, carries six traffic lanes across this part of the East China Sea. The 36km crossing, claimed to be the longest sea-crossing bridge in the world when it was opened, has a large service centre on an artificial island about half way along. The main structure, which is in the northern half, is a 908m-long cable-stayed bridge with a main span of 448m.

Hangzhou Old Bridges, Hangzhou, Zhejiang Province, China

There are two interesting old bridges in this city. The **Yudai (Jade Belt) Bridge** on the West Lake is a simple three-span stone slab structure, those over the side spans being slightly sloping, but it is capped by a magnificent double-tiered pavilion. The three-arched **Gongchen Bridge**, which was built in 1631, is the biggest of the ancient bridges here, being 992m long with a 16m-high main arch.

Haoshang Bridge, Leshan, Sichuan Province, China

Built in traditional style but only opened in 1994, this beautiful bridge links the area containing the 71m-tall Giant Buddha of Leshan to temples on a nearby island. It has three main segmental stone arches, the two side spans carrying a level deck which, when it meets

Haoshang Bridge

the central one, rises up on steps to the crown. Above each of the two piers an arcade of five smaller arches supports a pagoda which is linked to the bank by a covered walkway.

Harbour Footbridge, Zadar, Zadar County, Croatia

An earlier footbridge across Zadar Harbour was built in 1928 and destroyed by bombing in 1944. The present steel structure, which was opened in 1962, links the eastern side of the old city walls on Zadar peninsular to the newer parts of the city and fully encloses Jazine Bay to its south. This bridge is 152m long and 6m wide and has a low arching elevation over eight spans.

Hardinge Bridge, Paksey, Pabna District, Bangladesh

Eastern Bengal Railway's double-track railway bridge across the Padma River, opened in 1915, is 1,798m long. The main part consists of fifteen spans, each 105m long, of steel Petit-type through trusses on trestle piers. There are also three 23m-long land spans at each end. The exceptionally deep piers are founded 46m below low water level. The bridge, which was designed by A.M. Rendel, was damaged during the 1971–72 Bangladesh Wars.

Harp Motorway Bridge, Ljubljana, Province of Ljubljana, Slovenia

This 38m-wide bridge has two spans, each 41m long, which are supported by three parallel mast and cable-stay systems, one along each edge of the deck and one between the carriageways. The masts have six parallel cable stays on each side and the criss-cross effect in three-quarter views thus looks very dramatic. The bridge was opened in 1999.

Hartland Covered Bridge, Saint-John, New Brunswick, Canada

The first bridge here across the Saint John River was opened in 1901. It had seven Howe truss spans and its overall length was 1,282ft. Two spans were brought down by river ice in 1920 and, when these were rebuilt, the original timber piers were replaced by concrete ones and these now have ice-breaking vertically-angled upstream pier ends. The timber siding and roof were added at the same time, making it now the world's longest covered bridge. The separate external covered walkway was added on the south side in 1943 and the bridge has been a National Heritage Site since 1980.

Hartland Covered Bridge

Håverud Aqueduct, Håverud, Västra Götaland County, Sweden

The Dalsland Canal was built to provide a link for water-borne traffic between Lake Vänern and the lakes at Värmland. In order to transport boats over a series of rapids at Håverud, the engineer Nils Ericson designed an aqueduct consisting of two 34m-long wrought iron plate girders connected by a watertight deck which supports a 4.4m-wide and 1.8m-deep water-filled channel between the girders. Built between 1864

Haverud Aqueduct

and 1868, the aqueduct carries the Dalsland Canal diagonally over the Upperud River and is itself crossed by a higher-level arched railway bridge and a road bridge.

Hawkesworth Suspension Bridge, San Ignacio, Western Belize, Belize

A single-lane suspension bridge across the Macal River in San Ignacio, which was built by the UK firm Head Wrightson and opened in 1949, was named after a former governor of British Honduras. The bridge has a 280ft-long main span with 100ft-long back spans on each side. A new bypass structure, which has an overall length of 475ft, has recently been opened and a campaign has been started to save the deteriorating old bridge by listing it as a National Monument.

Hawkesworth Bridge

Hegigio Gorge Pipeline Bridge, Komo Station, Komo-Magarima, Papua New Guinea

At a height of 393m above the Hegigio River, this suspension bridge is the world's highest pipeline structure. Its span is 470m and it was completed in 2005.

Helgeland Bridge, Sandnessjøen, Nordland, Norway

The road bridge across Leirfjord is a cable-stayed bridge that was opened in 1991. It is 1,065m long and has nine approach and three main spans, the central one of these being 425m long. The cable stays are arranged in the semi-fan layout and the deck is 12m wide.

Hell Gate Bridge, New York, New York, USA

The famous arched Hell Gate Bridge, built between 1912 and 1916, originally carried four railway tracks – separate pairs for passenger and freight traffic – at a height of 135ft above the East River. The 100ft-wide and 978ft-long main span is visually (but not structurally) bookended by 250ft-tall monumental stone towers above the abutments, a concept that was later used in the Sydney Harbour Bridge (q.v.) of 1932. The bearings at the ends of the bottom chord of the two-pinned structure are housed in these abutments. The bridge, with approaches, is 1,700ft long overall and was designed by Gustav Lindenthal and Harold W. Hudson.

Hell Gate Bridge

Henderson Waves Bridge, Singapore

This steel cycleway and pedestrian footbridge over a six-lane highway, opened in 2008, is 284m long with six 24m-long spans and a 57m-long central span. Its timber deck, which is 8m wide and reaches a height of 36m above water level, is supported on one side only by a wave-like upstand which, in turn, stands on concrete piers.

Henderson Waves Bridge

Hercilio Luz Bridge, Florianopolis, Santa Catarina, Brazil

Linking Santa Catarina Island to mainland Brazil, this road bridge's 340m-long main span makes it the longest suspension bridge in Brazil. Together with approach spans the overall length of the bridge is 820m. The eyebar chain structure was designed by the American firm of Robinson & Steinman and built by the American Bridge Company, with all the material coming from America. At the central, lowest part of their span, these eyebar chains are integrated with the main longitudinal stiffening truss, resulting in the

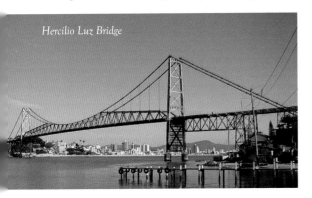

Hercilio Luz Bridge

non-uniform height of the truss and thus giving the bridge its unusual appearance. The towers are 74m tall and the mid-span clearance beneath the 14m-wide deck is 43m. The bridge, which was opened in 1926 and underwent major restoration between 1991 and 2019, was declared a National Artistic and Historic Monument in 1997.

High Bridge, Amsterdam, North Holland, Netherlands

This 93m-long footbridge across the entrance to a dock in Amsterdam's Eastern Docklands is more familiarly known as the **Python Bridge**. The red-painted latticework truss has a main span providing 9m of vertical clearance and a smaller side span. Built in 2001, it won an International Footbridge Award in 2002 for its designer Adriaan Geuze.

High Bridge Amsterdam

High Level Aqueducts, Caesarea, Haifa, Israel

The first masonry aqueduct at Caesarea (now called High Level Aqueduct I) was commissioned by King Herod in about 20BCE. The largest surviving section has eighty semicircular arches and is about 500m long. In 130CE, the Roman emperor Hadrian built High Level Aqueduct II alongside the earlier structure.

High Level Bridge, Edmonton, Alberta, Canada

This steel truss structure, which spans the North Saskatchewan River and was built between 1910 and 1913, has an overall length of 2,549ft over twenty-eight spans and its three central spans

High Level Bridge

are each 289ft long. The bridge has two separate 39ft-wide decks, with the upper one 156ft above water level carrying one railway line flanked by a tram track each side. The railway service stopped in 1989 but tourist trams still run during the summer. The lower deck carries the roads and footways. An LED lighting scheme on the bridge was opened in 2014.

Hiyoshi Dam Footbridge, Nantan, Kyoto, Japan

Also known as Friendship Bridge, this 3.5m-wide and 80m-diameter circular footbridge was designed as part of the landscaping around the Hiyoshi dam across the Katsura River. Designed by architect Norihiko Dan and spanning the spillway from the dam, the steel box girder bridge was built in 1998.

Hiyoshi Dam Footbridge

Hohenzollern Bridge, Cologne, North Rhine-Westphalia, Germany

The famous bridge over the River Rhine in Cologne, designed by Friedrich Dirksen, was built between 1907 and 1911 to replace the earlier two-track Cathedral Railway Bridge. With an overall length of 409m and a longest span of 168m, the three-span tied-arch structure originally carried four railway lines and a two-lane road. It was blown up by the German army in March 1945 to delay the Allies' advance and rebuilt by 1959. Further work between 1985 and 1988 increased the width of the bridge to 26m and it now carries six railway lines and a footway, the latter gaining fame as a 'love-lock' bridge. Hohenzollern Bridge is claimed to be Germany's busiest railway bridge, with more than 1,200 train movements a day.

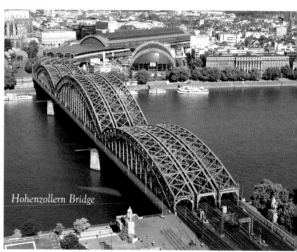

Hohenzollern Bridge

Hong Kong-Zhuhai-Macau Bridge, Hong Kong, China

This six-lane bridge (the name of which is usually abbreviated to HZMB) across the Pearl River Estuary links three major cities and consists of three cable-stayed structures and an undersea tunnel, all of which connect together at four artificial islands. HZMB has an overall length of 55km: the Hong Kong Link Road of 12km, the 29.6km Main Bridge and the Zhuhai Link Road of 13.4km. Work began in December 2009 and the formal opening was in October

2018. The largest of the bridges, Main Bridge, has a 22.9km-long viaduct, which includes the three cable-stayed structures over shipping lanes. The biggest of these, with a main span of 458m, is Qingzhou Channel Bridge, the others being Jianghai Channel Bridge and Jiuzhou Port Channel Bridge. The Qingzhou Bridge is distinctive because the connection between the tops of the twin towers on each of its two pylons looks like interlinked brackets. Total cost of the world's longest sea bridge was around US$20 billion.

Hongcun Moon Bridge, Huangshan City, Anhui Province, China

The Moon Bridge, a slightly stilted semicircular stone arch, is in the centre of a causeway that crosses South (Nanhu) Lake and appeared briefly in the background of the film *Crouching Tiger, Hidden Dragon*. The deck over the crown of the bridge is flat but the stepped approaches on each side are concave, giving the structure a distinctive profile. The village is a World Heritage Site.

Hongcun Moon Bridge

Howrah Bridge, Calcutta, West Bengal, India

This major steel cantilever bridge provides a railway link across the Hooghly River between the cities of Howrah and Calcutta and was opened in 1946. Its total length is 2,313ft, its main span is 1,500ft long and the main roadway deck is 71ft wide. The bridge was designed by the British civil engineering firm Rendel, Palmer & Tritton.

Howrah Bridge

Hradecky Bridge, Ljubljana, Province of Ljubljana, Slovenia

In 1867 a cast iron tubular arch bridge was built across the Ljubljanica River in the centre of Ljubljana to a design by Johann Hermann of Vienna. Named Hradecky Bridge after an earlier mayor, it spanned 30m with three parallel cantilevered ribs, each of these consisting of two half-arches connected at mid-span by a pin joint. This was a novel development as until then individual structural elements of bridges had always been rigidly connected together. In 1931 the structure was taken down, so that a new Cobblers' Bridge (q.v.) could take its place, and re-erected elsewhere in the city. Then, in 2011 it was moved again and is now considered to be the world's oldest surviving single-hinged bridge.

Hradecky Bridge

Huc Bridge, Hanoi, Red River Delta, Vietnam

The timber bridge starting from the northern shore of Hoan Kiem Lake provides access to the Jade Island of the Buddhist Ngoc Son Temple. An earlier bridge here, built in 1865, was burnt down in 1877 and its replacement collapsed in 1952 when too many people were on it. The present structure, designed by Nguyen Ngoc Diem, is a seventeen-span timber footbridge with a gently arched profile that stands on piers of circular timber columns. The bridge, which is 2.6m wide and 45m long, is painted vermilion red and decorated by flags. It is approached through a triumphal gateway.

Hunter Street Bridge, Peterborough, Ontario, Canada

The first two timber bridges that crossed the Otonabee River at Peterborough were succeeded by a cast iron structure that collapsed in 1875. This was followed by a wrought iron bridge that was itself replaced by the present bridge in 1921. The structure was designed by Frank Barber with architectural input from Claude Fayette Bragdon. It has a single main 234ft-span segmental arch over the river, which is believed to be the longest mass concrete arch in the world, and the 33ft-wide deck is supported by five subsidiary arches in each spandrel. Decorative features include tall columns above the main piers and glazed earthenware panels set in the concrete parapets. The bridge has an overall length of 900ft, with the main river arch being flanked on each side by five reinforced concrete land spans. These are grouped into two pairs of smaller arches that are separated from the main span, the abutments and each other by longer arches, giving the whole structure a distinctive and elegant appearance.

Hurricane Gulch Railroad Bridge, Anchorage, Alaska, USA

Built in 1921 by the American Bridge Company for the Alaska Railroad, this 918ft-long steel bridge carries a single rail track more than 300ft above the Hurricane Gulch tributary of the Chulitna River. The main structure is a parabolic arch spanning 384ft between abutments, the deck being supported above this on N-braced trusses.

Hussaini Hanging Bridge, Passu, Upper Hunza, Pakistan

Claimed to be one of the most dangerous bridges in the world, this structure was built on the orders of the country's president Ayub Khan and completed in 1968 after an earlier bridge had been washed away. There are five parallel wire cables spanning 900ft through which 4in thick planks about one foot apart are laced to form a very basic deck. Often, many of these planks are broken or missing. Higher up above each side of the deck there is a further wire suspension cable between which U-shaped loops of wire hang (and are also laced through the deck cables) to provide some rough side protection.

Hussaini Hanging Bridge

Hvítá Bridge, Hvanneyn, Western Region, Iceland

Árni Pálsson was the Icelandic engineer who designed this 106m-long reinforced concrete bridge across the Hvítá River that carried the main road connection between northern and southern Iceland. Opened in 1928, it has two low-rise semi-elliptical arches, each spanning 51m, that support the 3m-wide gently-arched deck on slim, square concrete posts. Considered to be one of the country's most photogenic bridges, it was declared in 2002 to have been the country's most outstanding engineering project of the 1920s.

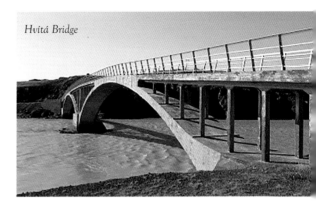

Hvítá Bridge

I

Ikitsuki Bridge, Nagasaki, Japan

Built between 1983 and 1991, this steel truss road bridge connects Ikitsuki and Hirado islands. The central span, which is 400m long, continues on each side onto a 200m-long side span, making this three-span structure the world's longest continuous truss. Together with shorter approach spans at each end, the bridge's overall length is 960m and it is 6.5m wide. The concrete piers have two columns supporting a crosshead beam.

Inca Bridge, Machu Pichu, Cuzco, Peru

The western entrance to Machu Pichu is along a pathway on top of a masonry wall that clings to the cliffside above a 2000ft drop into a canyon. At one point the path crosses over a wooden log bridge spanning about 20ft, which originally could be removed if the Inca settlement was at risk of being attacked.

Inca Bridge

Inner Harbour Bridge, Duisburg, North Rhine-Westphalia, Germany

This 3.5m-wide and 74m-span footbridge is suspended between 20m-tall twin-column pylons at each end, these columns being 420mm-diameter steel cylinders. The arched deck consists of fourteen independent sections which are joined together by hinges. In order to allow yachts to pass beneath the bridge, hydraulic cylinders are used to shorten the back stays, thus reducing the sag in the main suspension cables and thereby raising the 150-ton deck by 4.5m. The bridge was designed by Schlaich Bergermann Partners and opened in 1999.

International Bridge, Valenca, Viana do Castelo, Portugal

Built between 1884 and 1885, this double-decked bridge across the Minho River on the border between Portugal and Spain was designed by the Spanish engineer Pelayo Mancebo. It is a 318m-long, five-span, wrought iron, lattice box girder with the N-551 road on the lower deck and a railway above. The structure, which now also carries the Portuguese Way part of the pilgrims' route to Santiago de Compostela, underwent major rehabilitation in 2012. A new prestressed concrete relief bridge to the south was completed in 1993.

International Railroad Bridge, Sault Ste. Marie, Ontario, Canada

This 5,580ft-long structure carries a single railway track across Saint Mary's River and Canal to link Canada and the USA and was built by the Canadian Pacific Railway. It consists of several separate sections, which are effectively different structures, that have been rebuilt at different times in the bridge's lifetime. The oldest parts, built in 1887, are the approach spans which consist of nine camelback through trusses each spanning 239ft and a series of plate girder deck spans. The most interesting parts of the bridge are the opening spans. The double-leaf heel-trunnion bascule bridge, designed by Joseph Strauss and built in 1913, is 23ft wide and spans 336ft. An unusual feature is that, when the leaves are down, both top and bottom booms are locked together in order to give the structure the additional strength needed to carry heavy freight trains. The southern leaf was damaged in 1941 when it collapsed under a heavy load. The vertical lift bridge, which was built in 1960 to replace an earlier swing bridge built in 1895, has a 369ft-long span and its deck is 21ft wide.

Irene Hixon Whitney Bridge

International Railroad Bridge (foreground)

an overall length of 118m and its deck is 4m wide. Structurally, the bridge consists of an arch section and a suspension section end to end, although these partly overlap each other.

Irene Hixon Whitney Bridge, Minneapolis, Minnesota, USA

This through truss bridge for cyclists and pedestrians crosses sixteen lanes of the I94 Freeway and provides access between the Minneapolis Sculpture Garden and Walker Art Center and, on the other side of the freeway, Loring Park near Downtown West. Designed by Siah Armanjani, it was opened in 1988, has

Ivancice Railway Viaducts, Ivancice, South Moravia, Czech Republic

The first viaduct over the Jihlava River on this site was built between 1868 and 1870 to carry the Vienna to Brno railway line. Designed by Karl von Ruppert, it was 374m long and 42m high. A huge lattice girder of double-height X-braced panels separated by vertical members was supported on latticework piers with cast iron columns. These columns were replaced in 1892. In 1978 a new viaduct was opened alongside and the old one was dismantled in 1999. The new structure is a six-span welded steel girder standing on A-shaped piers.

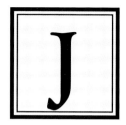

Jaargalant Wooden Bridge, Jaargalant, Khövsgöl Province, Mongolia

Built in 1940, the 128m-long rackety old timber bridge across the Ider River, which carried the road between Shine-Ider and Jargalant, was considered to be one of the most dangerous bridges in the world. Roughly built of timber poles and planks, it had about ten spans but the deck was neither level nor straight. A replacement structure has been due for completion for some years.

Jaargalant Wooden Bridge

Jacques Chaban-Delmas Lifting Bridge, Bordeaux, Nouvelle-Aquitaine, France

Inaugurated in 2013, this bridge across the Garonne River, has a main lifting span that is 117m long, the longest such structure in Europe, which is 32m wide and weighs 2,600 tonnes. When lifted, the clearance beneath is 53m. Overall, the length of the bridge is 575m and it is 45m wide. This allows it to carry four road lanes and two monorail tracks. Footways are cantilevered off the main deck and pass outside the slim towers, which are 87m high. These towers are illuminated with lights that change colour. The bridge was designed by Nabil Yazbeck and is named after a former mayor of Bordeaux (for forty-eight years) and Prime Minister of France.

Jacques Chaban-Delmas Lifting Bridge

Jacques Gabriel Bridge, Blois, Loir-et-Cher, France

An earlier bridge across the Loire here was recorded as having twenty arches in 1679. Following its collapse in 1716 the current structure was opened in 1724 and named Jacques Gabriel Bridge after its architect. This bridge has eleven semi-elliptical arches, is 283m long and 15m wide and its central span is 26m long. A commemorative 15m-high obelisk was placed on the parapet at the centre of the bridge. The structure has been deliberately damaged on four occasions: in 1793 during the Revolution an arch was blown up to prevent royalist troops entering the town; in 1870 another was destroyed to delay an invasion by the Prussians; in 1940 an arch was brought down to hinder the advance by the Germans, who themselves then destroyed three other arches during their retreat in 1944. The bridge was not re-opened until 1948.

James Joyce Bridge, Dublin, County Dublin, Ireland

The deck of this 40m-long, single-span, steel road bridge over the River Liffey is supported from two tied arches that lean outwards at an angle of 20°. The footways are cantilevered outside these arches and include seating areas. The bridge, which was designed by Santiago Calatrava, was named after the local author James Joyce and opened in 2003.

Japanese Covered Bridge, Hoi An, Quang Nam, Vietnam

This pagoda bridge, probably built in 1595, is the town's main symbol. The structure is tee-shaped in plan. The head of the tee is 60ft long and consists of five spans across a canal leading to the Thu Bon River – a centre flat beam span flanked on each side by two narrow stone arches. The stalk of the tee on the north side of the bridge contains a small 20ft square Buddhist temple above the canal, which is entered from the bridge passageway. The timber roof is clad with tiles and the whole structure was renovated in 1986. The bridge is within the area of the town's UNESCO World Heritage Site.

Japanese Covered Bridge

Japan-Palau Friendship Bridge, Koror, Palau

The first bridge between the Pacific islands of Koror and Babeldaob had an overall length of 385m with a deck width of 9m and was an elegant variable-depth box girder structure built of prestressed concrete. It consisted of two cantilever arms, meeting at a central hinge, that formed a main span 240m long. When it opened in 1978, it was the largest bridge of its type in the world and, as well as road traffic, the bridge also carried electrical and water mains. However, it slowly developed an increasing central sag and, despite being declared safe, collapsed without warning in 1996 killing two people, injuring four others and breaking the water and power connections. Its failure seems to have been caused by a combination of loss of prestress and consequent mid-span sagging due to 'creep' (long-term changes to a structure under continuous load) and structural alterations to deal with these problems. The replacement structure, built with aid from Japan, has a similar prestressed-concrete box girder profile but this is now supported by cable stays from a tall concrete pylon at each end.

Japoma Bridge, Edea, Littoral Province, Cameroon

The 160m-long steel arched railway bridge over the Sanaga River near Edea was manufactured and test-assembled in Germany before being shipped out to Africa and erected on site, being completed in 1911. Its use is now limited to cyclists and pedestrians only. During October 1914 local German troops damaged the bridge before retreating.

Jiaozhou Bay Bridge, Hongdao, Shanxi, China

The road crossing over the nearly-enclosed Jiaozhou Bay off the Yellow Sea has a total length of 16 miles. It is one of the longest bridges in the world and is nearly all over water. However, this includes passing over three navigable channels, each of which requires a separate major structure, and includes an oversea road junction. Throughout its length the bridge is 35m wide and has six traffic lanes and two hard shoulders. The longest span is the 260m cable-stayed bridge over the Cangkou Channel in the west of the bay. Daguhe Channel in the east is crossed by a self-anchored suspension bridge with a single 160m-tall concrete pylon supporting two 260m-long spans and Hongdao Channel in the north has two 120m-long spans supported by a central concrete pylon 160 high. The bridge was built between 2007 and 2011 and, at the 30th International Bridge Conference in 2013, the bridge won the George Richardson Award.

Jiemei (The Sisters') Bridges, Miamyang Xiaobei, Sichuan, China

This timber pavilion bridge over the Xiaocha River, built in 1873, has two slightly unequal spans separated by a rocky outcrop which supports the inner end of each span. Both walkways are covered with tiled roofs with uplifted corners and there is a second smaller and higher roof over each entrance.

Jihong (Rainbow in the Clear Sky) Bridge, Yongpin, Yunan, China

A chain suspension bridge over the Lancang River was built in 1470, replacing an earlier timber structure said to have been crossed by Marco Polo in the late thirteenth century. (Britain's first chain suspension bridge was built in 1817.) The bridge was 57m long and 4m wide and was supported by eighteen chains anchored to stakes driven into the river banks. Although the Japanese tried and failed to destroy the bridge by bombing in World War II, it was brought down by floods in 1986 but has since been replaced by a new structure located higher up the steep valley sides.

Jingxing Qialoudian Bridge, Jingxing, Hebei, China

The segmental stone arched bridge at Jingxing spans 15m across a 70m-deep chasm and supports the two-storied Bridge Tower (Qiaolou) Hall. The structure is believed to have been built at the end of the sixth century.

Jisr Jindas (Baybars) Bridge, Zeitan, Tulkarm Governorate, Palestine

The Jisr Jindas Bridge across the Ayalon River was built by Sultan Baybars in 1273, possibly on Roman foundations, and now carries the country's Route 434. It is 30m long and has three slightly pointed stone arches. The parapet on each side of the bridge has a carved inscription that is flanked by two lions.

John A. Roebling Suspension Bridge, Covington, Kentucky, USA

The suspension bridge over the Ohio River in Covington was built by Roebling between 1856 and 1866 and, when it was opened, the 1,057ft-long main span was the world's longest. The structure had particularly elegant masonry towers topped by decorative pinnacles. In 1896 the bridge was upgraded with a wider deck and additional cable stays. It became a National Historic Landmark in 1975 and a National Historic Civil Engineering Landmark in 1983, when it was given its present name in honour of its designer.

Roebling Suspension Bridge

John T. Alsop Jr Bridge, Jacksonville, Florida, USA

Work started on the steel vertical lift bridge across the St John's River, also known as Main Street Bridge, in 1938 and was completed in 1941. The bottom of the 365ft-long trussed girder

Jacksonville Main Street Bridge

lifting span is 35ft above water when the bridge is closed and can be raised by 100ft for water traffic. There are sixteen approach spans and, overall, the bridge is 1680ft long. The 58ft-wide deck carries four traffic lanes for Highway 90 and US1 as well as two footways. The top booms of the lifting span and the side span on either side of the lifting towers have a continuous gently-arched profile when the deck is not raised.

Jökulsá á Dal Canyon Bridge, Egilsstaðir, Eastern Region, Iceland

The polygonal reinforced concrete arch bridge, which carries National Road 1 over the Studlagil Canyon, is 125m long with a 70m-long main span and was built in 1994. The deck is supported on slim slightly-tapered spandrel walls. The bridge was designed by Línuhönnun Consulting Engineers.

Jokulsa a Dal Canyon Bridge

Juba Nile (Freedom) Bridge, Juba, Central Equatoria, South Sudan

The first bridge across the White Nile River in South Sudan was built between 1972 and 1974 at the end of Sudan's first civil war and consisted of twin Bailey bridges, each with six 43m-long spans. During the second civil war, between 1983 and 2005, one of the structures was destroyed and in 2006 two replacement Mabey Compact 200 bridges were installed. However, one of these was subsequently damaged and construction began in 2013 on a major new Japanese-funded bridge although, when further violence erupted the following year, the Japanese left the country. In 2019, construction resumed

on what is to be called the Freedom Bridge, with completion expected in 2021. This will be 560m long and 13m wide and will have four bowstring deck arches of steel. Together with access roads at each end, the total length of the new work will be 3,700m.

Juba Nile (Freedom) Bridge (2006)

Jules Wijdenbosch Bridge, Paramiribo, Suriname

Opened in 2000, this is the lowest bridge over the Suriname River and carries the country's East-West Link road along Suriname's north coast between Paramiribo and Meerzorg. The prestressed concrete box-girder bridge, which is 1,504m long and 9m wide, rises to reach a height of 52m above water level in the 155m-long main span over the river channel.

Jules Wijdenbosch Bridge

Jundushan Aqueduct, Yanqing, Beijing, China

This very industrial boxy-looking reinforced concrete aqueduct was built in 1988. The waterway is supported on a 276m-long cable-stayed beam that has a main span of 126m between two ladder-like pylons standing on square post-and-beam towers and this is claimed to be the longest span of any aqueduct.

Jungfern Footbridge, Berlin, Brandenburg, Germany

This, the tenth structure on the site, is claimed to be the oldest bridge in Berlin and was first built in 1701 and rebuilt in 1798. With a total length of 28m, it has three spans: a central twin-bascule 9m-long opening span that is flanked by short segmentally-arched masonry spans. Above the intermediate piers are four cast iron posts supporting the suspension chains and the handwheels by which the deck leaves can be opened. East Germany issued a postage stamp in 1985 showing the bridge.

Juscelino Kubitschek Bridge, Brasilia, Central-West Region, Brazil

Opened in 2002, this road bridge crosses Lake Paranoa to connect Brasilia's city centre to its international airport and is named after the president who inaugurated the building of the new city. The structure was awarded a medal at the 2003 International Bridge Conference in Pittsburgh. It is distinctive for its three main 60m-high steel arches from which the 1200m-long and 24m-wide curved bridge deck is suspended. These arches each span 240m and cross diagonally above the roadway from right to left, then left to right, and finally right to left again. The bridge was designed by architect Alexandre Chan and engineer Mário Vila Verde.

Jungfern Footbridge

Juscelino Kubitschek Bridge

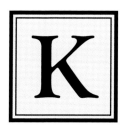

Kafue Railway Bridge, Kafue, Lusaka Province, Zambia

Built in 1906 to carry the railway from Livingstone to Lusaka over the Kafue River, this 1,400ft-long bridge is the longest in the country and has thirteen 108ft-long steel girder spans. The structure featured on some of Zambia's postage stamps in 1968.

Kaidatsky Bridge, Dnipro, Dnipropetrovsk, Ukraine

The six-lane Kaidatsky bridge, opened in 1982, carries the European E50 highway across the Dnieper River along a 6,000km-long route that has existed since the seventeenth century. The segmentally-arched concrete structure, which has an overall length of 1,732m and is 26m wide, is cranked in line as it crosses Namistanka Island.

Kakum Canopy Walk, Wawase, Ashanti, Ghana

Kakum National Park, established in 1992, includes 145 square miles of dense tropical jungle. This treetop walkway, which was opened in 1995, is 330m long and crosses seven wire rope suspension bridges at about 30m above ground level.

Kalte Rinne Viaduct, Breitenstein, Lower Austria, Austria

Completed in 1854, the Semmering Railway was deliberately designed by its engineer, Carl von Ghega, to harmonise with its landscape and this was recognised in 1998 by its addition to the list of UNESCO World Heritage sites. The railway has sixteen viaducts and more than 100 stone arch bridges but its signature structure is the 184m-long curved, two-tiered stone Kalte Rinne Viaduct. This has five segmental arches in the shorter lower tier with a further ten semicircular arches above and has a maximum height of 46m.

Kalte Rinne Viaduct

There is now a 23km-long walking trail along part of the railway route.

Kanoh Bridge, Kalka, Haryana State, India

Kanoh Bridge (No 493) on the Kalka-Shimla Railway was built in 1898 and, at nearly 2,000m above mean sea level, is the world's highest railway bridge. It is 53m long, 23m high with thirty-four arches in four tiers and carries a curved section of the 2ft-6in narrow-gauge single line railway track. The complete railway, together with two others, is now part of the Mountain Railways of India UNESCO World Heritage Site.

Kantara Moulay Bridge, Ismail, Khenifra, Morocco

The bridge here, probably built around 1700, has a main pointed stone arch with two much smaller flanking arches behind the main abutments. The structure was originally hump-backed, but a level deck was built in modern times and this work incorporates on one side an additional high-level rectangular opening to cope with flood waters.

Kanzler-Dollfuss Footbridge, Stams, Tyrol, Austria

The footbridge at Stams, which was built in 1935, is suspended from two wire ropes spanning 94m between tapering stone towers each topped by a pitched roof. Wire hangers from the cables support timber cross beams beneath the 1.1m-wide deck, with every fifth or sixth one of these extending outward, their ends being connected by a horizontal cable to limit torsional deflections and increase stability.

Kasakh Bridge, Oshakan, Yerevan, Armenia

Built across the Kasakh River in 1706 after an earlier bridge here had been swept away, this unsymmetrical structure has three main pointed stone arches, the piers protected by heavy round cutwaters. There are two smaller semicircular flood arches at one end.

Kasari Old Bridge, Matsalu, Lääne County, Estonia

When this road bridge over the Kasari River was completed in 1904 it was the longest reinforced concrete bridge in Europe and remained in use until 1990. It was then refurbished and re-opened as a footbridge. The bridge has three ribs and stands on masonry piers. Its deck is 6.5m wide and, with thirteen 21m-long shallow segmental arches, it has an overall length of 308m.

Kassuende Bridge, Tete City, Tete Province, Mozambique

The main and highest section of this reinforced concrete bridge across the Zambezi River has three full and two-half arches and these were built by balanced cantilever construction. Overall, the 15m-wide bridge is 715m long with three approach spans on one side and a viaduct of a dozen spans over the old flood plain on the other. The bridge was built between 2011 and 2014.

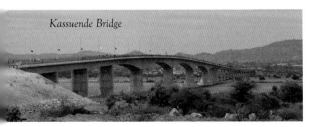

Kassuende Bridge

Katima Mulilo Bridge, Sesheke, Western Province, Zambia

The bridge here over the Zambezi River carries the Trans-Caprivi Highway, which is part of the trade route known as the Walvis Bay Corridor. Opened in 2004, the road bridge is 900m long and has nineteen spans. These were all built on the western bank in 13.4m-long segments and, after each successive one had been completed, the structure was push-launched out over the river at a speed of 10 metres per hour. Interestingly, a complication here is that the whole bridge is curved in plan, with a radius of 1600m. For the final push, the completed structure had a total mass of 13,000 metric tonnes.

Kattwyk Bridges, Hamburg, Lower Saxony, Germany

Although Hamburg is 100km from the North Sea it is one of Germany's principal seaports. As part of its infrastructure a major vertical lift bridge was built in 1973 to carry both road and rail traffic across the Süderelbe River. The 96m-long lifting span weighing 1919 tonnes can be raised by 46m up the 81m-high flanking towers, which contain electric motors and counter-balancing weights. The towers are flanked on each side by an 84m-long fixed approach span, with all three spans consisting of Warren girders. Overall, the bridge is 287m long and 11m wide. A new similar structure with a main span 108m long was opened in 2020. This is devoted to rail traffic only.

Kattwyk Bridge

Kawarau Bridge, Queenstown, Otago, New Zealand

A suspension bridge across the Kawarau River, designed by Harry Higginson, was built in 1880 with the road deck 43m above the river. This deck is supported by X-braced side trusses which hang from chains spanning between four masonry pylons. After a new steel arch road bridge was built nearby in 1963, the old structure became the first commercial bunjee-jumping site in the world and there is now a zip-ride as well. The bridge also carries the Queenstown Trail.

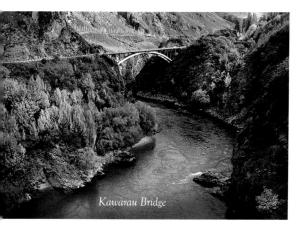

Kawarau Bridge

Kawazu-Nanadaru Loop Bridge, Shimoda, Shizuoka, Japan

The traffic on this double-decked helical bridge passes round two successive full 360° circles, moving 45m vertically, as Highway 414 carries vehicles up or down the steep valley between two mountains. (The structure is sometimes wrongly

Kawazu-Nanadaru Loop Bridge

called a spiral bridge, but the radius is constant rather than having the increasing or decreasing radius of a true spiral.) The structure, which was built in 1982, has a radius of 40m and is supported by six radially-oriented frames, each of which consists of a pair of inverted Y-shaped columns connected by a transom beam below each deck. Including approach viaducts, the overall length of the bridge's road deck is 1100m.

Kazarma Bridge, Arkadiko, Peloponnese, Greece

The triangular-corbelled arch here, made from massive cyclopean stonework and dating from about 1300–1200BCE, is one of the oldest bridges in the world still in use. It was built to provide a military link between Tiryns and Epidauros and spans about 1m across a stream at the bottom of a small gulley. The overall length of the deck is 10m and its width of about 3m would have allowed chariots to cross.

Kazarma Bridge

Kazungula Bridge, Kazungula, Southern Province, Zambia

The recently-built four-span bridge linking Zambia and Botswana crosses the Zambezi River at the point where the Chobe River joins it and where the borders of four countries meet at one point. It has a gentle horizontal curve to avoid crossing over the territories of either Zimbabwe or Namibia. It is 923m long and 19m wide and there are six cable-stayed spans, the longest being 129m. Completed in 2020, it carries a single narrow-gauge rail line flanked on each side by a single traffic lane and a footway.

127

Kazurabashi Vine Bridge, Shikoku, Ehime Prefecture, Japan

The first vine bridge over the Iya River here is said to have been built in the twelfth century. The present one, which is rebuilt every three years around a permanent core of steel cables, is a big tourist attraction, with a 500 yen toll charged to cross it. The unusual structure spans 45m and is 14m above the river.

Kazurabashi Bridge

Kelefos (Tzelefos) Bridge, Agios Nikolaos, Famagusta, Cyprus

The attractive, single-span, pointed stone arch across the River Diarizos is considered to be the largest medieval bridge in Cyprus. Known as a Venetian bridge, because it was built when the Venetian Republic governed Cyprus between 1489 and 1571, it was used by camel trains carrying copper to local ports for export to Venice. The arch spans about 11m and the deck is about 2½m wide.

Kellams Bridge, Stalker, Pennsylvania, USA

David Kellam built this 384ft-long bridge across the Upper Delaware River in 1890. It is a rare underspanned suspension bridge, in other words, one where the central lowest part of the suspension cables is beneath the level of the bridge deck, which therefore has to be supported on posts above the cables rather than from hangers below the cables. The bridge is 379ft long and 20ft wide and was renovated in 1959 and 2018.

Keshwa Chaca Bridge, Huinchiri, Canas Province, Peru

This bridge over the Apurimac River is the last remaining Inca fibre suspension bridge and its reconstruction in June every year represents a 600-year-old part of the country's cultural heritage. The bridge is 33m long and 1.2m wide and, at the centre of the span, the footway is about 15m above river level. It consists of four floor cables and two handrail cables, with each cable being braided from three ropes. These ropes are handwoven in advance, by up to one thousand people, from Peruvian feathergrass and it then takes two to three days to build the bridge. The cables have to be replaced in June every year.

Keshwa Chaca Bridge

Khndzoresk Swinging Bridge, Tatev, Syunik, Armenia

The suspension footbridge here is 160m-long and 1½m-wide and was built in 2012. Its purpose was to provide an access route for tourists to the Old Khndzoresk village in which most of the people used to live in caves carved out of the rock.

Khudafarin Bridges, Jabrayil, Hadrut, Azerbaijan

Two ancient arch bridges cross the Aras River a short distance downstream from a modern dam. The higher one was probably built in the eleventh or twelfth century and originally had eleven

Khudafarin Bridge

arches, but most of these have been reduced to pier stumps. An artist's depiction of this bridge appeared on an Azerbaijani postage stamp. The bridge further downstream is complete with fifteen pointed arches. Most sources consider it to date from the thirteenth century but one records a date of 1786, which is possibly the date of a major rebuilding.

Kiel-Hörn Folding Bridge, Kiel, Schleswig-Holstein, Germany

This unusual 5m-wide opening footbridge crosses Kiel Fjord. It is approached over a fixed span, at one end of which a pinned joint connects to three further pin-linked spans that are cable-stayed from two portal frames also hinged at the end of the fixed span. Together, these folding spans are 26m long. The bridge is opened by winching back the stays to shorten them, thus retracting the free end of the bridge as the three spans concertina together. The bridge was designed by Gerkan, Marg & Partners and completed in 1997.

Kiel-Horn Folding Bridge

Kigamboni Bridge, Kurasini, Dar es Salaam, Tanzania

Built between 2012 and 2016, this 680m-long cable-stayed road bridge crosses Kurasini Creek with a main span of 200m. Its deck, which is 32m wide, carries a six-lane dual carriageway highway with a cycle track and footway on each side.

King Fahd Bridge, Bamako, Koulikoro, Mali

The first modern bridge over the Niger River at Bamako, the capital and largest city of Mali, was built in 1957 and this one, 500m upstream and sometimes called New Bridge, was sponsored by the Saudi Fund for Development and opened in 1992. It is a multi-span column-and-beam structure that carries dual two-lane carriageways with a cycle track and footway on each side.

King Fahd Causeway Bridge, Al Khobar, Eastern Province, Saudi Arabia

Built between 1981 and 1986, this 16-mile-long crossing over the Gulf of Bahrain links Al-Aziziyyah in Saudi Arabia to Al-Jasra on the main island of Bahrain. There are five principal bridges: two on the eastern Bahraini side and three on the Saudi side, one of which includes a raised structure over a shipping channel. A $3.5 billion expansion project, called the **King Hamad Causeway**, is due for completion in the late 2020s.

King Fahd Causeway Bridge

Kingdom Centre Skywalk, Riyadh, Riyadh Region, Saudi Arabia

The upper third of the 302m-tall Kingdom Centre skyscraper block splits into two separate towers with the space between them in the shape of an inverted parabolic arch. The tops of the towers are linked by an enclosed steel footbridge spanning 65m. The tower and bridge were completed in 2002.

Kingdom Tower

Kintai Bridge, Iwakuni, Yamaguchi Prefecture, Japan

Early timber bridges over the Nishiki River at Iwakuni were regularly brought down by floods. Finally, in 1674, a five-span structure standing on stone piers was successfully completed. In this, the lower outer arches included intermediate trestle supports but the bridge was famous for its high segmental timber arches built without nails. This stood until washed away in 1950 and a copy was completed in 1953. The bridge, which is 193m long and 5m wide, has been a National Treasure since 1922.

Kintai Bridge

Kinzua Viaduct, Westline, Pennsylvania, USA

The railway trestle bridge over Kinzua Creek was first built in wrought iron in only ninety-four working days in 1882. Eighteen years later it was dismantled and rebuilt in steel so it could carry heavier trains. Its total length was 2,052ft and its 10ft-wide deck had a maximum height above the creek of 301ft – the fourth highest in the country. After the railway line was closed the structure became the focal point of a state park and was being reconstructed when in 2003 it was struck by a tornado with wind speeds of more than 90mph. Twenty-three of the forty-one spans and eleven of the intermediate piers were brought down. In 2011, the remaining length of the viaduct was re-opened as a skywalk with a glass-floored observation deck. The bridge is included in the ASCE list of historic bridges.

Kinzua Viaduct

Kirkgoz Kemeri Bridge, Limyra, Antalya Province, Turkey

The 360m-long late-Roman bridge over the Alakir Cayi River is believed to have been built to link Limyra to Antalya on the coast. Considered to be one of the earliest bridges to have been built with segmental masonry arches, it originally had twenty-six or twenty-seven although most are now buried. These all had a significantly flat profile, with a span-to-rise ratio ranging from 5.3 to 6.4:1, and span lengths between 11.6m and 15m.

Kızılçullu Aqueducts, Smyrna, Izmir Province, Turkey

Two Roman aqueducts, one containing two tiers of slightly pointed arches (three over the river itself), were built by the Romans over the River Meles. They probably date from the second century CE. They are chiefly known now through the pictorial records of the structures made by William James Müller and Dr Moritz Busch during the nineteenth century.

Kochertal Viaduct, Schwäbisch Hall, Baden-Württemberg, Germany

This prestressed concrete viaduct over the Kocher River, which was completed in 1979, was the highest bridge in the world at 185m, until the Millau Viaduct (q.v.) was opened in 2004. There are seven 138m-long main spans with the outer span at each end being 81m long. The structure consists of a single cell reinforced concrete box girder 6.5m deep supporting a road deck 31m wide. The edges of the cantilevered wings are supported by raking struts from the bottom of the box girder. The viaduct carries the A6 motorway.

Koh Pen Bamboo Bridge, Kampong Cham, Kampong Cham Province, Cambodia

The bamboo bridge linking the river island of Koh Pen to the city of Kampong Cham has been built by the islanders every year when low water levels in the Mekong River stop the ferry service. Considered to be the longest bamboo bridge in the world, the 1km-long structure is estimated to contain 50,000 bamboo sticks and takes a month to build. The deck is made from strips of bamboo which are attached to bamboo posts by twisted wire. Every year in May, at the start of the rainy season, the bridge is taken down and stored. Since a new 779m-long concrete bridge was completed in 2017 about 2km to the south of the bamboo bridge site, the bamboo structure has been made narrower just for pedestrians and cyclists and it is uncertain how long the old custom can be maintained.

Koh Pen Bamboo Bridge

Kohukohu Bridge, Panguru, Northland, New Zealand

The simple masonry arch bridge in the village of Kohukohu was built at some time between 1843 and 1851 across the mouth of the Waihouuru Creek and is considered to be the oldest bridge in New Zealand. The structure, which spans about 10ft, is made from stone blocks thought to have been brought in as ballast on ships taking felled timber back to Australia. Reclamation of the harbour has left the bridge surrounded by dry land.

Kohukohu Bridge

Koishikawa Korakuen Garden Bridges, Tokyo, Japan

These world-famous gardens were started by Mito Yorifusa in 1629 and include two bridges. **Engetsu-kyo (Full Moon) Bridge** across the South Pond is a semicircular structure springing from the water level and **Tsuten-kyo Bridge** is a timber structure crossing a rocky ravine.

Konitsa Bridge, Konitsa, Epirus, Greece

The nearly semicircular stone arch bridge across the Aoos River was built in 1870. At nearly 20m high it is considered to be the biggest arch in the Balkans. A bell on the bridge warns when the wind is high and it can be dangerous to cross.

Koornbrug, Leiden, South Holland, Netherlands
The first bridge here over the Nieuwe Rijn was built in the fifteenth century and became a place where grain was traded, hence its name (Corn Bridge in English), but the present three-arched structure dates from 1642. In 1824, in order to keep the grain dry, it was given its two elegant colonnaded roofs over the sidewalks, which were designed by the city architect Salomon van der Paauw.

Koornbrug

Krasnoluzhsky Bridges, Moscow, Russia
There are now two Krasnoluzhsky bridges across the Moskva River. The first, originally called the Nicholas II Railway Bridge (but renamed the **Krasnoluzhsky Railway Bridge** after 1917), was designed by the structural engineer Lavr Proskuryakov and built in 1907. This was a crescent steel arch structure spanning 135m, which was replaced in 2001, the original one then being re-erected and

Krasnoluzhsky Railway Bridge

renamed the Bogdan Khmelnitsky Pedestrian Bridge (q.v.). The replacement rail bridge is a steel arch similar to its predecessor. The second of the current bridges is the **Krasnoluzhsky Road Bridge**, which was built between 1997 and 1998 to carry the Third Ring Road over the Moskva River. This is a 413m-long concrete structure consisting of twin box girders with a main span of 145m. The deck, which is 40m wide, carries ten traffic lanes, a cycle way and a footpath.

Kristiansand to Trondheim Proposed Submerged Bridge, Kristiansand, Agder County, Norway
Every other entry in this book deals with one or more bridges that are about to be built or have already been built, some of them more than 2,000 years ago and, together, the story they tell is one of continuing development in civil engineering. It seems fitting, therefore, that there should be one entry that looks to the future and covers what some people call 'blue skies thinking' (although in this case 'blue seas thinking' might be a more apt description). The idea of a Submerged Floating Tube Bridge (SFTB) was first proposed in 1886 by UK naval architect Sir James Edward Reed. Further development has concluded that the concept would probably be most suitable where a fixed crossing was needed of a strait 2000-3000m wide with water depth of several hundred metres or more and where a fully serviced under-water tube could be anchored at a depth of 20-50m. The Norwegian government is investigating using the concept for the E39 road in Norway between Kristiansand and Trondheim.

Krk Bridge, Rijeka, Primorge-Gorski Kotar, Croatia
Krk Bridge links the island of Krk to the Croatian mainland. When it was completed in 1980, this bridge's main span of 390m was the longest reinforced concrete arch in the world. Separated from it by a section of four post-and-beam spans, there is a second arched span of 244m and fourteen further post-and-beam spans. The bridge deck is 11m wide.

Kubelbrücke, Herisau, Appenzell Ausserrhoden, Switzerland

After an earlier bridge here across the Urnasch River was destroyed by floods in 1778, this bridge was built by Hans Ulrich Grubenmann in 1780 and it is his only surviving bridge. The main structure is a pentagonal bridge arch inside the covered bridge housing and the bridge is just under 3m wide and spans 30m. Grubenmann was the youngest of three bridge-building brothers who often worked together and whose bridges combined timber arches and trusses. Their best-known was the 1757 Schaffhausen Bridge over the Rhine (q.v.).

Kuldiga Bridge, Kurzeme, Courland, Latvia

Built in 1874 across the River Venta, this brick bridge is claimed to be one of the longest brick road bridges in Europe. It has an overall length of 164m, with seven 17m-long segmentally-arched spans between stone piers, and is 26ft wide. Two spans which were blown up in 1915 were rebuilt in 1926 and the bridge was restored in 2008.

Kuldiga Bridge

Kurilpa Bridge, Brisbane, Queensland, Australia

The 470m-long bridge for cyclists and pedestrians over the Brisbane River is a multi-masted cable-stayed structure with a span of 420ft. The main deck is 6.5m wide but there are also two seating areas and all these spaces are covered by a full canopy. The design is based on the principles of tensegrity, first put forward by Buckminster Fuller, in which individual non-connected compression members are contained within a framework of tension members. There are twenty structural steel masts and sixteen horizontal boom members, these being connected together in a complex

Kurilpa Bridge

system by nearly 7km of high-strength steel cables. Solar-powered LED lighting illuminates the bridge at night. The bridge, with engineering design by Ove Arup & Partners working with Cox Rayner Architects, was opened in 2009.

Kurishima-Kaikyō Bridge, Imabiri, Ehime Prefecture, Japan

The bridge here, which links Oshima Island to Shikoku across the Seto Inland Sea, was built between 1988 and 1999. It has a total length of 4,015m and consists of three separate end-to-end suspension bridges with two common anchorages where the different structures meet. The lengths of the main spans of the three bridges are successively 600m, 1,020m and 1,030m and the deck width for all three is 27m.

Kushma-Gyadi Bridge, Kushma, Parbat, Nepal

Opened in 2010, this 1½m-wide footbridge is 135m above the Modi River Gorge and is claimed to be able to support 1,800 people. Spanning 344m, it is the longest suspension bridge in Nepal and is stabilised by below-deck cables anchored lower down the gorge sides. There is a similar bridge nearby at Katuwachaupan.

Kutai Kartanegara Bridge, Tenggarong, East Kalimantan, Indonesia

The original road crossing over the Mahakam River at this site was by a steel suspension bridge built between 1995 and 2001. Its overall length was 710m with a 270m-long main span. In 2011 this structure collapsed following the failure of a hanger-to-deck connection during maintenance work that was in progress as a result of horizontal movement of the foundations. More than sixty people were killed or injured. A replacement steel arch bridge was completed in 2013 and formally opened in 2015.

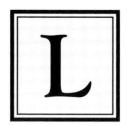

L

La Cassagne Bridge, Planès, Pyrenees-Orientales, France

The Little Yellow Train Railway tourist line, which started running between Villefranche de Conflent and Latour de Carol in 1911, crosses 75m above the River Têt on a rare 241m-long railway suspension bridge with a main span of 156m. Built to a revolutionary design by Albert Gisclard (1844–1909), which was later licensed by Ferdinand Arnodin, it consists of a combination of diagonal stays near the steel support pylons and hangers from suspension cables for the central part of the span. Gisclard and five others were killed in an accident during the final test run of a train in 1909. The bridge was classified in 1997 as a Historical Monument.

La Cassagne Bridge

La Marca Suspension Bridge, Villa Azurduy, Chuquisaca, Bolivia

This 361ft-span suspension bridge in rural southern Bolivia was built in 2019 by the not-for-profit organisation Bridges to Prosperity (B2P) which, since 2001, has built more than 200 bridges for isolated communities in twenty countries.

Lacey V. Murrow Memorial Bridge, Seattle, Washington, USA

The first bridge here, originally called the Lake Washington Floating Bridge, was opened in 1940 and given its present name in 1967. It closed in 1989 and during reconstruction in 1990 the supporting pontoons sank. The replacement structure was opened in 1993. This is a 6,620ft-long pontoon bridge that carries the four lanes of Interstate 90's eastbound traffic across the lake (named after the Washington's director of highways), while westbound vehicles use the **Homer M. Hadley Memorial Bridge** (named after a well-known Seattle bridge engineer). The Lacy V. Murrow Bridge is included in the ASCE list of historic bridges.

Homer M. Hadley Memorial Bridge (left) Lacey V. Murrow Memorial Bridge (right)

Laguna Garzón Bridge, José Ignacio, Rocha, Uruguay

Best known for its unusual shape, with traffic on the double-lane approaches having to slow down as they enter the one-way central circular section across the Garzón lagoon, this prestressed concrete bridge was designed by Uruguayan architect Rafael Viñoly and opened in 2015 to replace a car-carrying raft. Circular columns

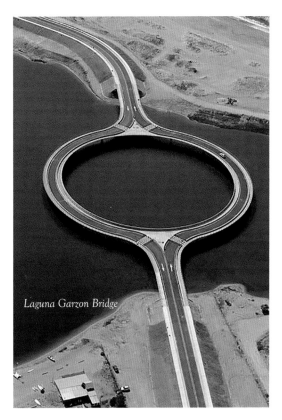

Laguna Garzon Bridge

world's longest continuous bridge over water. All the structural elements for this bridge are of concrete. These were precast and then prestressed off site in a purpose-built factory. Pairs of hollow concrete piles were driven into the soft ground and then connected by cross-head beams which, in turn, support longitudinal beams. A raised section allows water traffic to pass beneath the bridges. The American Society of Civil Engineers has designated the structure a National Historic Civil Engineering Landmark.

Lake Vranov Footbridge, Znojmo, South Moravia, Slovakia

Designed by Strasky, Husty & Partners, this prestressed concrete suspension bridge was completed in 1993. A-shaped pylons 252m apart carry twin suspension cables that support the slim stressed-ribbon waisted deck. This deck varies in width from 10m at the pylons to 6.5m at midspan and consists of precast units only 0.42m thick. The suspension cables are anchored beyond the end of the deck, these back spans being only 30m long.

Lake Vranov Footbridge

support the sixteen 20m-long spans of the bridge which has a walkway along each edge, the inner one being covered.

Lake Pontchartrain Crossings, New Orleans, Louisiana, USA

The first bridge across Lake Pontchartrain was built near New Orleans in 1883 by the Southern Railway and, at 21½ miles long, was then the world's longest continuous bridge over water. However, since then many sections have been filled in and only about 6 miles of trestle now remain. The bridge includes a rolling lift bascule that can be lifted for water traffic to pass. The first road crossing over the lake, opened in 1956, was the Lake Pontchartrain Causeway, a low-level trestle structure, also with a bascule span for water traffic. In 1969 a second parallel bridge was added with an 84ft-wide gap between the two structures and this is (by about 50ft) the longer one and, at 23.8 miles, was then the

Lambézellec Viaduct, Brest, Finistère, France

The steel railway bridge over a stream called the Spernot (a tributary of the River Penfeld) was built between 1891 and 1893 to connect the towns of Ploudalmézeau and Brest and is also known as the Brasserie Viaduct. The structure, which was

Lambézellec Viaduct

Land Bridge, Banff National Park, Alberta, Canada

The first land bridge over the Trans-Canada Highway in Banff National Park was completed in 1997 and there are now some half dozen such bridges (as well as nearly forty underpasses) that allow wild animals to cross the road safely. The crossing consists of a long semicircular archway over each carriageway, above which earth is placed and planted with grass and small shrubs to create a wide natural land corridor between the open ground on each side.

Land Bridge

designed by the engineers Louis Harel de la Noë and Armand Consider, has a 3.6m-wide and 109m-long deck with eight 13.5m-long plate-girder spans. These are supported by inverted V-shaped steel piers consisting of pairs of triangular legs, with each leg tapering to a pin joint at its base. The maximum pier height is 17m.

Lanaye Bridge, Liège, Walloon, Belgium

The overall length of this 13m-wide cable-stayed road bridge over the Albert Canal is 232m. It has a single 68m-tall inverted Y-shaped pylon supporting a fan of ten pairs of stays along the 177m-long main span. These are balanced by ten back stays that are fixed to ground anchorages arranged at right angles to the bridge axis. The bridge was designed by René Greisch and completed in 1982.

Langeais Bridge, Langeais, Indre-et-Loire, France

The D57 road bridge here over the River Loire, built between 1935 and 1937, is an unusual tied-tower suspension bridge. The overall length of the bridge is 358m and it includes three consecutive 90m-long main suspension spans between four brick piers as well as shorter end spans. The ends of the piers are corbelled out into circular brick towers either side of a large pointed arch opening. The towers are topped by vertical openings through which pass the suspension chains and the horizontal wire

Langeais Bridge

rope ties. Above the main pier arches, through which the traffic passes, the towers are linked by a decorative arcade of small pointed arches. The bridge was destroyed in 1940 to hinder the German army's advance and was rebuilt in the late 1940s.

Langebrug, Haarlem, North Holland, Netherlands

There had been timber drawbridges here for generations but the present structure replaced an iron drawbridge dating from 1932. This new bridge, completed in 1995, was designed by city architect Thijs Asselbergs and is a modern version of the typical Dutch drawbridge. It has a single pylon at the top of which is hinged an elongated X-frame. The two longer arms of this support each side of the outer end of the 10.9m-wide lifting roadway while a huge cylindrical counterweight is supported between the ends of the shorter inner arms. The appearance of this counterweight has led to the bridge being nicknamed the 'Paint-roller'.

Langebrug

Langkawi Sky Bridge, Langkawi Island, Kedar, Malaysia

Completed in 2005, this cable-stayed footbridge is 660m above sea level at the peak of the mountain called Gunung Mat Cincang. The 1.8m-wide curved footway is 125m long and this is claimed to be the longest curved bridge

Langkawi Sky Bridge

deck in the world. It is supported by eight cables from an inclined steel latticework pylon that is 82m high.

Langlois Bridge, Arles, Bouches-du-Rhône, France

The double-leaf timber drawbridge at Arles, with its distinctive balancing framework above, crosses the canal between Arles and Port-de-Bouc. and was built during the early nineteenth century. In 1930 the bridge was replaced by a concrete structure and this was blown up by the Germans in 1944. The bridge is famous for the paintings and drawings made of it in 1888 by Vincent van Goch and a modern reconstruction has been built by the Arles tourist authorities.

Langlois Bridge

Lansdowne Bridge, Sukkur, Sindh Province, Pakistan

The first bridge at this site, built between 1887 and 1889, is a huge structure designed by Sir Alexander Rendel (1829–1918) to carry the East India Railway's link between Lahore and Karachi across the Rohri channel of the Indus River. It consists of two 170ft-tall anchored cantilever trussed frames (each a half-hexagon) and weighing 3,300 tons, that are 310ft long with a 200ft-long suspended span between their inward ends. The bridge originally carried a railway that was flanked by timber walkways for pedestrians and cyclists. A second railway bridge, the Ayub Bridge (q.v.), was built 100ft away and opened in 1962, since when the deck of the Lansdowne Bridge has been rebuilt to carry light road traffic.

Lascellas Bridge, Lascellas-Ponzano, Aragon, Spain

The first bridge here over the Alcanadre River, built by Mariano Royo in 1860, was a partly underslung suspension bridge spanning 94m, the middle third of the deck being propped on wooden posts above the cables. This structure became famous through the work of the British photographer Charles Clifford, who was appointed by Queen Isabel II to record her opening of the bridge as part of a public relations exercise. The current bridge, which was completed in the 1990s, is a five-span prestressed concrete structure that is 225m long and carries the N-240 highway.

Latin Bridge, Sarajevo, Bosnia and Herzegovina

The first stone bridge here over the River Miljacka, replacing an earlier wooden structure, was built in 1565 and rebuilt following a flood in 1795. It has four stone arches and there are flood relief arches above cutwaters on two of the piers. The footways are cantilevered out on each side. The bridge is famous for being very near where Gavrilo Princip assassinated Archduke Franz Ferdinand on 28 June 1914, thereby precipitating the First World War, and the bridge was therefore sometimes called the Princip Bridge.

Lego Bridge, Wuppertal, North Rhine-Westphalia, Germany

The concrete beam bridge crossing Schwester Street at a slight skew was originally built to carry a railway, but this was decommissioned in 1991. In 2011 the bridge was painted by artist Martin Heuwold to look as if it were made from Lego bricks, the work being awarded the Deutscher Fassadenpreis Advancement Prize in 2012.

Lego Bridge

Leipheim Bridge, Leipheim, Bavaria, Germany

The bridge at Leipheim, which carries the A8 autobahn across the Danube, was completed in 2001 to replace an earlier structure opened in 1936. It is a six-span prestressed concrete box girder with an overall length of 375m.

Lejonströmsbron Bridge, Skellefteå, Västerbotten, Sweden

This bridge across the Skellefte River was completed in 1737 and is Sweden's oldest timber

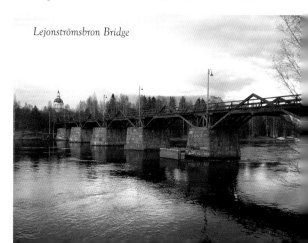

Lejonströmsbron Bridge

bridge. Designed by Carl Spennare, it is 207m long and 5m wide. There are six river spans consisting of A-frame trusses spanning between large stone piers with pointed cutwaters. There was a battle here in 1809 when the Russians crossed thawing ice to capture Skellefteå, the war resulting in Sweden ceding Finland to Russia.

Lekki-Ikoyi Bridge, Lagos, South West, Nigeria

The main part of this 1358m-long road crossing over Five Cowries Creek, opened in 2013, is a cable stay bridge with a main span 110m long with 9m clearance beneath its deck. The distinctively-shaped pylon is 91m tall. The approach viaducts have thirty-eight spans.

Lekki-Ikoyi Bridge

Leonardo da Vinci Bridge, Ås, Viken County, Norway

The artist Vebjørn Sand's conceptual design of this footbridge is based on an unexecuted design by Leonardo da Vinci for a huge masonry bridge across the Golden Horn in Constantinople, and the engineering design was by Reinhert Structural Engineers. The 2.8m-wide deck, which crosses 5.8m above the E18 road, is a 109m-long doubly-curved steel-reinforced stressed-laminated member. This is supported by three separate glued and laminated timber arches: a 40m-span central arch beneath the deck with two outer arches leaning inwards to provide lateral support. The bridge was opened in 2001.

Léopold-Sedar-Senghor Footbridge, Paris, Ile-de-France, France

This unusual footbridge across the River Seine in Paris has three single-span arches: two outer arches linking the higher riverbank terraces on each side and a central one between the lower terraces, with all three sharing the same apex level above mid-river. It was designed by engineer:architect Marc Mimram and was opened in 1999. The bridge is 106m long and its decks, with a total width of 15m, have been surfaced with Brazilian timber. It is the third structure to cross the river at the site, the previous ones being called the Solferino Footbridge.

Lerez River Bridge, Pontevedra, Galicia, Spain

Designed by Leonardo Fernandez Troyano, the Tirantes road bridge over the River Lerez was completed in 1995. It is a cable-stay structure with the cables supporting the 125m-long main span in a semi-fan layout. The anchorage lines for the two sets of back stays are at right angles to the backward-leaning pylon, the cables thus forming a hyperbolic paraboloid shape. The single pylon is 140m high.

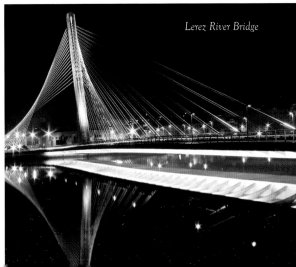

Lerez River Bridge

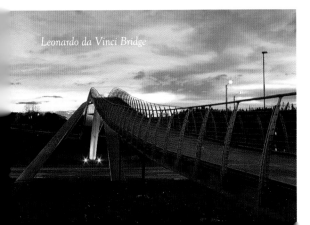

Leonardo da Vinci Bridge

Lethbridge Viaduct, Lethbridge, Alberta, Canada

The huge Lethbridge Viaduct across the Oldman River, completed by the Canadian Pacific Railway in 1909, is one of the biggest trestle bridges in the world. It has an overall length of 5328ft with a total of sixty-seven spans, all but one of these being steel plate girders that are either 20m or 30m long. The structure has a maximum height of 314ft and its greatest width at the level of the footings is 101ft.

Lethbridge Viaduct

Lézardrieux Bridge, Lézardrieux, Côtes-d'Armor, France

The first bridge here across the Trieux River was a suspension bridge built in 1840 and this was replaced by a new bridge completed in 1925 and restored in 1993 that now carries the D786

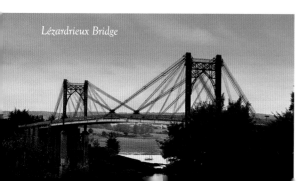

Lézardrieux Bridge

road. With masonry approach viaducts on each bank, this Gisclard stayed structure has a main span of 118m and 20m-long back spans. The two legs of the 20m-high steel portal-frame pylons are connected by an intricate pointed arch and themselves stand on masonry piers pierced by two pointed arches.

Liberty Bridge, Budapest, Central Hungary, Hungary

The first structure here was opened in 1896 as part of the World Exhibition celebrating one thousand years of Hungarian history. It was called the **Franz Josef Bridge** after the emperor of the time and spanned the Danube River to link the twin cities of Buda and Pest. Designed by Janos Feketehazy, the three-span girder structure was 334m long and the deck was 20m wide. The central span was blown up by the retreating Germans in January 1945 but this was rebuilt and the bridge then re-opened in 1946 as the Liberty Bridge. The spires above the two intermediate piers are decorated with bronze statues of falcon-like Tural birds, mystical symbols in Magyar mythology.

Lika River Bridge, Kosinj, Zadar, Croatia

The stone bridge here was built in 1936 to replace earlier timber structures that were frequently washed away by spring floods. It is 70m long and 5½m wide with three semicircular stone arches each spanning 18m that support a level roadway. There is a large circular tunnel-like aperture above each of the two mid-river piers.

Lima Bridge, Ponte de Lima, Norte, Portugal

The original bridge over the River Lima at this site was built by the Romans and had seven arches, which are now buried under a field. By the

Lima Bridge

time the existing town was founded in 1125 the course of the river had moved southward and is now crossed by a bridge dating from 1368. This originally had seventeen slightly pointed stone arches (three of which are now buried) with tall arches in some of the massive piers. The bridge originally had defensive fortifications but these are also now gone. It is 277m long and 4m wide and has been a National Monument since 1910.

Linn Branch Creek Bridge, Parkville, Missouri, USA

This structure was first built as a railway bridge in 1898 to a patented design by the engineer John Waddell. It was dismantled in 1981 and re-erected a short distance away in Parkville, where it is now a registered Historic Place footbridge. The bridge consists of two steel frames meeting at an apex to form a 12m-high 'A' truss, with vertical and diagonal members supporting the 30m-long deck at its quarter-span points.

Lionel Viera Bridge, La Barra, Maldonado, Uruguay

The first of two identical three-span stressed ribbon road bridges over the Maldonado River was built in 1963 to carry Route 10 and connect Punta del Este to La Barra. It was designed by the engineer Lionel Viera, and an identical copy followed in 1999. The line of the deck follows a gentle sinusoidal curve, the two high points being over the intermediate piers which consist of a five-ribbed V-frame. The deck is 10m wide and the span lengths are 30m, 90m and 30m.

Lipcani-Rădăuţi Bridge, Lipcani, Bessarabia, Moldova

This road bridge over the River Prut connects Lipcani in Moldova with Romania and was first built in 1937. It was destroyed twice in WWII: in 1941 it was blown up by the retreating Red Army during the German invasion of the Soviet Union and, after the Germans rebuilt it, it was destroyed again by Allied bombing in 1944. A replacement steel girder bridge, opened in 2010, is 246m long and 10m wide.

Lispole Viaduct, Dingle, County Kerry, Ireland

The narrow-gauge Tralee & Dingle Railway, which opened in 1891, crossed the Owenalondrig River valley on this structure. It consists of five semicircular masonry arches, four at one end and one at the other, with two intermediate metal lattice girder spans.

Llinars Bridges, Sant Celoni, Catalonia, Spain

The two adjacent structures at Llinars del Vallès, which carry the high-speed, twin-track railway between Barcelona and Perpignon, have an overall length of 574m and were opened in 2006. The first, a 267m-long prestressed concrete viaduct across the Mogent River, has a maximum span of 48m. The other, which crosses the AP-7 highway on a sharp skew, has a central main span of 75m flanked by two 71m-long spans and smaller end spans of 45m. The supporting piers extend above the deck and, from their tops, a large-section curved member radiates downwards

Llinars AP-7 Highway Bridge

Lionel Viera Bridge

and outwards to support the centre of the steel and concrete deck span on each side. Because of the unusually large size of these tension members, the above-deck elevation of the structure looks somewhat like four consecutive king post trusses with upward-curving rafters.

LNG Pipeline Bridge, Bioko, Bioko Sur Province, Equatorial Guinea

An unusual structure was completed in 2006 on the shoreline of the Gulf of Guinea. It was built by Bechtel Corporation to support a 76cm-diameter pipeline carrying liquid natural gas between a 60m-high clifftop gas plant and a jetty from where the LNG can be directly pumped onto sea-going ships for export. The structure is a suspension bridge, 350m-long and 9m-wide, that has one short pier standing within the plant area at clifftop level and a tall one founded on the sea bed.

Løkke Bridge, Sandvika, Greater Oslo, Norway

This cast iron bridge over the Sandvikselven River in a village a few miles from Oslo was first built in 1829 and is the oldest such structure in Norway. It spans 23m and has ten diminishing circles in the spandrels. The bridge, which was renovated and moved to a new location in the village in 1977, appears at the bottom of Monet's 1895 painting 'Sandviken Village in the Snow'.

Løkke Bridge

London Bridge, London, England

The history of London Bridge is inextricably tied up with that of London itself. The first bridge over the Thames here was probably a simple timber structure built at an early stage during the Roman occupation between the first and fourth centuries and estimated to have been 620ft long and 40ft wide, although no written records have ever been found. No bridge can have existed in 993 when King Olaf I of Norway sailed with a Viking fleet to Staines, but later there was a wooden bridge wide enough for two wagons to pass each other. This featured in other attacks on London by the Danes when Olaf II, now supporting the Saxons under King Aethelred, attacked the Danish-held bridge in about 1014 and pulled it down by attaching cables to it from his war galleys. Successor bridges were destroyed by a storm in 1091 and by fire in 1136, being replaced after the latter disaster by a bridge in elm that was built by Peter, the chaplain at the parish church of St Mary Colechurch. Since then there have been three further bridges.

In 1176 Peter of Colechurch began to build a new bridge in stone just to the west of his earlier wooden structure. From 1202 he was assisted (or even, perhaps, directed) by Isembert of Saintes in France and, following Peter's death in 1205, Isembert completed the work. Old London Bridge was finished in 1209 and it served the city for 622 years until Rennie's replacement was opened in 1831. For the first five centuries of its existence, until Fulham Bridge was opened in 1729, it was the only crossing of the Thames below Kingston Bridge, about 20 miles upriver. Old London Bridge consisted of nineteen pointed arches of Kentish ragstone and a wooden drawbridge span, with each pier protected by a massive starling. The lengths of the arch spans varied between about 12ft and 33ft and the thicknesses of the piers between about 17ft and 36ft, the overall length being about 900ft with a deck about 20ft wide supporting a roadway about 12ft wide. The original bridge piers were founded on three rows of timber piles, mainly of elm, within which loose stones were placed supporting massive oak sleepers on which the stone superstructure was then constructed, progressing across the river at the rate of roughly one completed arch span every eighteen months. The protective starlings outside the piers also consisted of rows of timber piles enclosing loosely dumped stonework, each subsequent repair increasing the size of starling. The reduced width of waterway caused by the starlings resulted in a considerable difference in water level between each side of the bridge, the maximum recorded being more than 4½ft in 1736. In order to take advantage of this source of energy, mill wheels were built under the two southern arches in 1559 and in 1580 waterwheels were installed at the north end to pump water into the City. By 1767 there were seven wheels and these remained until removed by Act of Parliament in 1822. There were also buildings on the bridge, including at first a chapel (where Peter was buried), which was demolished in 1553, and two gateway towers and, later, many houses.

Between 1758 and 1766, under George Dance the Elder, all the remaining buildings were demolished, one of the central piers was removed and the Great Arch spanning about 50ft was built, some of the stone coming from the newly-demolished Moorgate. The bridge roadway was also widened to a standard width throughout by extending it out over the starlings. The bridge remained in this state for another sixty-five years until 1831, when it was completely demolished after the Rennies' replacement had been opened, one of the old stone alcoves from the bridge being re-erected at Dropmore in Buckinghamshire, another in the grounds of Guy's Hospital and two more in Victoria Park, Bow.

The detailed design of the new bridge and supervision of its construction was undertaken by Sir John Rennie, working to an outline design submitted shortly before his death by his father, also John. It was located about 100ft upstream of the old structure and consisted of five semi-elliptical arches: a central arch of 152ft span and 30ft rise, with side spans of 140ft and 130ft. The overall width was 56ft incorporating a 35ft carriageway and two 9ft footways. The

foundations, constructed in timber cofferdams about 45ft below high water level, consisted of beech and elm piles driven into the riverbed supporting the stonework bases of the piers. Three types of granite facings were used: red-brown from Peterhead for the arches, grey from Devon for the upstream side, and a light purple from Aberdeen for the downstream face. The overall appearance was of classic simplicity without undue ornamentation. The bridge was opened on 1 August 1831. Between 1902 and 1904 it was widened to 68ft by cantilevering out granite corbels from the top of the side walls to support extended footways, with a lighter open balustrade replacing the original solid parapet.

Apart from some problems of settlement, Rennie's bridge, like its predecessor, also proved to be inadequate to meet the ever-increasing demands of vehicle and pedestrian traffic and, in 1965, a decision was taken to rebuild. The replacement bridge, literally constructed around its predecessor, is a prestressed concrete box girder structure designed by consulting engineers Mott, Hay & Anderson and built by John Mowlem & Co. The three spans are 260ft, 340ft and 260ft and are faced with polished granite. The four parallel box girders were each assembled from eighty-nine precast sections supported from a temporary steel gantry and then post-tensioned together to form a single integral structure. The Queen opened the bridge in 1973. As part of the reconstruction project, the granite facing stones of Rennie's bridge were carefully dismantled and the front four inches cut off into slips to clad a replica reinforced concrete bridge at Lake Havasu in Arizona. (See AEBB.)

Innumerable paintings of the bridge include those by Claude de Jong showing the medieval structure.

Long Biên Bridge, Hanoi, Red River Delta, Vietnam

This major steel bridge over the Red River, originally called the **Paul-Doumer Bridge**, was built between 1898 and 1903 and at the time was one of the longest bridges in Asia. It has an overall length of 1,683m and nineteen spans, the

Long Biên Bridge

longest being 106m. The bridge carries a single railway line flanked by motorcycle, cycle and pedestrian lanes. Some of the central part was brought down by bombing during the Vietnam War in 1967 and again in 1972.

Longeray Viaduct, Leaz, Ain, France

The reinforced concrete railway bridge, which crosses diagonally above the River Rhone, was built between 1941 and 1943. It is mainly recognisable because of its three vertically parabolic and tapering main arches that span 56m, 69m and 51m. These are separated by vertical piers that similarly taper inwards from bottom to top. The bridge deck is supported by the crown of each arch and by flanking vertical piers on each side. Overall, the viaduct is 70m high and, with its short semicircular approach spans, its total length is 290m.

Longeray Viaduct

Longteng Bridge, Sanyi Township, Miaoli County, Taiwan

The bridge here, built in 1907 to carry a railway, consisted of a central steel truss flanked on each side by four semicircular brick arches. An earthquake in 1935 caused all the arches to collapse, leaving just the T-shaped remains of the piers. The line was re-opened in 1938 with a new steel structure a short distance away. Following a further earthquake in 1999, one of the remaining piers from the original bridge collapsed. The ruins are now preserved as a monument to one of Taiwan's worst-ever earthquakes and are included on the country's list of Cultural Heritage Assets.

Longteng Bridge

Los Caros Bridge, Bahia de Caraquez, Manabi, Ecuador

Opened in 2010, this bridge, which is the second longest bridge in Ecuador, is the lowest crossing of the Rio Chone and carries the E15. A feature of the 1,980m-long structure is the use of special seismic isolators that saved it from damage in the earthquake of 2010.

Loto Samasoni Bridge, Apia, Samoa

The Mabey Bridge Company has built five of its proprietary Compact 200 structures, developed from the original Bailey Bridge of the Second World War, in Samoa. The Loto Samasoni Bridge, completed in 2019, carries a two-lane road and single footway across the Vaisigano River and is 37m long.

Lotus Lake Footbridges, Kaohsiung, Kaohsiung County, Taiwan

The footbridge across the south-west corner of Kaohsiung's artificial Lotus Lake, opened in 1951, has seven main zigzag lengths leading to the seven-storey octagonal Dragon and Tiger Pagodas. Other footbridges on the lake lead to the Wuliting and Zuoying Yuandi Temples.

Lotus Lake Footbridges

Lowen Bridge, Berlin, Brandenburg, Germany

This decorative suspension bridge in Glienicker Park was designed by the engineer Ludwig Ferdinand Hesse and built in 1838. The metal suspension rod system on each side of the deck is anchored in the mouths of cast bronze lions standing on the abutments.

Lower Volta Bridge, South Tongu, Volta, Ghana

Also known as **Sogakope Bridge** and completed in 1967, this structure carries the N1 highway across the Volta River between Abijan and Lagos. Consisting of multiple prestressed concrete T-section beams, its overall length is 650m. It has not been well maintained and potholes in the roadway led to deaths in 2019.

Loyang (Ten Thousand Peace) Bridge, Chuanchow, Fujian, China

The forty-seven-span stone slab bridge across the mouth of the Loyang River was built between 1053 and 1059 and was one of the five great ancient bridges of China. It was 1,200m long and 5m wide and was founded on an underwater causeway with each span between the boat-shaped stone piers consisting of seven stone slabs 7m long by 60cm wide and 50cm thick.

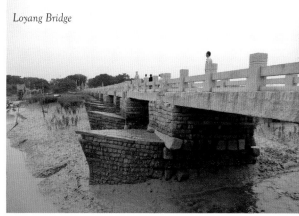
Loyang Bridge

After being badly damaged in 1937 during the Sino-Japanese War, the bridge was rebuilt with a reduced length of 730m.

Luanfeng (Mythical Bird Peak) Bridge, Xiadang, Fujian, China

The Luanfeng Bridge was built in 1800 and restored in 1964. Its clear span of 37m is the longest timber bridge span in China. The bridge's overall length is 48m and width is 5m.

Lucan Bridge, Dublin, County Dublin, Ireland

An earlier bridge here over the River Liffey was built in about 1200 but it and later replacements were destroyed by floods in about 1730 and 1786. The present segmentally-arched structure, designed by the architect George Knowles, was built in 1814. Its 110ft span makes it the largest stone arch bridge in Ireland and it has very attractive cast iron balustrades.

Lucan Bridge

Lucano Bridge, Villanova, Padua, Italy

The old Roman bridge that crosses the Anio River here is thought to have been built in 2BCE, possibly by Marcus Plautius Silvanus, and originally had five small arches. It is known to have been destroyed and rebuilt in ancient times but underwent a major restoration in 1835 and now has seven low arches. In the mid-twentieth century the road was diverted onto a new structure to the north. Adjacent to the bridge is a circular castellated tower, known as the Mausoleum of the Plautii, and the bridge and tower feature in a well-known painting by Pietro della Valle and engravings by Giambattista Piranesi and others. The bridge was placed on the World Monuments Watch in 2010.

Lucano Bridge (print c1840)

Lucky Knot Bridge, Changsa, Hunan Province, China

This most unusual bright red steel footbridge across Dragon King Harbor River has three interlinked 185m-long footways and reaches a maximum height of 24m above water level so boats can pass under it. In elevation the bridge looks roughly like overlapping sine curves, with the apex of one curve coinciding with the trough of another, and there are interconnections

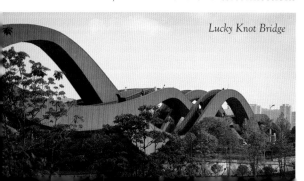

Lucky Knot Bridge

between the footways at the five points where they overlap. The footbridge, which was designed by NEXT Architects, was opened in 2016.

Luding Bridge, Garze, Sichuan, China

This chain suspension bridge, built during the period 1705–1710 and spanning 100m across the Tatu River, was an important link on the road between Sichuan and Tibet. The wooden deck is supported by nine chains, with two further chains on each side of the deck acting as handrails. The Red Army forced a crossing over the bridge in 1935 during the Long March and it is now a major tourist destination.

Lugard Footbridge, Kaduna, Kaduna State, Niger

This little suspension bridge was first built by Sir Frederick Lugard in 1904 at Zungeru and was then re-erected at Kaduna in 1920. Its wooden deck is 14m long and 1.8m wide. Sir Frederick, later Lord Lugard, is famous for having effectively created the modern Nigeria when in 1914 he consolidated the Protectorates of Northern and Southern Nigeria.

Lumberjack's Candle Bridge, Rovaniemi, Lapland, Finland

Rovaniemi's signature cable-stayed bridge carries the E75 road over the junction of the River Ounasjoki with the River Kemijoki. Designed by the Finnish civil engineering firm Suunnittelukortes Aek Oy (later part of WSP Group) and completed in 1989, it has an overall length of 223m and has five spans, the biggest being 126m long. The prestressed concrete deck is 26m wide. The bridge's name derives from the lights around the tops of the twin columns of the H-shaped main pylon, 47m above the deck, which are adorned with two large representations of a lumberjack's candles.

Lumberjack's Candle Bridge

Lupu Bridge, Shanghai, China

The Lupu Bridge is named to mark its role connecting the Luwang and Pudong districts of Shanghai across the Huangpu River. It is a half-through tied-arch all-welded steel bridge consisting of two inward-leaning box ribs. Overall, it is 3,900m long with a main span of 550m and side spans of 100m, and its 29m width accommodates six traffic lanes with flanking footways. When it opened in 2003 it was the world's longest arch bridge and there was originally a sight-seeing platform above the arch crown, but this is now closed.

Lupu Bridge

Lusitania Bridge, Mérida, Badajoz, Spain

Santiago Calatrava's bridge over the Guadiana River, opened in 1991, was built to take road traffic from the nearby old Roman Bridge (q.v.). The new structure has an overall length of 465m with three 45m-long approach spans on each side of a tied arch spanning 189m. This arch is in three parts: the lower part on each side is in concrete and these act as abutments for a central

Lusitania Bridge

steel trussed arch. The 24m-wide bridge deck is supported by twenty-three pairs of hangers, one hanger to each side, and there is a central raised footway. Each of the concrete supporting piers beneath the deck splits into two separate tapered columns.

Luzancy Bridge, Luzancy, Seine-et-Marne, France

An earlier bridge over the River Marne at Luzancy was a chain suspension bridge spanning about 54m that was completed in about 1833. The later prestressed concrete bridge, designed by Eugène Freysinnet and built between 1941 and 1946, was one of the first of this new type of structure. It has been regarded as either a very flat segmental arch or as a two-hinged rigid portal frame, but it is considered to be a great demonstration of the elegance and lightness offered by the new technology. Spanning 54m, it consists of six parallel longitudinal units, each made up from three precast sections, and is stressed transversely as well as longitudinally.

M

Ma Kham Bridge, Kanchanaburi, Thailand

An earlier timber bridge over the Mae Klong River near here was originally built by Allied PoWs between October 1942 and October 1943 as part of the Burma-Siam Railway and was the inspiration for Pierre Boule's novel *The Bridge on the River Kwai*, later filmed by David Lean. The present 300m-long steel structure consists of eleven spans, nine of which are segmental arched trusses from an early rebuilding after the bridge had been bombed in 1945. The two parallel boom girder spans are replacements from the late 1950s. The film itself was made near Kitulgala in Sri Lanka using a temporary mock-up timber cantilever structure.

Ma Kham Bridge

Maameltein Bridges, Maameltein, Tyre, Lebanon

The arch ring of a single-span segmental arch is all that remains of the original Roman bridge and a campaign exists to try to save it. A reinforced concrete deck arch structure nearby now carries a road across the gulley. This bridge was damaged in 2006 when it was bombed by the Israeli air force and then had to be rebuilt. The modern bridge features on a 10 piastre Lebanese postage stamp issued in 1971.

Mackinac Straits Bridge, Mackinaw City, Michigan, USA

The 'Big Mac' suspension bridge, designed by David B. Steinman and opened in 1957, carries the four-lane Interstate 75 across the Straits of Mackinac on its 54ft-wide deck. Together with approach spans its total length is 26,372ft and the central suspension span is 3,800ft long. The height of the towers is 525ft and the maximum clearance below the bridge is 155ft. Because of the exceptionally challenging weather in the Straits, with violent winds and thick ice-drifts, the steel and concrete pier foundations are fixed to bedrock more than 200ft below water level and weigh almost one million tons. The bridge, which is included in the ASCE list of historic bridges, is closed on Labor Day every year for the Mackinac Bridge Walk.

Mackinac Bridge

MacNimir Bridge, Khartoum, Sudan

Named after an 1822 tribal leader and opened in 2007, this bridge crosses the Blue Nile to link Khartoum city centre with Khartoum North. The cable-stayed main span is 80m long and there are a further seven river spans and more land approach spans resulting in the bridge having an overall length of 1165m.

Magdalena Footbridge, Pamplona, Navarre, Spain

The twelfth-century bridge in Pamplona carries the Camino Frances Pilgrimage Route to Santiago de Compostela over the River Arga.

Magdalena Footbridge

It has three main slightly pointed arches with small flood relief arches through the piers above the cutwaters.

Magdeburg Water Bridge, Magdeburg, Saxony-Anhalt, Germany

This, the longest navigable aqueduct in the world, crosses the Elbe River to connect the Elbe-Havel Canal to the Mittelland Canal. Construction began in 1998 and the canal crossing opened in 2003. The main structure over the river is an N-braced steel girder spanning 106m, but the approach spans use the canal trough as a beam structure to span between flat U-shaped reinforced concrete supports. The overall length of the aqueduct is 918m with a longest span of 106m, the width 34m and the water depth 4.3m.

Magdeburg Water Bridge

Mağlova Kemer Aqueduct, Istanbul, Marmara Region, Turkey

The first masonry aqueduct across the Alibey River was completed in 1563 but was destroyed by a flood a few months later. The replacement structure, which was built by the famous Ottoman architect Mimar Sinan between 1554 and 1562, is 258m long and 36m high. It has an unusual arrangement of arches. In the central part, there are two tiers, with each tier alternating between a wide arch and a smaller bay, the smaller upper bay consisting of twin arches arranged vertically. The intermediate piers are stiffened laterally on one side by buttresses and, on the approach spans at each end, the bottom tier is higher. The lower part of the aqueduct is now partially submerged by water in the Alibey reservoir.

Main Viaduct, Nantenbach, Bavaria, Germany

This 374m-long railway bridge, which was built in 1994, crosses the River Main on both a curve and a gradient and leads directly into the Schönrain Tunnel. Its 208m-long central span is a Warren girder segmental arch made of steel. This is flanked by two half-arch side spans that are each 83m long. The reinforced concrete deck is fully integrated with the upper boom of the steelwork.

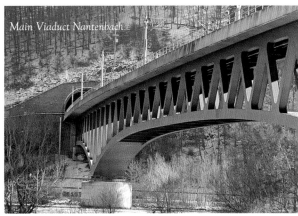

Main Viaduct Nantenbach

Makira Bridges, Makira, Makira-Ulawa Province, Solomon Islands

Two identical new bridges across the Magoha and Maepua Rivers were opened in 2012 as part of a project to upgrade the province's roads and twenty-five of its bridges. Each structure consists of a 72m-span network arch supported on concrete piers and carries a single carriageway flanked on each side by footways.

Mala Rijeka Viaduct, Podgorica, Central Region, Montenegro

The single-track railway bridge crossing 200m above the Mala Rijeka River was the highest railway bridge in the world until the Beipan River Shuibai Bridge (q.v.) was completed in 2001. The five-span steel truss viaduct, which is 499m long with a 151m-long main span, was built between 1969 and 1973.

Mala Rijeka Viaduct

Malabadi Bridge, Silvan, Diyarbakir, Turkey

Believed to have been built between 1146 and 1155, this is often considered to be the most beautiful bridge in Turkey. With a total length of about 150m, its main opening over the Batman River is a single, tall, pointed stone arch. This spans about 39m, is 19m high and carries a stepped footway 7m wide. There are other arches in the approaches as well as a small arch immediately behind the abutment

Malabadi Bridge

on the right bank. Above this abutment a simple triumphal gateway crosses the deck and both abutments include vaulted brick chambers. The bridge has been nominated for World Heritage status. A new concrete arched bridge was built a short distance upstream in 1955.

Malan Bridge, Herat, Herat Province, Afghanistan

The Herai Raud River, also known as Pul e Malan, is crossed at Herat by the Malan Bridge. This was built in the early twelfth century in the architectural style called Khorasanian. It has twenty-two pointed brick arches, the piers between them being fronted by very large cutwaters with tiered capping. Instead of a capping, one pier near each end of the bridge has a circular tower on each side of the bridge deck. This has tall lancet openings and is called the Kandahar gateway. The whole bridge was rebuilt in 1995 by the Danish Committee for Aid to Afghan Refugees.

Malan Bridge

Malir River Bridge, Karachi, Sindh, Pakistan

This multi-span concrete structure carrying the N5 road over the River Malir was the largest bridge in Pakistan when it was opened in 2009. With an overall width of 24m, it is 5,000m long and the longest span is 30m. The structure consists of steel beams spanning between twin-column concrete piers and cross-head beams.

Malleco Railway Viaduct, Collipulli, Araucania, Chile

Painted a striking bright yellow, this steel railway viaduct crosses the deep gorge in which the Malleco River flows. Designed by the Chilean engineer Victorino Aurelio Lastarria and built between 1886 and 1890 from parts fabricated in France, the 348m-long viaduct had five 70m-long spans and, with its highest ten-tier support tower being 76m tall and the rails 102m above the bottom of the gorge, it was then the tallest railway bridge in the world. The French firm Schneider et Cie did the detailed design and fabrication before the 1500 tons of wrought ironwork were then exported to Chile for erection. The structure originally had continuous 7m-deep lattice deck girders supported on doubly-tapering latticework towers but, in 1927, it was strengthened by the construction of an additional tower at each end and the introduction of steelwork haunches between the tops of the original towers and the underside of the lattice girders. Now painted yellow, the bridge was declared a Chilean National Monument in 1980 and an ASCE Historic Civil Engineering Landmark in 1995. It suffered some slight damage during an earthquake in 2010. The **Malleco Road Bridge** now crosses the gorge a short distance away from the railway bridge.

Malleco Viaduct

Manhattan Bridge, New York, New York, USA

Designed by Leon Moisseiff, the 120ft-wide double-deck Manhattan Bridge is a suspension bridge that carries seven road lanes and four subway lines to connect Manhattan Island to Long Island. The overall length of the bridge is 6,855ft with a 1,470ft-long main span between towers that are 336ft tall. Foundation work started in 1901 and the bridge opened in 1909. Although not strictly part of the bridge itself, mention must be made of the associated civic plaza that was built between 1910 and 1915. The entrance onto the bridge from the Manhattan side was through a grand triumphal archway, based on the Porte St Denis archway in Paris, that was flanked on each side by a colonnaded quadrant. The bridge is included in the ASCE list of historic bridges.

Manning Crevice Bridge, Riggins, Idaho, USA

The first bridge here, 14 miles east of Riggins, crossed diagonally over the Salmon River and was built in 1934. After it had reached the end of its working life a new structure was built alongside and the old bridge was demolished once the new one had been opened in 2018. Because of difficulties in bringing construction plant and materials to the south side of the river, the structure was designed as a single-pylon suspension bridge, this 110ft-tall pylon being located on the north bank. The pylon's two legs lean together slightly and are linked by double X-braces. The span is 300ft long and the bridge deck is 16ft wide. In 2020 the structure became a prize winner in the National Steel Bridge Alliance's Long Span Bridge category.

Maputo-Katembe Bridge, Matola, Maputo Province, Mozambique

The bridge across Maputo Bay, built by the China Road & Bridge Corporation and opened in 2018, is part of the 187km-long highway link to the South African border and includes the longest suspension bridge span in Africa. Its cables are supported by 140m-tall towers, the main span is 680m long and the deck is 60m above water level. The S-shaped northern approach viaduct is 1,097m long and the southern 1,264m, giving a total crossing length of 3,041m. Although the whole scheme was criticised for its high cost of more than US $786 million, the bridge won a 'Global Best Project' award in 2019.

Marcelo Fernan Bridge, Cebu City, Central Visayas, Philippines

The Mactan Channel that separates Cebu and Mactan Islands is crossed by this prestressed concrete bridge with extradosed cables that was opened in 1999. The main structure has a 185m-long central span and the bridge's overall length, including approach spans, is 1237m.

Marcelo Fernan Bridge

Marco Polo (Lugou) Bridge, Beijing, China

Marco Polo saw this bridge, one of the five great ancient bridges of China, on his visit to the country during his travels between 1271 and 1295, writing that it was 'perhaps unequalled by any other in the world', and it is sometimes called the Marco Polo Bridge. It crosses the Yongding River with eleven segmental stone arches and is 267m long and 9m wide. Work on the bridge started in 1189 and was completed three years later. The parapets consist of flat stone slabs set on their long narrow edge between smaller parapet columns, each of which supports a large decorative stone lion, sometimes with smaller lions hiding under it. Originally there were 627 lions altogether but only 485 now survive. Four

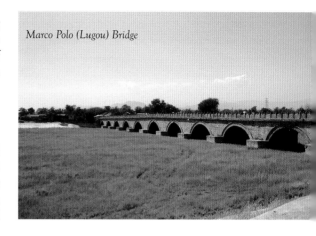

Marco Polo (Lugou) Bridge

5m-tall columns mark the corners of the bridge. The bridge was reconstructed in 1698 following flood damage and fully restored in 1986. In 1937 the bridge was the place where the Sino-Japanese war broke out, which lasted until the Japanese surrender in 1945.

Margaret Bridge, Budapest, Central Hungary, Hungary

This bridge across the Danube, which is broadly Y-shaped in plan, links the southern tip of Margaret Island to both river banks. Built between 1872 and 1876, it is 638m long and 25m wide. Although it was damaged in 1944 by an accidental explosion, it was quickly rebuilt and then reconstructed again between 2009 and 2011.

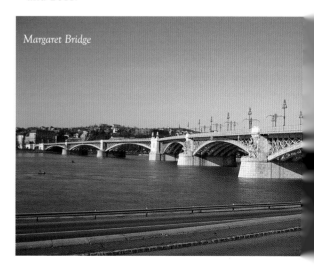

Margaret Bridge

Maria Pia Bridge, Porto, Norte, Portugal

This wrought iron structure, designed by Gustave Eiffel and similar to but less well-known than his later Garabit Viaduct (q.v.) of 1884, was completed in 1877 and named after the king of Portugal's wife Queen Maria Pia. A two-pinned parabolic arch spanning a then record-breaking 160m supports a 344m-long X-braced truss that carries a single-track rail line 60m over the Douro River. In 1991 a replacement structure was opened and Eiffel's structure is no longer in use. The bridge is included in the ASCE list of historic bridges and its north-eastern end is adjacent to the Porto World Heritage site.

Marina Cristina Bridge, San Sebastian, Gipuzkoa, Spain

In 1893 a temporary timber bridge was built over the Urumea River to link San Sebastian's town centre to its railway station, and the permanent reinforced concrete structure was completed in 1905. This is 88m long, with three 24m-span, low segmental arches, and is 20m wide. It is, however, one of the most exuberantly decorated bridges ever built, with a monumental 18m-high stone obelisk at each of the four corners. These stand on three twin-columned bases which support an arched superstructure beneath a short tapering pylon that, in turn, is crowned by a golden casting of a horse. The pointed cutwaters are topped by elaborate carved winged figures, behind which the pier ends are faced by twin columns that support viewing balconies, while the stone balustrades include cast iron panels and support tall cast iron lighting columns with further decoration. The bridge was the work of engineer José Eugenio Ribera and architect Julio Maria Zapata.

Marina Cristina Bridge

Maronne Bridge, Saint-Geniez-ô-Merle, Corrèze, France

This most unusual structure, completed in 1999 to replace an earlier suspension bridge dating from 1852, carries the D13 road at a height of 30m over the Maronne River. The 58m-long reinforced concrete deck is supported on five parallel glued-and-laminated (glulam) timber beams. Beneath these beams are six inclined support frames, each of which has five glulam posts, the posts of the central frames crossing each other beneath the deck, the span between their bases being 27m. The bridge was designed by a firm called SOGELERG.

Martyrs Bridge, Bamako, Koulikoro, Mali

Opened in 1957, this bridge carries the RN7 road and links the Bamako city centre to its suburbs on the southern bank of the Niger River. The multi-span structure was renamed the Martyrs Bridge after Mali gained its independence from France in 1961.

Mary McAleese Boyne Valley Bridge, Drogheda, County Louth, Ireland

The bridge here stands in the sensitive area of the 1690 Battle of the Boyne and its 170m-long main span eliminated the need for river piers. It is an asymmetrical cable stay bridge, which is

Mary McAleese Boyne Valley Bridge

supported by fourteen pairs of stays from a single inverted Y-shaped pylon 95m high. The bridge deck, which was incrementally launched, is 35m wide and has an overall length of 370m. The bridge was designed by Roughan & O'Donovan and opened in 2003.

Maslenica Bridges, Zadar, Zadar County, Croatia
Two bridges have been built near Maslenica to carry road traffic across a deep inlet off the Adriatic Sea. The one nearer the mouth of the inlet was built in 1997 to carry the A1 Zagreb-Split motorway. It consists of a reinforced concrete box-section arch spanning 200m that supports eight of the piers for a twelve-span prestressed concrete beam structure. About a mile further up the strait was an earlier bridge that was built in 1961 to carry the D8 road. Following its destruction in 1991 during the Croatian War of Independence, a replacement was opened in 2005. Similar to its predecessor, this consists of a steel arch spanning 155m that supports the central part of a seventeen-span reinforced concrete beam-and-slab structure, which is now also the site of Croatia's highest bunjee jump.

Maslenica Bridge

Matadi Bridge, Matadi, Central Province, Democratic Republic of Congo
When this Japanese-sponsored bridge across the Congo River was completed in 1983, its main span

of 520m made it the longest suspension bridge span in Africa until the Maputo-Katembe bridge (q.v.) was opened in 2018. Overall, Matadi Bridge is 722m long and 12m wide. (It is hoped that work will begin soon on the Brazzaville-Kinshasa Bridge – q.v. – that will provide a second connection between the country and its smaller independent neighbour, the Republic of the Congo.)

Matadi Bridge

Mauricio Báez Bridge, San Pedro de Marcoris, Dominican Republic
The Mauricio Báez Bridge over the Higuamo River was completed in 2007 to replace an earlier suspension bridge. It has an overall length of 606m with a main cable-stayed span of 390m – the longest of any bridge in the Caribbean. The inclined supporting legs of the inverted Y-shaped pylon are in reinforced concrete and the upper mast is of steel. The main span, which is a composite structure of reinforced concrete and structural steelwork, has a 25m-wide deck with aerodynamic fairings made of aluminium to cope with wind speeds up to 150mph during tropical hurricanes. The potential effect of earthquakes is also limited by built-in damping mechanisms.

Mayfly Footbridge, Szolnok, Jász-Nagykun-Szolnok, Hungary
This elegant footbridge across the Tisza River was opened in 2011. Its 390-ton steel structure was fabricated and assembled in Budapest

Mayfly Footbridge

Medjez-el-Bab Bridge

and then, stretched between two barges, was transported 500km to its final location. This structure consists of twin tubular steel arches spanning 120m, splayed out 30° from the vertical, from which tie rods within the plane of each arch support a truss along either side of the 5m-wide deck. At night, the bridge is illuminated by LEDs on the outside of the arches and within the handrails. The total length of the bridge including approach spans is 450m and, when built, it was the longest footbridge in Hungary.

Mears Memorial Bridge, Nenana, Alaska, USA

The single-track Mears Bridge across the Tanana River is named after Colonel Frederick Mears, chairman and chief engineer of the Alaska Engineering Commission that built and operated the 470 mile-long Alaska Railroad. The structure was designed by Ralph Modjeski and opened in 1923, when its 700ft-long main span became the longest truss span in America.

Medjez-el-Bab Bridge, Béja, Béja Governorate, Tunisia

The masonry bridge across the Medjerda River at Béja was completed in 1677 and has been a Tunisian listed monument since 1920. It has eight main semicircular arches with tall narrow intermediate flood relief arches set higher up within each of the piers and abutments. After two arches were destroyed during the North

African battles of World War II, a temporary Bailey bridge was quickly erected across the gap in 1942 and this scene was captured in a well-known picture by Henry Carr. The arches were rebuilt after the war.

Megyeri Bridge, Budapest, Central Hungary, Hungary

The longest bridge in Hungary is the cable-stayed Megyeri Bridge over the Danube, with an overall length of 1800m and a 300m-long central span. The structure has twin 100m-tall A-shaped pylons, one of which is located on Szentendrei Island. There is a total of eighty-eight stays spreading out in a semi-fan arrangement on both sides of these pylons. The bridge, which carries the M0 motorway, opened in 2008.

Megyeri Bridge

Mehmed-Pasha Sokolovic Bridge, Visegrad, Republika Srnka, Bosnia-Herzegovina

The bridge here over the Drina River was designed by the famous Ottoman architect Mimar Sinan (1489–1588) and built between 1571 and 1577. It has eleven slightly pointed arches, with spans of from 11m to 15m between pointed piers, and is 180m long. Above the cutwaters at each end of one of the central piers are decorative additions. On one side is a tall, flat inscribed monumental panel which, being offset outside the line of the parapet, backs a small public open space. Opposite this there is another open space where the 6m-wide bridge deck widens out above a semicircular tower standing on the cutwater at the other end of the same pier. The bridge was damaged in both World Wars and in 1992, during the 'ethnic cleansing' of the Bosnian War, it was also the site where several thousand Bosniaks were murdered by Bosnian Serbs. The bridge, which is now limited to pedestrians only, was included on the 2006 World Monuments Watch, has been on the UNESCO World Heritage List since 2007 and underwent a major renovation programme between 2010 and 2013. It is the subject of the novel *The Bridge on the Drina* by Ivo Andrić (1892–1975), published in 1945. This chronicles the facts and myths associated with the bridge across nearly four centuries of Bosnian history. Andrić was awarded the Nobel Prize for Literature in 1961.

Mehmed-Pasha Sokolovic Bridge

Mekong River Crossing, Steung Trang, Kampong Cham, Cambodia

This road bridge across the Mekong River between Krouch Chhmar and Steung Trang is being built by the Shanghai Construction Group. Work began in 2017 and completion is expected in 2021, when it will become one of only three bridges over the 500km-long Mekong River in Cambodia. The concrete beam structure will be 1,131m long and 14m wide.

Menai Straits Bridges, Menai Bridge, Isle of Anglesey, Wales

Telford's **Menai Bridge**, probably his greatest memorial, was the final link in his London to Holyhead road. Work started in 1819 and was not completed until 1826. The suspension chains of the bridge have a main span of 579ft between stone piers, and side spans of 280ft. Although there are deck support hangers on these side spans, the approaches to the main span consist of 52ft-span stone arches, three at the mainland end and four at the Anglesey end. Originally there were four main groups of wrought iron eyebar chains supporting the 30ft-wide timber deck, with one group on each side of the two separate carriageways, and each group consisting of four chains placed one above another. Lack of proper stiffening resulted in the bridge suffering frequent storm damage and the first deck collapsed in 1839, its heavier timber replacement being itself replaced by a steel deck in 1893. The whole bridge was reconstructed in 1940 when the sixteen wrought iron chains were replaced by two steel eyebar chains on each side of the deck, which was again rebuilt, this time with cantilevered footways that pass outside the towers. It is listed Grade I and is included in the ASCE list of historic bridges.

Completed in 1850, nearly quarter of a century after Telford's nearby suspension bridge, the magnificent **Britannia Bridge** was built by Robert Stephenson to carry the Chester & Holyhead Railway over the Menai Strait. The rail tracks ran through twin separate rectangular box girders prefabricated on shore from riveted wrought iron plates, floated out and jacked up

Menai Straits Bridge

Merchants' Bridge Erfurt

vertically into their final positions. There were four spans – two central spans each of 460ft and two half-length side spans – and these were fully connected together to operate as the first fabricated continuous beam structure. The main spans remained Britain's longest railway bridge spans until the Forth Bridge (q.v.) was completed in 1890. The three intermediate masonry towers extended well above the top of the box girders to make provision for supplementary suspension chains, although these were never installed.

A fire in 1970 left the tubes sagging irreparably and a double-deck structure, with a new upper deck road relieving traffic on the Menai Bridge, was constructed between the existing towers. This new structure consists of two steel arch main spans, while each of the 230ft-long side spans was rebuilt as a three-span reinforced concrete column and beam viaduct. (See *AEBB*.)

Merchants' Bridge, Erfurt, Thuringia, Germany

Earlier timber bridges here are known from 1117, when the burning of one was recorded, and the present structure was completed in 1325 to carry the Via Regio Pilgrims' Way across the Breitstrom channel of the Gera River. It has six segmental stone arches, with spans ranging from 5.5m to 8m, an overall length of 125m and the width of the roadway is 5.5m. It is an inhabited bridge and there were originally sixty-two separate timber-framed houses that had been completed by 1486, the longest row of buildings on any European bridge. With these buildings the overall width of the bridge is 26m. Some houses were destroyed in 1945 but had been rebuilt by 1954. Most are four-storeyed and there are about eighty inhabitants at present. Originally there had also been a church at each end of the bridge and St Agidien's bridge chapel is known to have existed in 1110. The bridge was restored between 1985 and 1986.

Merritt Parkway Bridges, Greenwich, Connecticut, USA

The bridges on what was the first limited-access highway with two separate carriageways were designed by George L. Dunkelberger in a variety of then-modern architectural styles. Typical features include castellated parapets and sgraffito panels. Originally there were sixty-eight bridges, the older ones dating from 1935 and later reconstructions up until 1997. They are all listed on the National Register of Historic Places and in the Historic American Engineering Record and were included in the 2010 World Monuments Watch.

Mes Bridge, Shkodër, Shkodër County, Albania

Mes Bridge (sometimes called Ura E Mesit Bridge) over the Kir River is considered to be the largest and best-preserved Ottoman bridge in Albania. The 12½m-high main arch and one other were originally built in 1770, a further

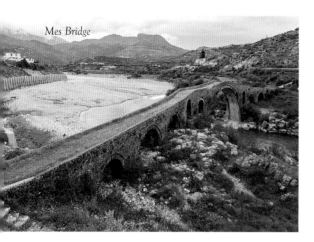

Mes Bridge

eleven arches some years later. The bridge, which is 108m long and 3½m wide, is now bypassed by a modern structure.

Metro Bridge, Kiev, Kievshchyna, Ukraine

The story of the bridges at this site encapsulates the sad fact that the strategic importance of such major structures often leaves them badly damaged or destroyed as a result of warfare. The first permanent bridge across the Dnieper River at Kiev was the **Nicholas Chain Bridge**. Charles Blacker Vignoles (1793–1875) was commissioned in 1846 by Tsar Nicholas I of Russia to build the structure and, when it was opened in 1853, it became the largest suspension bridge in Europe. Its overall length was about 2,260ft, with four main end-to-end chain suspension spans each 440ft long, two 225ft-long end spans and a 50ft opening section. The chains were supported on five elegant arched stone piers, making it the first multi-span suspension bridge in Europe, and the structure was widely considered to be one of the continent's most handsome bridges. It was destroyed in 1920 by the Polish army retreating after the Soviets counter-attacked a Polish-Ukrainian alliance. A new deck was built between the original piers but the whole structure (now called the **Bosch Bridge**) was blown up by the Soviet army when it withdrew from Kiev in 1941. A replacement temporary pontoon bridge was then built by the occupying German forces but they destroyed it when they abandoned the

city in 1943. The current **Metro Bridge**, built beside the line of the old structures, was opened in 1965. This is double-decked with rail tracks above the road. It has an overall length of 700m with seven arched spans and has been a National Heritage bridge since 2008.

Meuse Footbridge, Chooz, Ardennes, France

The 110m-long, three-span footbridge over the Meuse at Chooz was designed by Marc Mimram and opened in 2002. Each side of the deck of the 52m-long central span is hung on stainless steel rods from two tubular steel arches. A feature of the structure is that the main span's outer arch at the piers becomes the inner arch at mid-span, the two tubes having crossed at about the span's quarter-points. The side spans, which are 28m and 30m long, are supported by tubular steel half-arches beneath the deck.

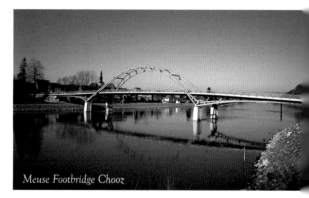

Meuse Footbridge Chooz

Miho Museum Bridge, Shigoraki, Shiga, Japan

This unusual 120m-long suspension bridge has no cable towers or pylons. The ninety-six stay cables that support one end of the bridge deck are anchored into the rock face, from

Miho Museum Bridge

which the 7.5m-wide footway emerges out of a tunnel to cross the bridge, and – instead of a pylon – pass over a 19m-high parabolic arch. A simple inverted cable stay system beneath the deck, with a 2m-long kingpost, supports the remaining length of deck to the opposite abutment. The bridge was designed by Leslie E. Robertson Associates with architect I. M. Pei and built in 1997.

Mile into the Wild Walkway, Keenesburg, Colorado, USA

The Wild Animal Sanctuary was established in 1980 and, in 2012, it opened an elevated pedestrian footbridge through its 790-acre parkland. This bridge is 4,800ft long, can simultaneously accommodate more than 4,000 people and its height above ground ranges from 18ft to 42ft. The simple structure consists of twin X-braced steel columns supporting N-braced edge trusses.

Millau Viaduct, Millau, Aveyron, France

Designed by Michel Virlogeux with architectural input by Sir Norman Foster, this is one of the world's great modern bridges. Opened in 2004 and carrying the A75 motorway over the Tarn River valley, the viaduct is 32m wide and 2,460m long with eight cable stay spans, the longest being 342m. The deck is a maximum of 245m above the valley floor and the pylons in which the cable stays are anchored reach a further 87m above the deck, the highest pylons in the world. Among its many honours the bridge won the 2006 Outstanding Structure Award from the International Association for Bridge and Structural Engineering.

Millennium Bridge, Ourense, Spain

The unusual combination stay and suspension bridge, built over the River Minho in 2001, was designed by engineer Juan Calvo and architect Alvaro Varela. Above each of the two main river piers, which are 120m apart, the pylons supporting the stays lean in towards each other, with a harp-arrangement of stays and back-stays between the pylons and the centreline of the deck. A walkway, elliptical in plan, loops round the top of each of the inclined pylons and then passes outside the edges of the main deck. In elevation, this loop follows a catenary curve, passing beneath deck level at approximately the third points of its span, and then acting as an under-spanned suspension cable supporting the central one third length of the deck span on raking props. The bridge carries a four-lane road with a footway each side.

Millennium Bridge, Ourense

Millau Viaduct

Millennium Bridge, Podgorica, Central Region, Montenegro

The cable-stayed bridge over the Morača River is 173m long with a main span of 140m. Its deck, which carries dual two-lane carriageways with flanking footways, is supported from a single 57m-tall pylon by twelve cables along its centreline. These are balanced by twenty-four back stays, anchored either side of the road, which form a kind of framed entrance. The bridge, which opened in 2005, was designed by Professor Mladen Ulićević of Podgorica University.

Millennium Bridge, Podgorica

Milo Bridge, Kankan, Guinea

The 1,900m-long, reinforced concrete bridge across the Milo River, a tributary of the Niger, was completed in 1949 to carry Guinea's N2 road. The bridge has eleven segmentally-arched spans, with an arcade of seven narrow round-headed arches over each pier.

Mindaugas Bridge, Vilnius, Vilnius County, Lithuania

The opening of this bridge across the Neris River in 2003 was part of the festivities celebrating the 750th anniversary of the coronation of Lithuania's King Mindaugas. The 101m-long structure has twin flat segmental steel arch ribs on each side of its 20m-wide deck.

Mirejovice Bridge, Mirejovice, Central Bohemia, Czech Republic

The interesting feature about this 1904 bridge is that it was constructed as part of a river control

system. The piers that support the bridge also act as the supports for a mechanical weir that crosses the river beneath the bridge and controls the river's water levels. In addition there is a lock between two of the piers. The bridge itself, which consists of steel trusses with an overlapping N-braced design, is 288m long, 7m wide and has seven spans.

Mitava Footbridge, Jelgava, Zemgale, Latvia

Built in 2012 to provide a crossing for cyclists and pedestrians over the River Driksa, this cable-stayed bridge has an overall length of 150m. Its main span is 75m long between two 24m-high pylons which support the deck on twenty-eight stays. The structure is the longest pedestrian/cycle bridge in the country.

Mizen Head Bridge, Crookhaven, Munster, Ireland

The bridge here, designed by Noel Ridley, was built in 1909 to connect Cloghan Island to the Irish mainland and thus provide access to the Mizen Head lighthouse. Considered to be one of the oldest reinforced concrete

Mizen Head Footbridge

structures in Ireland, the bridge consists of two inclined arch ribs that were precast in sections at the site. These were progressively built out from the abutments, being tied back by cables anchored to the rock face above the sea inlet. The 172ft-long completed span supports a 4½ft-wide cast in-situ footway 150ft above sea level.

Mkapa Bridge, Mkapa, Ikwiriri, Tanzania

The 970m-long Mkapa Bridge across the Rufiji River is named after Benjamin Mkapa, Tanzania's third President. Replacing a ferry crossing and opened in 2003, it is the longest bridge in the country and its structure consists of beams spanning between piers.

Moa Bridge, Moa, Pujehun District, Sierra Leone

The new 162m-long, four-span concrete beam bridge over the River Moa, completed in 2019, is part of a major EU-funded project to improve road links between southern Sierra Leone and Liberia.

Mohammed VI Bridge, Rabat, Rabat-Salé-Kénitra, Morocco

The deck of this cable-stayed bridge, opened in 2016, crosses 100m above the Bouregreg River as part of the motorway link between Rabat and Sale. Its two 200m-high concrete pylons, which are each shaped like the eye of a needle, are said to represent doorways to the two cities. Twenty pairs of cables hang from these to support the 30m-wide concrete deck. The bridge is 950m long with a central main span of 376m and back spans of 287m.

Mohammed VI Bridge

Mohlapiso Bridge, Mohlapiso, Qacha's Nek District, Lesotho

The 215m-long four-span Mohlapiso Bridge was built between 1989 and 1991 to carry Lesotho's A2 highway over the Senqu River. It consists of post-tensioned beams, precast on site, that stand on circular columns up to 15m high to support a reinforced concrete deck. The approach roads consist of bitumen-surfaced reinforced earth.

Mombasa Gate Bridge, Likoni, Mombasa, Kenya

Construction of a new cable-stayed bridge across the Mtongwe Passage west of Likoni is due to start in 2021. Its overall length will be 1,400m with a 660m-long main span and side spans of 330m and the main pylon will be 69m high. The bridge will carry four traffic lanes. The project is to be funded by concessionary loans from Japan.

Monorail Suspension Bridge, Putrajaya, Malaysia

The bridge carrying a monorail line across Putrajaya Lake was completed in 2003 as part of a Malaysian government project to improve communications within the city of Putrajaya. There were to have been two monorail lines and twenty-six stations along a route length of

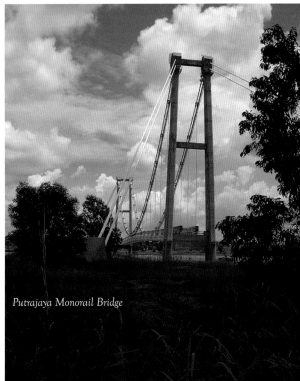

Putrajaya Monorail Bridge

20km but work was abandoned the following year because funding had run out. The structure is a 240m-long and 10m-wide suspension bridge and has been cynically called the 'suspended suspension bridge'.

Montauban Bridge, Montauban, Tarn-et-Garonne, France

The 205m-long brick bridge over the River Tarn at Montauban, built between 1311 and 1335 by Etienne de Ferrières and Mathieu de Verdun, has seven pointed 21m-span arches separated by piers with large pointed cutwaters on each side. In order to deal with the destructive flooding to which the Tarn is prone, there are additional tall high-level openings above the piers. An early bridge chapel existed until 1828 and a triumphal gateway, built in 1701, was demolished in 1870 because it was restricting traffic.

Montauban Bridge

Montignies-St-Christophe Roman Bridge, Mons, Hainaut, Belgium

Although it is believed there had been a Roman bridge across the River Hantes in this village, the current structure probably dates from the eighteenth century and was rebuilt in 1999. It is 3m wide and has an overall length of 21m containing thirteen semicircular stone arches, the piers standing on the crest of a small weir. Immediately after passing under the road bridge, the river makes a right-angled turn to the left and runs beside the road.

Montjean-sur-Loire Bridge, Montjean-sur-Loire, Maine-et-Loire, France

This site has an interesting history of tied suspension road bridges across the River Loire. An earlier structure, which was brought down by a storm in 1935, was replaced in 1937 and dynamited in June 1940 to delay the Germans' advance. It was rebuilt between 1948 and 1949 with four 90m-long main spans and two back spans supported by five steel portal frame pylons on masonry piers. The bridge carries the D6 road and is within the Loire Valley UNESCO World Heritage Site.

Montjean-sur-Loire Bridge

Moresnet Viaduct, Plombières, Liège, Belgium

Between 1915 and 1916 German forces occupying Belgium built a twenty-two-span curved railway viaduct across the Geul River to enable swift movement of their artillery between Aachen and Antwerp. The Belgians demolished one span in 1940 but this was repaired by a new German occupying force. Most of the bridge was then destroyed by the retreating Germans in

Moresnet Viaduct

1944 but had been repaired by 1950. It was then reconstructed in the early 2000s as a steel truss structure 1,107m long and 52m high.

Mosta Bridge, Mosta, Northern, Malta

Work on the first bridge across the Wied il-Ghasel River valley began in 1844 but was not finished until 1871. The structure consisted of an arcade of four masonry arches. Later, the British built a higher-level metal structure above the original masonry to improve access to the fort on one side of the valley. The condition of this deteriorated over the years, latterly being used for bunjee jumping and, following its removal, the earlier masonry bridge was raised to provide a new road link at cliff-top level.

Mouhoun Bridge, Boromo, Balé, Burkino Faso

The bridge here carrying the RN1 road over the Mahoun River was opened in 2018 to replace two older bridges and provide a modern link between the country's capital Ouagadougou and the neighbouring countries of Côte d'Ivoire and Mali. The three-span crossing is a steel ladder beam structure with integrated concrete deck and is 107m long and 12m wide.

Mountgarret Bridge, New Ross, South Leinster, Ireland

A timber bridge here across the River Barrow was built by Lemuel Cox in 1795. It was 64m long and 5½m wide. This was replaced in 1930 by a 64m-long reinforced concrete bridge with a steel, rolling-lift bascule opening section giving a clear width for navigation of 40ft.

Movable Bridges, Buenos Aires, Buenos Aires Province, Argentina

There are two movable bridges across the River Riachuelo at the entrance to the main port of Boca at Buenos Aires. The first of these is a transporter bridge, the **Transbordador del Riachuelo Nicolas Avellaneda**, and was named after a former president of Argentina. Work began on fabricating the structure in England in 1908, it opened in 1914 and was taken out of service in 1960. The travelling platform was

8m by 12m. In 2017 the original transporter was reopened and, since 1995, has been one of the city's Sites of Cultural Interest. A second movable bridge, **Puente Nicolas Avellaneda**, is a vertical lifting railway bridge that was built about 100m from the earlier structure in 1940. It is claimed to be the world's only lifting bridge with an additional transporter gondola, although this is now only used when the normal roadway is closed for maintenance. The overall length of the bridge is 1,650m but it has a central 60m-long vertical lifting section which enables the clearance beneath the structure for passing vessels to be increased from 21m to 43m.

Transbordador del Riachuelo Nicolas Avellaneda (in front), Puente Nicolas Avellaneda (behind)

Moving Bridges, Chicago, Illinois, USA

It is claimed that Chicago has more than sixty bridges, many of which are swing, bascule and vertical lift types of moving bridges, and one representative of each of these types is noted here. The Chicago, Madison & Northern Railroad built its **Chicago Sanitary & Ship Canal Swing Bridge** in 1899. This is a Pratt through truss, built in 1899, which is the canal's longest swing bridge with an overall swinging length of 479ft. The **Wells Street Bridge** is a double-deck plate girder bascule lift bridge that carries road traffic and, on the deck above, a line of the city's elevated 'L' railway system. Built in 1922, the distance

Wells Street Bridge, Chicago

Mudeirej Bridge

between its trunnion axles is 268ft. The **Canal Street Bridge**, opened in 1914, is a vertical lift railway structure that was built by the Pennsylvania Railroad to replace an earlier swing bridge across the Chicago River. The tracks are supported by a 1,500-ton main span consisting of hog-backed N-braced girders. This is 273ft long and can be raised a maximum of 111ft up the 185ft-high towers.

Moyola Park Bridge, Castledawson, Ulster, Northern Ireland

The footbridge that was built in this park by James Dredge in 1846 was one of a pair, the second of which was destroyed by floods in 1929. It has two 66ft-long chain suspension spans supported on a single, central, four-column frame, with the other ends of the chains anchored in the ground at the abutments. The 6ft-wide deck hangs between these chains supported from them by inclined stays rather than vertical hangers.

Mudeirej Bridge, Sawfar, Mount Lebanon, Lebanon

The first bridge here was opened in 1998 to carry the Beirut-Damascus Highway and, with its deck 80m above ground level, was then the country's highest bridge. It was destroyed in 2006 during the war between Israel and Lebanon and rebuilt

in 2008. The structure consists of steel girders supported on twin reinforced concrete columns.

Murinsel Bridge, Graz, Bundesland, Austria

Murinsel (or Mur Island) Bridge on the River Mur is a 47m-long floating platform resembling a half-open seashell that is anchored along the middle of the river and linked at each end by a Warren girder footbridge to the river banks. In the centre of the island is a small amphitheatre seating about 350 people. This is flanked by a glass dome-covered café and a play area for children. The structure was designed by Vito Acconci to commemorate Graz being the 2003 European City of Culture.

Murinsel Bridge

Muskauer Park Bridges, Bad Muskau, Saxony, Germany

The park here, created in 1815 by Prince Hermann von Pückler-Muskau, eventually extended from its original centre at Muskau Castle across the River Neisse into Poland. Four bridges linking the two parts were all destroyed in 1945. The rebuilt English Bridge, a simple four-span steel joist structure with masonry river piers, was opened in 2011. In 2004 the park was given UNESCO World Heritage status.

Musmeci Bridge

English Bridge, Muskauer Park

Musmeci Bridge, Potenza, Basilicata, Italy

The highly unusual reinforced concrete bridge at Potenza was designed by the Italian engineer Sergio Musmeci and built between 1971 and 1976 to provide a road connection between the town centre and the nearby motorway. It crosses both the Basento River and railway lines. The doubly-curved shell structure is 560m long and consists of four main 70m-span arches between abutments and three intermediate piers, which themselves have low arched openings. All these arches are contiguous and are made of a 30cm-thick membrane of reinforced concrete. Each of the main arches is surmounted by what has been described as a 'finger-like' structure, these upstand 'fingers' supporting the sides of the 16m-wide road deck. A planned internal footway within the 'hand' above the main arches and linked across the crown of the pier arches to form a sinusoidal shape was never completed. In 2003 the Italian government declared the structure to be a monument of cultural interest.

N

Nahr al-Kalb Bridge, Mazraat El Ras, Aintoura, Lebanon

The Nahr al-Kalb River was anciently called the Lycus River and later the Dog River. It is claimed that a stone bridge was built over the river by Sultan Selim in the seventeenth century, but the present structure is believed to date from 1892. It has two differently-sized large arches and one smaller one. A stamp showing the bridge was issued in 1950.

Nakabuta Bridge, Sigatoka, Nadroga-Navosa, Fiji

The Fijian government commissioned the Mabey Bridge company to replace an earlier bridge here with this 29m-long proprietary Compact 200 structure, developed from the original Bailey Bridge of the Second World War. It was opened in 2017.

Nanjing Yangtze River Bridge, Nanjing, Jiangsu, China

The double-decked steel trussed structure over the Yangtze River in Nanjing was built between 1960 and 1968 and is the world's longest road:rail bridge. The 20m-wide upper road deck, which carries National Highway 104, is flanked by 2m-wide footways and the 14m-wide lower deck carries the Beijing-Shanghai Railway. The

Nanjing Yangtze River Bridge

overall length of the bridge is 1,576m and it has ten main spans, the longest being 160m, at the ends of which are four 70m-high decorative masonry corner towers topped by statuary. The whole structure was closed between 2016 and 2018 for renovation.

Nanpu Bridge, Shanghai, China

Nanpu Bridge across the Huangpu River, which was built between 1988 and 1991, has an overall length of 8,346m and the main cable stay part is 760m long with a central span of 423m. The H-shaped reinforced concrete pylons are 150m high, each with twenty-two pairs of steel cable stays supporting the deck, and the deck itself is a composite steel and concrete beam structure that has 46m of clearance beneath it. At one end of the bridge, the roadway descends in a true spiral (with decreasing radius) onto the road network. The slightly larger sister structure, **Yanpu Bridge**, was completed two years later and has a 602m-long main span.

Naras Bridge, Manavgat, Antalya Province, Turkey

The bridge here across the Kargicayiri Creek is believed to have been built in the Seljuk era (1037–1260) on top of the remains of an earlier Roman structure, probably an aqueduct. There are five pointed arches with spans ranging from 9m to 11.2m long and the bridge is cranked slightly in plan. It was repaired in 1960 and further restored in the 1990s, when it was given a concrete deck and timber parapets.

National Post Office Bridge, Guatemala City, Guatemala Department, Guatemala

This stylish bridge carries a link over a street between two parts of the city's main Post Office building. Built between 1937 and

National Post Office Bridge

back from the vertical so its own dead weight helps to balance the force from the cable stays. The bottom of the pylon incorporates openings through which riverside traffic and pedestrians can pass.

Neckar River Footbridge, Stuttgart, Baden-Württemberg , Germany

The twin suspension cables for this 114m-span footbridge across the Neckar River pass over single tubular masts at each end of the 164m-long S-shaped deck but, in plan, bow out towards mid-span as a result of the stainless steel hangers being connected to the outside edges of the deck. The bridge was designed by Sclaich Bergermann Partner and opened in 1989.

Nelson Mandela Bridge, Johannesburg, Gauteng Province, South Africa

Completed in 2003, this 284m-long structure carries a road over forty-two separate tracks in a railway yard and is the largest cable-stayed bridge in South Africa. The taller X-braced steel pylon is 42m high and the main span is 176m long.

1940, it consists of a semicircular archway supporting a five-arched covered arcade with a cupola above.

National Unity Bridge, San Pedro Garza Garcia, Nuevo León, Mexico

The cable-stayed bridge carrying four lanes of traffic on highway 410 across the Rio Santa Catarina was designed by Daniel Tassin and completed in 2003. It has an overall length of 304m and its main span is 185m long. The deck is suspended from thirteen pairs of cable stays anchored into a massive reinforced concrete pylon, which is 134m tall. This has only two large diameter vertical back stays but also leans

Nesenbachtal Bridge, Stuttgart, Baden-Württemberg, Germany

This Stuttgart bridge, designed by Schlaich Bergermann Partner and completed in 1999, carries a bypass road over the Neenbach River valley linking two tunnel sections of the bypass. The 151m-long structure is a trapezium-shaped hollow box steel girder with three spans of 25m, 50m and 36m. Tubular steel arches span across the 7.5m-wide roadway supporting both a 3.5m-wide footway above it and translucent side screening that acts as a noise barrier.

National Unity Bridge

Nesenbachtal Bridge

Neuilly Bridge, Paris, Ile-de-France, France

The famous French civil engineer Jean-Rodolphe Perronet, first director of the prestigious École des Ponts et Chaussées, built a bridge across the Seine in 1774 to replace an earlier timber one. His structure was 219m long and had five stone arches, each spanning 39m between piers that were only about 4½m thick, which included the first use of cornes de vache splays. This is the name given to the tapering soffit between the main barrel of the arch and a larger diameter arch on the face of the bridge. Perronet's bridge was demolished when the present structure was built between 1936 and 1942. This has a total of twelve spans, these including crossings of an island and a small back channel, the main opening being a two-hinged steel arch spanning 82m. This bridge now carries a wider roadway and a metro line.

New Bridge, Bratislava, Slovakia

The 432m-long cable-stayed steel road bridge over the Danube in Slovakia's capital Bratislava was built between 1967 and 1972 to a design by the local Dopravoprojekt Company. The main feature is its 85m-high A-shaped pylon, which is topped by an enclosed circular observation deck nicknamed the Flying Saucer accessed by a lift in one of the pylon legs. The backwards-leaning pylon supports the 303m-long main span of the 21m-wide bridge deck through three main stays and these are balanced by five back stays to an anchorage included in the southern abutment. The footways on either side are below the main deck.

New Bridge

New River Gorge Bridge, Fayetteville, West Virginia, USA

When opened in 1977, this steel bridge's arched span of 1,700ft made it the world's longest single-span bridge. Its overall length is 3,030ft, it is 69ft wide and its maximum height above the ground is 876ft. The original structure included a 2ft-wide service catwalk beneath the deck but this can now be used for assisted walks.

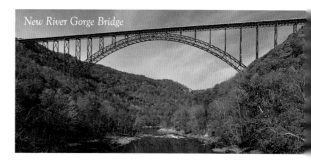

New River Gorge Bridge

Nine Arch Bridge, Demodara, Uva, Sri Lanka

A somewhat unusual tourist destination in Sri Lanka is this single-track mixed-gauge railway bridge, opened in 1921. The curved masonry structure, which is 300ft long and has nine spans, is 80ft high and consists of stone arches spanning between brick piers and is 9,600ft above sea level. The tapering piers have tall openings through them and at their tops are the corbels on which the arch centring was constructed. The parapets are relatively low.

Nine Arches Bridge

Nine-Hole Bridge, Hortobágy, Hajdú-Bihar, Hungary

A timber structure dating from 1697 over the River Hortobágy in what is now the National Park here, was replaced by this bridge built between 1827 and 1833. It was designed by Ferec Povolny and is still Hungary's longest stone road bridge. There are nine segmental arches over a 92m length between the abutments. An interesting feature is that the width of the roadway decreases through the approaches to the actual river crossing, and it is supposed that this was to make it easier to drive animals across. The park and bridge are now a UNESCO World Heritage Site.

Nine-Hole Bridge

Noefefan Bridge, Oecusse, Timor Timur, Timor-Leste

The three-span steel bridge across the Tono River, built between 2015 and 2017, has a structure of tied ladder arches. The overall length is 380m and the bridge deck is 10m wide.

Nogat Bridge, Malbork, Zulawy Region, Poland

The river crossing here has an interesting back story. Completed in 1857, the original 280m-long railway bridge over the River Nogat at Malbork (then Marienburg in Germany) was a double-box girder structure consisting of three wrought iron lattice-work trusses. It had two 103m-long spans that were push-launched into position. The bridge was mostly noteworthy for its fortifications, which were designed by Johann Leopold von Brese-Winiary. These included tall castellated towers at each end of the bridge and on the mid-river pier, and there was also an equestrian statue standing above the tracks at each end. A second bridge, completed in 1891, had two steel lenticular trusses. The bridge was destroyed in March 1945 and there is now a modern replacement.

Noorabad Bridge, Marena, Gwalior, Pakistan

Noorabad Bridge is a sixteenth-century Mughal stone structure over the Sankh River and has seven pointed arches spanning between piers fronted by pointed cutwaters. Above the springing level of the arches these cutwaters are topped by semi-octagonal extensions that end at parapet level. Each of the pier extensions that flank the central arch supports a domed roof on a circular colonnade and the four corners of the bridge are marked by tall tapering spires each supporting a crown and dome. The old bridge is now bypassed by a new structure.

Noorabad Bridge

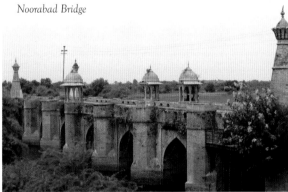

Normandie Bridge, Le Havre, Seine-Maritime, France

Designed by Michel Virlogeux (of Millau Viaduct fame – q.v.), the bridge here was completed in 1995 to carry the A29 autoroute over the River Seine estuary and its 856m main span was the world's longest cable stay span when opened. The overall length of the bridge is 2141m and its 24m-wide box girder deck is supported from 215m-high inverted Y-shaped pylons.

Normandie Bridge

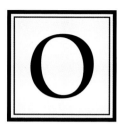

O

O'Connell Bridge, Dublin, County Dublin, Ireland

The first stone bridge over the River Liffey at this site was called **Carlisle Bridge**. Opened in 1794, it was designed by the architect James Gandon. It had three semicircular arches, was narrow and slightly hump-backed. In 1880 a reconstructed, widened and level bridge, now with three shallow semi-elliptical arches spanning 40ft, 49ft and 40ft, was opened. Unusually, at 155ft wide, it is wider than it is long (140ft) and provides a grand entranceway onto O'Connell Street.

Obel Railway Bridge, Molki, Gash-Barka, Eritrea

The first railway in this country, which was a single narrow-gauge (950mm) track between Eritrea's capital city of Asmara and the Red Sea port of Massawa, was later extended westward. Building work continued between 1887 and 1932 and the railway was then largely rebuilt in the 1990s after Eritrea had become independent from Ethiopia. The largest bridge on the line is the viaduct across the Obel River which has fourteen slightly-pointed stone arches. This structure featured on a 10 Nakfa Eritrean banknote issued in 1997.

Obelisk Bridge, Drogheda, North Leinster, Ireland

Obelisk Bridge is so called because of it being adjacent to the monument, commemorating the 1690 Battle of the Boyne, which was blown

Obelisk Bridge

up during the Irish Civil War. Built in 1868, the bridge is a fine example of the lattice girder type of structure, having five levels of flat bar X-crossings in each of the 128ft-long twin lattices on either side of the 15ft-wide deck.

Oberbaum Bridge, Berlin, Brandenburg, Germany

The Oberbaum Bridge, designed by Otto Stahn (1859–1930), must be one of the most elaborately decorated bridges ever built. The 150m-long double-decked structure over the River Spree was constructed between 1894 and 1896 and, because it links two parts of the city previously divided by the Berlin Wall, is now a symbol of German unity. Between early 1992 and late 1994, the bridge was fully rebuilt and restored to its previous appearance, although the reconstruction included a new 22m-long central span for the upper-deck twin-track Klosterstrasse Railway Viaduct, which was designed by Santiago Calatrava. Seen from the north-west the whole structure consists of a roadway bridge with three segmental masonry arches on either side of a central steel beam span. Behind this is a seven-part railway viaduct, the six main viaduct sections each consisting of seven pointed masonry arches with battlemented parapets, and the central section being the Calatrava span. Towering over the railway viaduct and on either side of this central span are 34m-tall masonry towers. Each of these has a square base that supports a corbelled octagonal enclosed gallery with a further battlemented tower section above topped by a spire.

Oberbaum Bridge

Octavio Frias de Oliveira Bridge, Sao Paulo, Southeast Region, Brazil

The main feature of this prestressed concrete cable-stayed bridge over the Pinheiros River, opened in 2008, is its 138m-high X-shaped pylon. This stands on, and parallel to, one bank of the river at a T-junction where two slip roads from a main road on the opposite bank curve round to cross each other under the pylon before crossing another riverside highway and then joining a road at right angles to the river. The two sections of curving road over both the highway beside the pylon and the river are supported by 144 stays from the pylon. The maximum individual span length is 150m. Oliveira was a Brazilian newspaper proprietor who led the campaign for direct presidential elections.

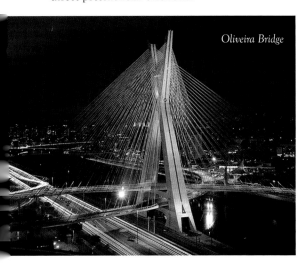

Oliveira Bridge

Ojuela Bridge, Mapimi, Durango, Mexico

The first suspension bridge on this site, completed in 1898, was designed by Wilhelm Hildenbrand and built by the Roebling's Sons Company to help in the exploitation of local gold and silver mines. It had a span of 316m and, at the time it was built, was the third longest suspension bridge in the world. In 1991 the original structure was scrapped and a replacement was built for use as a footbridge by tourists. The timber-trussed deck is about 275m long between its stone abutments and the cables span between twin four-legged steel pylons set slightly further back behind

Ojuela Bridge

these abutments. The maximum height of the 2m-wide bridge deck above ground level is about 110m. There is now a zipline beside the bridge.

Okavango River Bridge, Mohembo Village, North-West District, Botswana

Work began in 2016 on the Okavango River Bridge, located just to the north-west of the elephant lands of the Okavango Delta. The stand-out feature of this cable-stayed bridge is that each of the two main 55m-tall pylons which support the stays look like a pair of crossed elephant tusks. Overall, the steel structure over the main channel is 12m wide and 489m long with a central main span of 200m and 100m-long back spans. There are a further eighteen approach spans. Following delays, the bridge was due for completion in mid-2020.

Öland Bridge, Kalmar, Småland, Sweden

This prestressed concrete box girder bridge carries Route 137 linking the Baltic Sea island of Öland to the mainland. It is 6072m long – one

Öland Bridge

of the longest bridges in Europe – and has a total of 157 spans, with the high section over the main channel having five 130m-long spans giving 36m clearance for vessels below. The 13m-wide structure was built between 1967 and 1972.

Old Bridge (Karl Theodor Bridge), Heidelberg, Baden-Württemberg, Germany

Apart from short-lasting early timber and stone bridges built by the Romans over the River Neckar in Heidelberg, it was not until the thirteenth century that further timber bridges were recorded. One of these was brought down by an ice floe in 1284, and five more had fallen for the same reason by 1565. Another was demolished by the French during the War of the Grand Alliance in 1689, following which ferries and pontoon bridges were used until the timber **Nepomuk Bridge** was completed in 1708. After this collapsed in 1784 the present 200m-long crossing, built on the foundations of its predecessor by Elector Charles Theodore, was opened in 1788. It has nine masonry arch spans with large pointed cutwaters, above which are projecting balconies supported on corbelled brackets. The bridge was painted by Turner. Two piers and three arches, blown up by the Germans in 1945, were rebuilt by 1947 and all the arches were raised in 1969 to allow construction of an autobahn along one bank. The bridge was included on the 2002 World Monuments Watch.

Old Bridge, Heidelberg

Old Bridge, Albi, Tarn, France

The first crossing over the River Tarn in the Episcopal City of Albi, a stone toll bridge, was built between 1035 and 1042. It is 151m long

Old Bridge, Albi

and has seven pointed stone arches spanning 21m and two semicircular end arches. In the fourteenth century the town was protected by the addition to the bridge of a gate tower with drawbridge and houses were also built on the piers. In 1820 the roadway was straightened and widened by building shallow, segmental brick arches between the projecting ends of the piers. Just upstream from the bridge is a five-ribbed railway bridge with an arcade of three arches over each pier. Both bridges are included within Albi's UNESCO World Heritage Site.

Old Bridge, Béziers, Hérault, France

The twelfth-century stone bridge over the River Orb has a total of fifteen arches along its 242m length, but normally the river runs through only six. Four of the piers include high-level flood

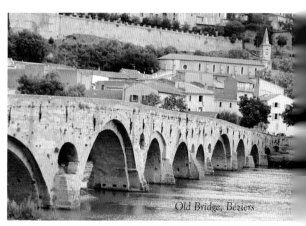

Old Bridge, Béziers

relief openings. Since 1945 the structure has been restricted to pedestrian use only and one arch had to be rebuilt in 1964. The bridge was listed as a Historic Monument of France in 1963.

Old Bridge, Pisek, South Bohemia, Czech Republic

The bridge over the Otava River at Pisek, built between 1250 and 1275 and forming part of a salt road, is the oldest bridge in the country. Originally, it had six arches but an additional wider arch was added during rebuilding following flood damage in 1768 and it is now 110m long and 6m wide. When built, the structure included two protective towers but one was swept away in the 1768 flood and the second was demolished in 1825. The bridge, which has been a national cultural monument since 1998, was damaged again by floods in 2002.

Old Bridge, Sospel, Alpes-Maritimes, France

Sospel's thirteenth-century bridge over the Bévéra River, less than 10 miles from the coast, was originally on the old salt road between Nice and Turin and, because of this importance, was protected by a gatehouse tower housing a customs post. The bridge has two segmental stone arches spanning 14m and 15m with a 5m-wide central pier on which the seventeenth-century three-storey tower stands. It is considered to be one of Europe's oldest toll bridges. The tower was bombed in October 1944 and has been faithfully rebuilt.

Old Bridge, Sospel

Old Main Bridge, Würzburg, Bavaria, Germany

The Old Bridge across the River Main links Würzburg with Marienberg. It was constructed between 1473 and 1543 to replace an earlier Romanesque structure that had been built between 1120 and 1133. There are eight arches (two now being recent reinforced concrete rebuilds), the intermediate piers having rounded cutwaters upstream and pointed cutwaters downstream. Since about 1730, the recesses above the cutwaters have housed twelve statues of saints or kings. The overall length of the bridge is 185m with the largest span being 18m. The bridge deck is 7m wide and since 1992 has been restricted to pedestrians only. Much of the city (including the bridge) was badly damaged in 1945 but was accurately rebuilt soon afterwards, mainly by the town's women.

Old Main Bridge, Würzburg

Old Railway Viaduct, Mogadishu, Benadir, Somalia

In the early twentieth century the Italians built a narrow-gauge railway in Somalia between Mogadishu and Villabruzzi (now called Jowhaar). The railway was dismantled by British troops during WWII but this multi-span viaduct is a distinctive survivor.

Old Rhine Bridge, Vaduz, Oberland, Liechtenstein

The covered bridge that crosses the Rhine to link Liechtenstein to Switzerland is less than a mile from Vaduz Castle, the private residence of the

Old Rhine Bridge, Vaduz

advance of the American army through the city. Most of the river traffic now bypasses the city on the Regen-Danube Canal. The medieval town centre, together with its buildings including the Old Bridge, is a UNESCO World Heritage Site.

Old Suspension Bridge, Mallemort, Bouches-du-Rhône, France

The first suspension road bridge across the Durance River at Malleport was built between 1844 and 1848 by the brothers Louis and Laurent Seguin, grandsons of Marc Seguin, and later improved by Ferdinand Arnodin. It has a total length of 300m and the deck, originally of timber, is 6m wide. There are twin supporting columns above each pier that are topped off by semicircular cable saddles. Most unusually, the main and both back spans were each hung from separate cables that, after passing over the saddles, came down the sides of the columns to low-level anchorages. There was also an additional lower cable spanning between metal masts on the outside face of each column. Parts of the bridge have been swept away by floods several times, the first in 1872, and it was bomb-damaged in 1940. It has been replaced by a new concrete bridge built alongside and access to the old structure is now closed off by a wall across the deck.

House of Liechtenstein royal family. An earlier bridge was built in 1871, but the present timber lattice girder structure was completed in 1901, although since then it has been raised twice to reduce the risk of flood damage. The 135m-long bridge is supported by five intermediate piers with four 20m-long river spans flanked at each end by a 26m-long shore span and is nearly 6m wide. In 1975 a new two-span road bridge was built a short distance away and the old bridge is now restricted to pedestrians and cyclists.

Old Stone Bridge, Regensburg, Bavaria, Germany

Regensburg, the oldest city on the Danube, dates back to Roman times and the present 309m-long bridge was built between 1135 and 1146 to replace an earlier timber structure, reputedly on the orders of Charlemagne. This bridge has sixteen semicircular stone arches. Originally, the bridge was protected by three gate towers standing on it but, during the Thirty Years' War in 1633, the Swedish army attacked the city, destroying two of the towers and one of the arch spans. These were all later rebuilt, although the middle tower was demolished in 1784 and the north tower after being damaged in 1809 when the French attacked the city. The bridge also once had watermills, the last demolished in 1784, and a chapel that was replaced by a tollhouse in 1829. In 1945 two arches were blown up to hinder the

Old Suspension Bridge, Mallemort

Old Yamuna Bridge, New Delhi, Punjab, India

The wrought iron railway bridge across the Yamuna River, built between 1863 and 1866 to carry a single-track railway line linking the centre of Delhi to Shahdara, is one of India's oldest and longest bridges. The double-decked structure includes a road below and has an overall length of 2,640ft with twelve spans, each 203ft long. It was widened to carry a second rail track in 1934. The bridge has a very low clearance above the river and the road is often closed when river levels are high.

Olstgracht Bridge, Almere, Flevoland, Netherlands

This 42m-long pedestrian and cycle bridge over part of Almere's harbour is a four-span, cable-stayed, inverted Fink truss supported by five 11m-tall masts. One mast stands on each abutment, its top tied back to an anchorage behind. A further diagonal cable from the top of each of these masts slopes down to support the base of another mast and, from the tops of these two masts, further cables support a fifth one. There are thus three intermediate masts above the water. Cross beams support separate 3m-wide twin walkways with anodised aluminium decks that pass either side of the masts. The bridge, which was designed by René van Zuuk Architects, was completed in 2004.

Olstgracht Bridge

Orb Aqueduct, Beziers, Hérault, France

The stone aqueduct that carries the Canal du Midi over the River Orb was opened in 1858 and is 28m wide and 240m long, the longest aqueduct on the canal. It has seven segmental arches, with the pier ends having heavily coursed rounded cutwaters topped by similarly-coursed

pilasters. The elevation on each side of the aqueduct is further enlivened by an arcade containing a footway that screens the canal trough (although the footways are no longer in use) with the towpath above.

Orb Aqueduct

Øresund Bridge, Malmö, Oresund Region, Denmark

This rail and E20 motorway crossing between Sweden and Denmark, which consists of a tunnel and an artificial island as well as an 8km-long bridge, was opened in 2000. The bridge is double-decked, with the motorway above the railway, and is carried throughout on deep steel girders. For the main part of the crossing these girders are supported by parallel cable stays from two 204m-high pylons, the central span being 490m and the side spans each 160m long. There are also forty-nine approach spans in which the girders span 141m between concrete piers. The engineering design was by Ove Arup & Partners.

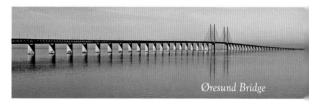
Øresund Bridge

Osten Transporter Bridge, Osten, Lower Saxony, Germany

The transporter bridge over the Oste River in Germany was built in 1909 and, since 1974, has operated solely as a tourist attraction. The

Osten Transporter Bridge

main structure is a two-pinned portal frame of latticed steelwork spanning 80m over the water. Unusually, the travelling platform, which has capacity for six cars or 100 people, is not suspended by wires beneath a rail-mounted carriage but the two are rigidly connected into a rectangular vertical framework.

Oversteek Bridge, Nijmegan, Gelderland, Netherlands

The bridge (its name meaning the 'Crossing' in English) commemorates 20 September 1944, the day when 700 American soldiers forced a crossing over the Waal River and forty-eight were killed. In their memory at sunset every evening forty-eight streetlamps on the bridge light up successively at about walking pace. The structure itself, which carries four main traffic lanes for the S100 road, a service lane and a two-way cycle track shielded by a glass wall, has a tied arch above the centreline of the deck that then forks into two separate struts just above the abutments. This arch, with a main span of 285m, was built on the river bank and then moved by barge to its final location where it was craned into place. The bridge, which with approach spans is 1600m long, was designed by Ney Poulissen Architects & Engineers and was opened in 2013.

Ozama River Bridges, Santo Domingo, Distrito Nacional Province, Dominican Republic

The first fixed crossing to be built over the Ozama River in Santo Domingo, the **Juan Pablo Duarte Bridge** is a suspension structure constructed between 1953 and 1955. The steel deck of its 176m-long main span is suspended between 55m-high steel pylons. Immediately adjacent and built between 1998 and 2001, the new **President Juan Bosch Bridge** has an overall length of 647m. The main cable-stay part has a central span of 180m supported by forty-eight double stays in a semi-fan layout from 63m-high reinforced concrete pylons. The 34m-wide reinforced concrete deck carries four lanes for road traffic and there is space for two future metro lines.

P

Padma Bridge, Mawa, Dhaka Division, Bangladesh

Construction of this steel truss bridge across the Padma River began in 2014 and it is due to open in 2021. The structure will have a roughly square cross section, within which a single railway line will run, and the upper deck with cantilevered extensions will carry four road lanes. Overall, it will be 6,150m long consisting of 41 spans each 150m long.

Padre Templeque Aqueduct, Hidalgo, Eastern, Mexico

This Mexican aqueduct, now a UNESCO World Heritage Site, was built by the Franciscan friar Father Francisco Templeque between 1554 and 1571 and has a total water channel length of 45km. It has three arched structures. The Main Arcade, which has sixty-seven semicircular arches standing on tall piers, is claimed to be the highest-ever single arcade aqueduct. This crosses the Papalote River ravine with a three-tiered 39m-high section of single arches and is decorated with an aedicule standing on the top of the middle-level arch. The two other arcades have forty-six and thirteen arches. A feature of the structures is the extreme slenderness of the arcades in elevation, although the lower sections of the piers for the taller arches thicken out from front to back.

Pakse (Lao-Nippon) Friendship Bridge, Wat Luang, Luang Praban, Laos

The first road bridge across the Mekong River was completed in 2000 and has an overall length of 1,380m. There are fourteen concrete spans consisting of precast units that were put into place using a 157m-long erection gantry before they were prestressed together to become self-supporting. For the 143m-long biggest span the prestressing cables are extradosed with the back stays supporting the adjacent spans. Laos issued a postage stamp showing the bridge to commemorate its opening. Since then, further crossings have been built and the Fifth Friendship Bridge is due to open in the mid-2020s.

Pamban Bridge, Mandapam, Tamil Nadu State, India

Pamban Bridge is claimed to be India's longest and oldest sea-crossing railway bridge. The 1m-gauge single-track structure was built over the Pamban Channel between 1911 and 1913 in order to connect the island of Rameswaram to the mainland. With 144 spans, it is 6,776ft long and its tracks are 41ft above sea level. The original Scherzer double-leaf rolling-lift opening span was replaced and the track gauge widened in 2007 and there are plans to build a new 63m-span lifting span. A road bridge alongside the original structure was opened in 1988.

Padre Templeque Aqueduct

Pamban Bridge lifting spans

Panam Bridge, Sonargaon, Dhaka, Bangladesh

The historic city of Sonargaon, an earlier capital city of the region, includes a seventeenth-century hump-backed brick bridge with an overall length of 53m. The structure has three narrow and steeply-pointed arches and is now a protected monument.

Panam Bridge

Panama Canal New Bridges, Colon, Paraiso, Panama

The first major bridge across the Panama Canal, opened in 1962, was the Bridge of the Americas (q.v.). Since then, two further bridges have been completed. The cable-stayed **Centennial Bridge** at Paraiso carries six lanes of the Pan-American Highway over the canal's Culebra Cut and was opened in 2004. The overall length of this bridge is 1052m and its longest span is 420m. The latest bridge is the prestressed concrete **Atlantic Bridge**, built between 2013 and 2019, which crosses the canal at its Atlantic end to link the port city of Colon with the north-western parts of the isthmus. The overall length of this bridge is 2,031m, which includes a 530m-long cable-stayed span with back spans of 260m. In 2018 work started on a fourth bridge over the canal just to the north of the existing Bridge of the Americas (q.v.). This is to be a cable-stayed structure with a 510m-long main span and will carry six traffic lanes as well as, on its northern side, two metro lines and a pedestrian/cycle way.

It is also planned to feature an observation deck at the top of one of the pylons.

Panzendorfer Bridge, Lienz, Tyrol, Austria

This covered timber bridge, its roof 66m long between the ends of its overhanging gables, crosses the Villgratenbach Stream below Heinfels Castle and is considered to be the most interesting bridge of its type in the Tyrol. Built in 1781, it has three spans of 19m, 21m and 21m and its roof structure, with three overlapping layers of cedar shingling consists of more than twenty bays. The bridge is now a protected historic monument.

Paradise Island Bridges, Nassau, Grand Bahama, Bahamas

The **East Bridge** that links Nassau's Paradise Island to the mainland, which was opened in 1967, is 1,560ft long and 36ft wide. It has fifteen low-level spans, each consisting of four reinforced concrete beams spanning between cross-head beams supported by twin circular columns, and three high-level spans over the main river channel. As a result of the warm weather and sea water the bridge had suffered badly from 'concrete cancer', in other words cracking and spalling of the concrete, and was restored in 2016. The **West Bridge**, opened in 1998, has beams supported on T-shaped column heads. Nassau is famous for having an hotel with a very expensive suite on a high-level bridge between two tower blocks. This is the **Royal Towers Hotel Bridge**.

Paradise Island Hotel Bridge

Paris Bridge, Andorra la Vella, Andorra

This stayed steel road bridge across the La Valira River has a span of 23m and was completed in 2006. On each side of the river two tubular struts share a common foundation, from which they lean out in a V-shaped end elevation. The open tops of these Vs meet over the centre of the river at two large spheroids to give an A-shaped side elevation. The spheroids themselves are connected by a tubular strut crossing from side to side above the deck. The deck, which is waisted, is supported by cross beams spanning between tubular edge beams, curved in plan, that are hung from stays anchored in the two spheroids, from where they radiate out in a fan arrangement.

Paris Bridge

Park Footbridge, Opatówek, Wielkopolska, Poland

Built in 1824 over an artificial moat surrounding a now-demolished palace, this is Poland's oldest cast iron bridge. It consists of four segmentally-arched beams, each made up of three sections bolted together, that now support a 3.5m-wide concrete deck that was probably originally of timber planks. The arches span 10.3m, the overall length of the structure is 13.8m and its restored ornate cast iron handrails consist of interlaced circles.

Parkovy Bridge, Kiev, Kievshchyna, Ukraine

The pedestrian-only suspension bridge spanning 400m over the River Dnieper was built in 1957 to connect the city to the park on Trukhaniv Island. The bridge is not centred on the river as one pier is located on the line of the right bank embankment wall and, as well as the back span, two further spans cross the river to the island. The bridge is now used for bungee jumping into the river and, nearby, there is also a 530m-long zip line from the right bank over the river to the island.

Passerelle des Deux Rives, Strasbourg, Bas-Rhin, France

The cycle and footbridge across the Rhine between Strasbourg and the German city of Kehl was opened in 2004 to mark the first transnational gardening show in the two cities. The cable-stayed bridge, which was designed by Marc Mimram, has two support pylons, one near each river bank, giving a main span length of 177m. There are two separate arched decks, which pass on either side of these pylons, with struts between them, and they then connect together in a single 100 square metre platform above mid-river.

Passerelle Marguerite, La Foa, South Province, New Caledonia

An earlier timber bridge built here by convicts in 1893 was replaced in 1909 by this unusual part-stay/part-suspension bridge designed by Albert Gisclard. The structure, which spans 48m and is 3m wide, has twelve stays supporting each end of

Passerelle Marguerite

the main span together with further suspension cables supporting the central part of the deck. These are balanced by six larger back-stays. The bridge was strengthened in 1927 to carry heavier vehicles but, after being bypassed by a new concrete bridge in 1989, remains in use as a footbridge. Pictures of the bridge featured on a postage stamp issued in 1985 and on a postmark in 2002.

Paul Sauer Bridge, Storms River, Eastern Cape, South Africa

The 100m-span reinforced concrete deck arch bridge over Storms River was designed by Riccardo Morandi and opened in 1956. The arch supports the level deck above on twelve radiating sets of triple columns. An interesting feature is that it was constructed in four half-span half-width sections. These were fabricated in a near-vertical position above each abutment using climbing formwork and then rotated outwards and downwards to link together and form two parallel arch rings. The bridge was renovated in 1986.

Peace Bridge, Calgary, Alberta, Canada

Structurally, this bridge is a steel truss consisting of a slightly-arched double helix tube spanning 126m. The deck is 20ft wide with separate cycle and pedestrian lanes and is covered by a glass roof. The bridge was designed by Santiago Calatrava and opened in 2012.

Peace Bridge, Calgary

Peace Bridge, Derry, County Londonderry, Northern Ireland

The new cycle and pedestrian suspension bridge over the River Doyle was opened in 2011. It effectively consists of a pair of inclined, single-mast, single-cable, self-anchored suspension bridges. These overlap at mid-river in a 'structural handshake' and support an S-shaped 4m-wide deck which is 235m long overall with a 101m-long main span. The bridge was designed by AECOM and Wilkinson Eyre Architects and funded by the European Development Fund for Peace.

Peace Bridge, Plauen, Saxony, Germany

The stone bridge across Syrabach Creek, built between 1903 and 1905, has an overall length of 133m with its single 16m-wide segmental arch spanning 90m and reaching a height of 18m. This is the longest stone arch bridge in Europe. The main spandrel at each end is pierced by three large-diameter weight-relieving tunnels. Behind these on one side there is a large semicircular arch for a river-side road which is balanced on the other side by a fourth tunnel. The 17m-wide deck is supported on a tall blind arcade of semicircular arches.

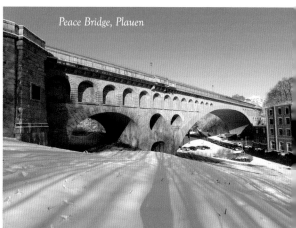
Peace Bridge, Plauen

Peacock Island Palace Bridge, Potsdam, Brandenburg, Germany

Peacock Island in the Havel Lake, which is near the Sanssouci parkland where Prussian royalty lived, was bought in 1793 by King Friedrich

Peacock Island Palace Bridge

Wilhelm II and he is believed to have designed the island's Pfaueninsel Palace. This consists of a mock front between twin four-storeyed towers with corbelled and turreted tops. These tower tops are connected by a wrought iron, pointed arch Gothic bridge that has very delicate parapet railings.

Peacock's Lock Viaduct, Tuckerton, Pennsylvania, USA

The viaduct here, named after a nearby lock on the Schuylkill River crossed by the structure, was built by the Philadelphia & Reading Railroad between 1853 and 1856 and has nine 46ft-long semicircular arch stone spans. However, it is chiefly significant for the large and unusual circular opening through the spandrel area immediately above each pier. Although common in European structures, this is a rare feature in an American bridge.

Pedestrian Living Bridge, Limerick, County Limerick, Ireland

Designed by Wilkinson Eyre Architects and Arup Consulting Engineers, this curved footbridge across the River Shannon connects two parts of the Limerick University campus. The 350m-long structure has seven equal spans, the intermediate piers being located on islands in the river. Beneath the side of each span a suspension cable supports the edge of the deck on posts that are

angled out from the vertical. Opened in 2007, the structure is the longest footbridge in Ireland.

Pedro & Inês Footbridge, Coimbra, Portugal

The bridge here, opened in 2007 to provide a walkway linking parkland on both banks of the River Mondego, consists of two double-cantilevered 4m-wide arms, each mounted on a single river pier. These two halves meet above mid-river making a 110m-long main span with 64m-long back-spans. However, the axis of each half-bridge is offset from the other, thus creating a short wider section of deck at their junction. There is also a shorter approach span at each end. The parapets are formed from colourful, glazed, fractal-shaped panels.

Pedro e Inês Footbridge

Pentele Bridge, Dunaujvaros, Fejer, Hungary

This 41m-wide major new road bridge, with an overall length of 1,682m, was opened in 2007 to carry the M8 motorway across the Danube. Its main span of 308m consists of two steel segmental tied-arches tilted inwards from the vertical. This structure was floated out to its final position on barges. The thirteen approach spans from the right bank of the river have an overall length of 1,065m and were incrementally push-launched from the river back to the shore, while the left bank approach has four 75m-long spans. Interestingly, the

Pentele Bridge

is 170m above ground level. It is propped at the centre by inclined struts from lower down the towers. This arrangement is designed to cope with any long-term differential settlement of the two towers. The building, which was completed in 1996, was designed by Pelli & Associates.

Pinnacle Sky Bridges, Singapore

This public housing project, completed in 2009, consists of seven fifty-storey towers that are connected at both the 26th and 50th floor levels by bridges with a total length of 500m. These incorporate a jogging track, playgrounds and gardens. The supporting structures, which consist of steel beams supporting an enclosed concrete deck, are all 20m wide and span lengths vary from 4m to 48m.

street lights across the bridge are inclined inwards at the same angle as the arches.

Pernstejn Castle Footbridges, Nedvedice, South Moravia, Czech Republic

The access road into this ancient and picturesque stronghold, which is one of the best-preserved castles in the country, crosses the main moat on a 'Gothic' bridge, and drawbridges cross smaller water barriers. There are also two timber covered bridges connecting the castle with an external tower which could be used as a last resort if the castle were captured. It was besieged during the Thirty Years War, but unsuccessfully.

Petronas Twin Towers Sky Bridge, Kuala Lumpur, Selangor, Malaysia

The double-deck footbridge that spans between the two 452m-tall skyscraper Petronas Towers provides short cuts and additional escape routes for office workers in the towers and also helps to form a grand gateway entrance for the country's capital city. The bridge structure spans 58m and

Pipeline Bridge, Dargan Ata, Lebap, Uzbekistan

This bridge across the Amu Darya River, previously known as the Oxus River, was opened in 1964. It is a suspension bridge with a main span of 390m that carries two pipelines and a single-lane road. (The area in which the bridge lies is perhaps best known from Robert Byron's justly famous travel book *The Road to Oxiana*, published in 1937.)

Pira Delal, Zakho, Kurdistan, Iraq

There was a bridge over the Khabur River here in Roman times but this structure, the name of which means beautiful bridge, may date from about 1000AD. It has five roughly semicircular stone spans, with a tall central arch flanked by three subsidiary arches on one side and one on the other, this having been rebuilt in the 1960s. The central pier has a large stepped buttress on its downstream face. The bridge is 115m long and 5m wide and its crown is about 16m above water level.

Petronas Twin Towers Sky Bridge

Pira Delal Bridge

Platjan Bridge, Mathathane, Central District, Botswana

An earlier crossing here of the Limpopo River between Botswana and South Africa consisted of a 100m-long single-lane low-level causeway built over a series of concrete pipes. Work started in 2017 on a new bridge consisting of a concrete slab deck spanning between trapezium-shaped concrete piers.

Ploče Gate Footbridges, Dubrovnik, Dubrovnik-Neretva, Croatia

Two similar semicircular stone arch footbridges at the entrance to the medieval Dubrovnik Old Town were both designed by Paskoje Milicevic and built in the second half of the fifteenth century. The Inner Bridge has a single arch and the Outer Bridge a double arch. The low parapet walls are pierced by quatrefoil openings, below which there is continuous stone bench seating for visitors.

Ploce Gate Outer Footbridge

Plougastel Bridge, Brest, Finistère, France

The striking bridge here over the River Elorn was built between 1926 and 1930 to a design by Eugène Freyssinet and, when completed, it was the world's largest bridge of its type. The structure consists of three segmental, hollow-box, reinforced concrete arches, each spanning 188m, and the total bridge length is 888m. The arches were constructed on the same centring, which was moved to each successive span on floating concrete barges. Spandrel piers on the arches supported two 9m-wide separate decks, one for road and one for rail. One arch was dynamited by the retreating German Army in August 1944 and rebuilt after the war. A parallel relief structure, the **Pont de l'Iroise** cable stay bridge, was built between 1991 and 1994 to carry all but cyclists and pedestrians. Plougastel Bridge now features in the annual Brest to Paris cycle race.

Plougastel Bridge

Polcevera Viaduct, Genoa, Lombardy, Italy

The first viaduct here, designed by Riccardo Morandi and built between 1963 and 1967, included a cable-stayed structure that carried Genoa's A10 motorway, part of the strategic road link between Italy and France. This had three main pylons about 90m high and an overall length of 1182m with a maximum span of 210m. It came down during a heavy rainstorm on 14 August 2018, killing forty-three and injuring thirteen people, when a corroded main stay suddenly gave way triggering the collapse of one pylon together with a 260m-long section of deck. The replacement **Genoa San Giorgio Bridge**, designed by Renzo Piano and opened in 2020, is a viaduct with three 100m-long and sixteen 50m-long spans. It has a hollow hybrid steel and concrete deck 5m deep and 30m wide. The new bridge is equipped with a dehumidifying system and four maintenance robots.

Pollaphuca Bridge, Britonstown, County Wicklow, Ireland

The bridge here carries the N81 road across a deep gorge through which the River Liffey used to fall precipitately before it was dammed upstream for a hydro-electricity scheme. It consists of a tall, slightly pointed stone arch spanning 65ft with a rise of 40ft and has triple mouldings to the voussoir ring. Its deck is nearly 150ft above where the waters once rushed. Pilasters rise up on either side of the abutments to end in castellated open-backed turrets above the parapets and each spandrel is decorated with a tall, blind, lancet arch. The bridge was designed by Alexander Nimmo (1783-1832) and completed in 1827.

Pompey's Bridge, Mtskheta, Mtskheta-Mtiancti, Georgia

The segmental stone arch here was given its name when it was built in 65BCE during Pompey's military campaign in what is now Georgia. It is also known as the **Magi Bridge**, a reference to the local rulers acquiring the area from Persia in the third century BCE. The bridge was rebuilt in the fifth century CE but, following the construction of a power station in 1923, it is now usually half

submerged. The ancient historical monuments of Mtskheta, including the bridge, have been a UNESCO World Heritage Site since 1994.

Pont de Cassagne, Planès, Eastern Pyrenees, France

Designed by Albert Gisclard, this bridge was built between 1905 and 1908, although the opening of the single-track line was delayed until 1910 following an accident during load tests that killed Gisclard and five others. The structure has an overall length of 253m, with a span of 156m between 30m-tall three-stage steel pylons standing on masonry piers, and the bridge deck is 80m above the River Têt. It is the only suspension bridge still in service on French railways and the suspension arrangement itself is most unusual, with the majority of the cables being stayed supports forming a triangular stiffening system rather than end-to-end catenaries.

Cassagne Bridge

Pont de la Margineda, Fontaneda, Andorra

The fourteenth-fifteenth-century bridge across the Gran Valira River in Fontaneda was built with voussoirs of pumice stone in order to keep the loading on the temporary centring as light as possible, but the abutments and parapets are of granite. Overall, the structure is 33m long and the crown is about 8m above water level. The bridge, which is the largest of the country's medieval bridges, is registered as part

Pompey's Bridge

of Andorra's cultural heritage and features on the 5-centime commemorative brass coin issued in 2013 and the 2.30 postage stamp of 1990.

Pont del Escalls, Escaldes-Engordany, Andorra

This thirteenth-century single arch stone bridge, now used by pedestrians only, crosses the Valira del Nord. It is considered to be one of the oldest bridges in the country and is registered in the Cultural Heritage of Andorra.

Pont des Amidonniers, Toulouse, Haute-Garonne, France

The fine, classical-looking modern bridge over the Upper Garonne River in Toulouse is a reinforced concrete structure clad in stone. It was built between 1903 and 1911 to a design by Paul Séjourné. On each side of the bridge there is a separate 3m-wide rib that consists of five semi-elliptical arched spans with cornes-de-vache splays. These spans vary in length from 46m at mid-river to 38m at the ends. The brick spandrels above the piers are pierced by flood relief tunnels which are topped by further semi-elliptical arches. The 22m-wide deck includes cantilevered footways.

Pont des Amidonniers

Pont des Arts, Paris, Ile-de-France, France

The first footbridge over the River Seine on this site was built between 1802 and 1804 and linked the main court of the Louvre Palace and the Institut de France. It had nine cast iron arches and was designed by Louis-Alexandre de Cessart and Jacques Dillon. At some stage it had very attractive cast iron lamp standards above each pier, but these are not always shown in old paintings of the bridge. The bridge was damaged by bomb blasts during both World Wars as well as by boat collisions. It was listed as a national historic monument in 1975 but, as a result of its deteriorating condition, had to be closed in 1977 and was then partly brought down when hit by a barge in 1979. The current steel structure, built between 1981 and 1984, is similar in overall appearance but has only seven arches. Its total length is 155m and it is 11m wide. In 1991 the bridge became part of the Central Paris Banks of the Seine UNESCO World Heritage Site.

Pont des Arts

Pont des Belles Fontaines, Juvisy-sur-Orge, Paris, France

The Antibes road out of Paris to the south (now the D77) was on an embankment when it reached the River Orge. This necessitated a relatively high arched bridge to carry the road across the river, with retaining walls to support the sloping sides of the embankments on either side which themselves had to be retained with seven narrow arched props that also crossed the river. During construction of these retaining walls and arched props a spring was discovered which was piped up into two large decorative fountains mounted on each of the road bridge's parapets, giving the completed structure a very handsome look. The work was completed in 1728 but in 1971 the fountains were taken down and re-erected in a local park.

Pont des Echavannes, Saint-Marcel, Sâone-et-Loire, France

Designed by Emiland Gauthey and Thomas Dumorey, this masonry bridge over the Sâone River was completed in 1790 and now carries the N73 road. It has seven semi-elliptical arches, the longest spanning 13m. The pier ends have cutwaters that are triangular in both plan and elevation and above which are large, vertically-oriented, elliptical flood-relief tunnels.

Pont des Trous, Tournai, Wallonia, Belgium

This bridge over the Scheldt River, built between 1281 and 1304, is considered to be one of very few medieval fortified bridges left in the world (others include those at Cahors in France – q.v., Monnow in Wales and Warkworth in England – see *AEBB*). Although the Tournai bridge contained a high-level walkway protected by tall parapets that passed between the medieval towers on each bank and was supported by three pointed river arches, the whole structure was more a defensive water gate for the city than a bridge. However, in what has been described as an act of state vandalism, the three arches were dismantled in 2019 with a view to being rebuilt later with a higher and wider central arch that would improve river navigation.

Pont du Gard, Nîmes, Gard, France

The magnificent 160ft-high, three-tier aqueduct over the River Gardon was built by the Romans in about 40–60AD. The bottom tier is 466ft long and contains six arches, the middle has eleven arches and the top tier has thirty-five arches extending over 902ft. The aqueduct is now a Unesco World Heritage site.

Pont du Gard

Pont Félix-Houphouët-Boigny, Abidjan, Ivory Coast

An earlier floating bridge had been built here in 1931 but the present road-over-railway double-decked structure was completed in 1957 to provide a modern link across the Ébrié Lagoon connecting the two halves of Abidjan City. This is a concrete box girder bridge with precast sections prestressed together. It has eight 47m-long spans and an overall length of 372m. In 2018 a major upgrade began which included reinforcing the foundations by installing new piles.

Pont Flavien, Saint-Chamas, Bouches-du-Rhône, France

The Roman bridge here over the River Touloubre was built, after his death, on the instructions of Lucius Donnius Flavos, probably to replace an earlier timber structure. Overall, it

Pont des Trous

Pont Flavien

is 25m long and 6m wide and has a 12m-long single masonry arch that is nearly semicircular. At each end of the bridge there is a 7m-high triumphal arch, with fluted Corinthian pilasters at the outside corners, that is topped by a pair of carved lions.

Pont Galliéni, Lyon, Rhône, France

The first bridge here across the Rhone, completed in 1849, was demolished in 1889. It was replaced in 1891 by a steel arched bridge with its deck supported on spandrel posts, but this was destroyed in 1944. A second replacement bridge lasted until 1962 when construction started on the present structure. This steel haunched girder bridge, which was opened in 1965, is 29m wide and has three spans along its overall length of 204m.

Pont Julien, Bonnieux, Vaucluse, France

The 85m-long Roman bridge over the Calavon River at Bonnieux has a main semicircular stone arch spanning 16m, with a flanking arch each side, and probably dates from about 3BCE. The road over the top of the arches slopes down noticeably on each side of the central arch crown and the airiness of the structure is further emphasised by its having edge railings rather than parapet walls. The two intermediate piers have, just above the springing points for the arches, a tall arched flood opening.

Pont Julien

Pont Levant Notre Dame, Tournai, Hainaut, Belgium

This unusual opening table bridge, built in the late 1950s, is similar to a normal vertical lift structure except that, instead of being hung from towers, the moveable section of deck stands on a column at each corner. These columns incorporate hydraulic cylinders allowing the deck to be raised so that small vessels can pass beneath.

Pont Levant Notre Dame

Pont Neuf, Paris, Ile-de-France, France

The two-part Pont Neuf, the oldest bridge in Paris, has separate arms over the Seine on either side of the Île de la Cité's downstream end, with seven spans to the right bank and five spans to the left. Work began in 1578 and, following interruptions by wars, was finally completed in 1607. The bridge is 22m wide – for a long time the widest bridge in Paris – and its overall length is 232m. The arches were originally nearly semicircular but were changed to semi-elliptical during major reconstruction in the mid-nineteenth century. The pointed piers between the arches are topped by circular pedestrian refuges and at the tip of the island is a bronze equestrian statue of Henri IV.

Pont Neuf, Toulouse, Haute-Garonne, France

The stone bridge across the Garonne River, the oldest bridge in Toulouse, was designed by the architect Nicholas Bachelier and construction lasted nearly a century between 1542 and 1632.

The structure has seven semi-elliptical arch spans, the longest spanning 30m, and it is 220m long. There are distinctively-shaped openings above the piers. A triumphal archway at one end, built by the architect Mansart in 1686, was demolished in 1860.

Pont Saint-Jacques, Parthenay, Deux-Sèvres, France

Entry into Parthenay from the north-west is across the River Thouet on the St-Jacques Bridge and then through the thirteenth-century gatehouse tower at the end of the bridge. The first permanent crossing here was probably built in the twelfth century but there have been many re-buildings since. The present bridge, which dates from 1721, has four semicircular stone arches each spanning 5m between 2m-wide piers.

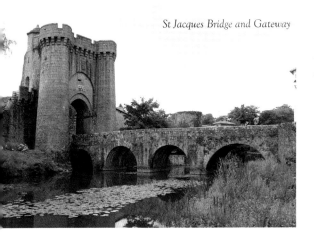

St Jacques Bridge and Gateway

Pont Vieux, Orthez, Pyrénées-Atlantiques, France

The fortified old bridge over the River Gave de Pau has three main pointed arches and a central octagonal tower that rises to about 13m above deck level. A second tower once stood on the river's right bank. It is not known when the bridge was built but a City of Orthez coin dated 1254 shows one of the arches with defensive towers. The bridge, which has an overall length of 46m and a main span 15m long, lies on the pilgrim route to Santiago de Compostela. In 1560 prisoners were cast to their deaths from

Pont Vieux, Orthez

the tower and in 1814, during the Peninsular Wars, the bridge was mined by the French to hinder Wellington's advance. Afterwards, the battle honour Orthez was awarded to about forty British regiments.

Ponte degli Alpini, Bassano del Grappa, Vicenza, Italy

An earlier wooden bridge over the Brenta River here, dating from 1209, was destroyed by floods in 1409, by fire in 1511 and washed away again in 1567. Andrea Palladio designed a replacement in the style of an ancient Roman structure, but this was rejected in favour of a new bridge similar to the original. This structure, also of timber, had five covered spans supported by four distinctive piers. The bridge was destroyed by floods in 1748, by fire in 1813 and by war in 1945, being rebuilt each time. Palladio's unused original design was copied by Henry Flitcroft for the park bridge at Stourhead in England (see *AEBB*).

Ponte degli Alpini

Ponte dei Salti, Lavertezzo, Ticino, Switzerland

An earlier bridge at Lavertezzo over the River Verzasca was built in the seventeenth century, was then partly destroyed in 1868 and rebuilt in 1960. The current attractive-looking structure consists of two segmental stone arches, each with a deep ring of voussoirs, the central pier of which is founded on a rocky outcrop that divides the river into two. The bridge's name means 'jump bridge' and it is now a popular bridge-jumping site.

Ponte della Maddalena

Ponte dei Salti

Ponte del Risorgimento, Rome, Italy

This bridge over the River Tiber, which was designed by Giovanni Porcheddu – the Italian licensee of Francois Hennebique's patent for reinforced concrete, was the first such structure in the world to achieve a span of 100m. It was completed in 1911 to mark the fiftieth anniversary of Italian unification and the proclamation of Rome as the country's capital. The 21m-wide bridge crosses the river with a single low-rise segmental arch.

Ponte della Maddalena, Borgo a Mozzano, Lucca, Italy

The bridge across the Serchio River near Borgo was probably built around 1100 for the Via Francigena Pilgrimage Route and was restored in the early fourteenth century. It has a high, nearly semicircular, central arch spanning 38m with two much smaller arches on one side. On the

other side the arches had to be demolished in 1889 and replaced by a higher one to make way for the railway between Lucca and Aulla.

Ponte delle Torri Aqueduct, Spoleto, Perugia, Italy

An earlier aqueduct at Spoleto is said to have been built in 604 for Theodelapius, third Duke of Spoleto. The current structure, which is 236m long and crosses 80m above the Tessino River, was probably built in the thirteenth century, possibly on Roman foundations, and is believed to have been designed by Matteo Gattapone of Gubbio. It has ten pointed arches and carries a narrow footway about 5m below the top of the enclosed water channel on the northern side of the structure. There are defensive towers at each end. The structure was damaged by an earthquake in 2016.

Ponte Novu, Castello-di-Rostino, Corsica, France

The seventeenth-century five-arched stone bridge over the Golo River was built by the Genoans when they held Corsica. The ends of the piers are fronted by massive buttresses. The bridge was later widened when the parapets and footways were rebuilt on semi-elliptical arches supported on corbelled brackets. Two of the main arches were destroyed in the battle of 1769 in which France defeated Corsica, thus leading to the French annexation of Corsica in 1770, and have never been rebuilt.

Ponte Salario, Rome, Italy

Parts of the original ancient bridge that crossed the River Aniene on this site, dating back to perhaps the first century BCE, can still be seen under the approaches to the existing bridge. This bridge was largely destroyed in the sixth-century Gothic War. The rebuilt structure was more than 6m wide and had a semicircular arch spanning 25m. With its eighth-century gate tower over the central arch, the bridge is best known for the picturesque depictions of it during its decrepitude, including a painting by Hubert Robert in about 1775. This tower was destroyed during the Napoleonic wars and the central arch was blown up in 1867. The bridge was rebuilt in 1874, with a single main segmental stone arch supporting a level roadway, and widened in 1930.

Ponte San Leonardo, Termini Imerese, Sicily, Italy

This decorative semicircular stone arch bridge was built in 1625 by Archbishop Agatino Daidone. The main arch is flanked on one side by a smaller arch and, on the other, by a sloping approach ramp along the bank of the river which contains two further small arches. Each abutment is strengthened by a sloping buttress which is topped by a blind aedicule containing an inscribed marble slab and the crown of the bridge parapet supports an elaborate carved upstand. The river bed is now dry, the flow having been diverted elsewhere.

Ponte Scaligero, Verona, Italy

This three-span, 120m-long bridge across the River Adige, which is also known as the **Ponte Castelvecchio (Old Castle Bridge)**, was built between 1354 and 1356 and it was claimed that its 49m segmental arch span next to the castle was then the longest in the world (the two other spans being 29m and 24m long). The arch rings and lower parts of the piers are in white marble and the remainder of the structure in red brick. The bridge's battlemented parapets maintain the swallow-tail feature of the merlons in the castle's fortifications. Unfortunately, the bridge was

Ponte Scaligero

completely destroyed in April 1945 but had been rebuilt by 1951. It is now on the World List of Tourist Attractions.

Pontoon Railroad Bridge, Wabasha, Minnesota, USA *

Between 1882 and 1951 the Milwaukee Road Railroad used a most unusual pontoon bridge to operate its services across the Mississippi River. The overall length of the main structure was 646ft and the pontoon part, which was 396ft long, consisted of a number of wooden boats linked side to side and on which short trestles were built to support the trackwork. One end of the floating section was pivoted to the end of the fixed bridge and the non-fixed end was then towed out by a tugboat so the whole pontoon could be aligned with the course of the river, leaving space for water traffic to pass. At either end of the main structure there were lengthy approaches of simple trestle spans giving an overall length of crossing of nearly 2,800ft. After the bridge was taken out of service it was later removed.

Port Mann Bridge, Vancouver, British Colombia, Canada

The first bridge here, built in 1964 to carry the Trans-Canada Highway across the Fraser River, was a three-span steel structure with a central main span of 366m flanked by 110m-long half

Port Mann Bridge

Poubara Liana Bridge, Franceville, Haut-Ogooué, Gabon

The first liana bridge across the Ogooué River was completed in 1915 and building a new one every year takes three months, after which the new link is floated in to replace the old one. The bridge is 53m long and its lowest part is 6m above river level.

Poubara Liana Bridge

arches, making it the longest arch bridge in Canada. Although the bridge was widened in 2001, increasing congestion led to the decision to build a replacement structure alongside the original one. The new structure was built between 2009 and 2015 and the old bridge was then demolished. This new cable-stayed bridge is 2,020m long with a main span of 470m. It has a 65m-wide deck carrying ten traffic lanes, which is the widest bridge deck in the world.

Port Nelson Bridge, Hudson Bay, Northern Manitoba, Canada

A seventeen-span steel truss bridge at the mouth of the Nelson River was built between 1915 and 1916 to link the mainland to a new artificial island that was intended to be developed as a new port. The scheme was abandoned when it was decided that the port of Churchill, 100 miles further north, would be more appropriate, although the bridge still stands.

Portuguese Bridge, Debre Libanos, Senafe, Ethiopia

Also known as **Ras Darge's Bridge**, this picturesque structure is probably a late nineteenth-century rebuild of an earlier crossing. It has three semicircular stone arches, with pointed cutwaters on the intermediate piers, and crosses a minor tributary to the Blue Nile.

Poughkeepsie Railroad Bridge, Poughkeepsie, New York, USA

This 6,768ft-long giant of a bridge, both physically and in terms of its heritage, was built between 1886 and 1888 to carry part of the New Haven Railroad across the Hudson River. Following fire damage in 1987, the steel cantilever truss bridge was decommissioned and in 2009 a new concrete deck was installed and it was re-opened to carry a footbridge called Walkway over the Hudson State Historic Park. It was the world's longest footbridge until 2016 (beaten by Mile into the Wild Walkway – q.v.)

Poughkeepsie Railroad Bridge

and since 1979 has been listed on the National Register of Historic Places. Its profile is unusual because it consists of deep anchor spans with cantilevered arms at each end that support trussed beam sections. Between approach viaducts on each side the 35ft-wide bridge consists of seven main spans, two of which are deep trusses 525ft long flanked on each outer side by a 548ft-long cantilever arm and suspension span, with another similar span between them, and with a shorter 201ft-long anchor span at each end of the bridge. The deck is 212ft above water.

Powerscourt Covered Bridge, Hinchinbrooke, Quebec, Canada

The two-span bridge across the Chateauguay River, built in 1861, is the oldest covered bridge in Canada. The internal structure is a McCallum arched truss, the only known surviving use of this kind of truss. The bridge's overall length is 167ft and its 20ft width allows it to carry a single traffic lane and a footway. Unusually, the roof is slightly arched along each span. The bridge is listed as an Historic Monument of Quebec.

Powerscourt Covered Bridge

Protville Bridge, Kantarat Binzart, Bizerte, Tunisia

The masonry bridge over the Medjerda River was built in about 1550. It has seven main semicircular arches with six further

Protville Bridge

tall and narrow flood arches set within the intermediate piers.

Puch Bridge, Ptuj, Styria, Slovenia

This bridge over the Drava River is named after the Slovene automotive inventor Johann Puch (1862–1914) and was completed in 2007. It is curved horizontally with a radius of 460m and has three main 100m spans and 65m end spans. Structurally, it is an extradosed cable-stayed prestressed concrete bridge in which the prestressing tendons appear at a low angle above deck level to pass through short, outward-leaning pylons.

Puente de la Mujer (Women's Bridge), Buenos Aires, Buenos Aires, Argentina

The elegant, cable-stayed, reinforced concrete footbridge in the Buenos Aires dockland area was designed by Santiago Calatrava and is similar to his Alamillo Bridge in Seville (q.v.). Overall, the 6.2m-wide deck is 170m long and has a 102.5m-long opening main span. This span is supported by cable stays from a cantilever arm inclined at 39° above the horizontal and both deck and arm rotate together on a bearing standing on a tapering cylindrical pier within the deck. The bridge was opened in 2001.

Puente Nuevo, Ronda, Malaga, Spain

An earlier badly-built bridge here over the 100m-deep gorge containing the Guadalevin River collapsed in 1741 killing fifty people. Its replacement, which has an overall length of 66m, was completed by the architect José Martin de Aldehuela in 1793 after forty-two years of construction. There is a small low-level pointed

Puente Nuevo Bridge

Puente Yayabo, Sancti Spiritus, Sancti Spiritus, Cuba

Built between 1817 and 1831 by Spanish conquistadores Domingo Valverde and Blas Cabrera, this slightly hump-backed bridge crosses the River Yayabo on five semicircular stone arches, the intermediate piers protected by pointed cutwaters. The four corners of the bridge are marked by elegant stone columns with streetlamps cantilevered off them. The bridge has been a national monument since 1995.

Putra Bridge, Putrajaya, Malaysia

The triple-decked concrete bridge across the 650-hectare artificial Putrajaya Lake, opened in 1999, uses Islamic architectural motifs similar to those of the seventeenth-century Pol-e-Khaju Bridge, one of the Zayanderud Bridges (q.v.) in Isfahan. It carries monorail lines, road carriageways and pedestrian promenades, and has a total length of 435m. There are five main spans with slightly arched soffits and the piers support half-octagon towers on each side of the bridge which house observation galleries and dining areas.

Putra Bridge

arch supporting a very tall rounded arch above which is a small aediculated feature window. This main arch is flanked on each side by buttresses topped by half domes and then by a smaller version of the tall central arch. The bridge is rumoured to be the one from which people were thrown to their deaths in the Spanish Civil War, a gruesome tale told to the hero Robert Jordan by his girlfriend Maria in Ernest Hemingway's 1940 novel *For Whom the Bell Tolls*.

Q

Quebec Bridge, Quebec, Canada

The story about the building of the bridges across the Saint Lawrence River at Quebec, with two catastrophic collapses taking the lives of eighty-eight workmen, is one of the most notorious in all construction history. Work on the first bridge, designed to carry a twin-track railway, started on site in 1904. This was to have two frameworks, each with a cantilever arm at both ends, that were linked by a central suspended span. In 1907, when this central span was being built outward from the end of one of the cantilever arms, the arm itself buckled and collapsed, killing seventy-five workmen. Work then started on a second bridge 65ft south of the first and, by 1916, all that remained to be done was to lift the now-prefabricated suspended span into place from pontoons anchored in the river. This span was 640ft long, weighed 5,000 tons and had to be lifted vertically 130ft into its final position. However, a complicated cruciform casting at one corner of the span failed, causing the whole unit to fall into the river, killing a further thirteen men. A new connecting span was then made and successfully lifted into place and the bridge finally opened in 1919. It is 95ft wide and its overall length is 3,238ft long, with a clear distance between the bases of the cantilever arms of 1,800ft (549m). It now carries one railway line and two road lanes, part of Route 175. Since 1995 the bridge has been a Canadian National Historic Site and is included in the ASCE list of historic bridges.

Queen Alexandrine Bridge, Stege, Zealand, Denmark

The steel arch bridge that carries Highway 59 across the Ulv Sound to the island of Møn was designed by Anker Engelund and built between 1939 and 1943. The overall length of the crossing is 745m, with the longest span being a 128m-long parabolic through arch flanked on each side by five smaller deck arches, and the deck is 11m wide. The bridge is shown on Denmark's 500 kroner banknote issued in 2011.

Queen Alexandrine Bridge

Queen Emma Bridge, Willemstad, Curacao, Venezuela

The floating bridge across Saint Anna Bay, which consists of a continuous wooden deck supported by sixteen pontoon boats, is 167m long and 10m wide and connects the Punda and Otrobanda areas of the city. Built originally in 1888, it was rebuilt and widened in 1939 and further restored in 2006. It is now limited to pedestrians only (with road traffic being carried by the Queen Juliana Bridge – q.v.). The bridge is lit at night from lamps supported on a series of hoops above the deck. The structure's key feature is that it is hinged at one end and, using two motor-driven propellers mounted on the pontoon at the other end, it can be swung open

Quebec Bridge

like a door to allow water traffic to pass. While this is happening, pedestrians can continue to cross the bay by ferry. The bridge is a major attraction in this World Heritage City.

Queen Juliana Bridge, Willemstad, Curacao, Venezuela

One of the first New World colonial towns, Willemstad was founded in 1527 and in 2010 it became a UNESCO World Heritage Site. An earlier structure across the narrow entrance into the harbour, the Queen Emma Bridge (q.v.), was built in 1888 and replaced in 1939. The current Queen Juliana Bridge, which was completed in 1974 after collapsing during construction in 1967 and killing fifteen workmen, provides a link between the Punda and Otrobanda areas of the city. Rising to 56m above water level in order not to restrict marine traffic in St Anna Bay, the four-lane bridge has an overall length of 500m, its main span being a rigid frame structure with inclined legs.

Queen Juliana Bridge

Queen Mary's Bridge, Füssen, Bavaria, Germany

This bridge is named after the wife of Maximilian II of Bavaria, whose husband gave it to her as a birthday present, and is chiefly known because there is an excellent view from it of the famous Neuschwanstein Castle. It was built in 1866 to replace an earlier timber bridge across the Pöllat River Gorge and consists of two separate N-braced arms, each about 60ft long, that are cantilevered out from the cliffs on each side of the gorge and bolted together where they meet. In 1984, these girders were renewed as part of a general restoration.

Queen's Bridge, Ourense, Navarre, Spain

Named after Muniadona, wife of Spain's King Sancho III, this 110m-long bridge carries the Pilgrims' Route to Santiago de Compostela over the Arga River and was built in the eleventh century. There are seven nearly semicircular arches, although the last arch on the eastern end is now buried, and the roadway is slightly hump-backed over the largest arch. The piers have pointed cutwaters and contain tall, pointed-arch openings.

Quepos Bridge, Quepos, Puntarenas, Costa Rica

Quepos Bridge over the Rio Cotos River was built in 1940 as a single-line rail bridge for trains carrying bananas from the plantations to the nearby port of Quepos. It now carries road traffic, including 30-tonne lorries, as well as cyclists and pedestrians but is so narrow that vehicles can only move in one direction at a time. As there are no traffic signals, driving over the bridge is dangerous and it has been called the Bridge of Death. The bridge itself is a steel girder structure. There are plans for a replacement bridge.

Querétaro Aqueduct, Santiago de Querétaro, Querétaro, Mexico

Built by the Spanish colonist the Marquis del Villa del Alguia between 1726 and 1738, this masonry aqueduct brought water to the old city from a spring nearly 2km away. The structure, which has an average height of 23m, is almost 1,300m long and has seventy-four slim semicircular arches each spanning 20m between large square piers. The aqueduct is part of the city's UNESCO World Heritage Site and is included in the ASCE list of historic bridges.

Querétaro Aqueduct

Rabat Bridges, Rabat, Rabat-Salé-Kénitra, Morocco

Although the French had occupied Algeria in 1830, it was not until 1912 that the Treaty of Fez made the adjoining territory of Morocco a French Protectorate, which lasted until 1956. During this period the French greatly improved the country's transportation system, an early example being the construction of the **Oued Sherrat Bridge**. This suspension road bridge over the River Sherrat, which was completed in 1917, has masonry piers with Moorish-style arched openings through which the roadway passes and the main suspension span is 102m long. Later strengthening of the bridge included the addition of six new pairs of stayed supports at each end. A more recent structure, completed by the Moroccan authorities in 2015, is the cable-stayed **Bouregreg Bridge**. This carries three lanes in each direction and has an overall length of 745m with a 375m-long main span. The deck is supported from two 200m-tall concrete piers.

Rainbow Bridge, Niagara Falls, Ontario, Canada

The first of a series of Whirlpool Rapids Bridges (q.v.) across the Niagara Gorge downstream from the Falls was built in 1848 but the current Rainbow Bridge dates only from 1941. After the **Upper Steel Arch Bridge**, opened in 1897, had collapsed as a result of an ice jam in 1938 there were extensive deliberations about a replacement structure and work eventually began in 1940 on this new structure. The road bridge has an overall length of 1,450ft with a single 960ft-span steel arch. Its 58ft-wide reinforced concrete deck, which is supported off the arch ribs by slim spandrel posts, is 202ft above water level. The whole area around the Canadian bridge end was later developed with the Rainbow Tower and Plaza that included a custom post and a

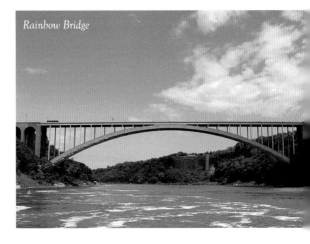

Rainbow Bridge

fifty-five-bell carillon. The Steel Bridge Institute declared the bridge to be 'the most beautiful steel bridge of 1941'.

Rama VIII Bridge, Bangkok, Krung Thep Maha Nakhon, Thailand

Built between 1999 and 2002, the cable-stayed bridge here across the Chao Phraya River has a 300m-long main span, a back span of 175m and, with the approach spans, its overall length is 1½ miles. The deck is supported by a harp-shaped layout of eighty-four cables anchored in a 160m-tall Y-shaped single pylon standing on the Thonburi bank. This pylon also carries a glass observation deck that is topped by a golden

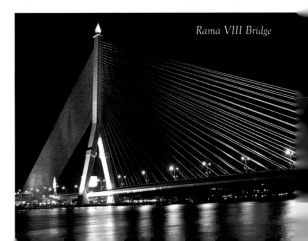

Rama VIII Bridge

spirelet shaped like a lotus bud and its base is surrounded by octagonal pavilions that represent an elephant's feet. The bridge, which carries four traffic lanes, a cyclist lane and a footway, is the world's longest asymmetrical cable-stay structure. It was designed by the Canadian firm Buckland & Taylor and is now an important national symbol for Thailand.

Ramstor Bridge, Nur Sultan, Central Kazakhstan, Kazakhstan

Ramstor Bridge is a steel network arch structure that carries the 50m-wide P3 road across the Ishim River. The overall length of the structure is 180m and the arch itself has a span of 120m diagonally over the 50m-wide river. The bridge was designed by the IHI Corporation and was built between 2007 and 2008.

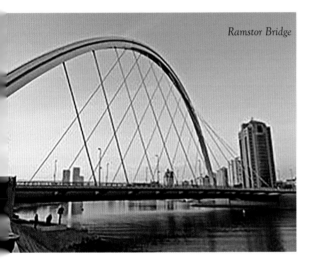

Ramstor Bridge

Reading-Halls Station Bridge, Muncy, Pennsylvania, USA

Built in 1846 with a span of 59ft, this is the USA's oldest metal truss bridge still in use. It was designed by Richard B. Osborne, a civil engineer with the Philadelphia & Reading Railroad, to carry a farmer's access road over a Norfolk Southern Railway line and in the late nineteenth century, when it was no longer needed there, was shortened and re-located to its present position. It now consists of a fifteen-panel truss with elaborate cast iron diagonal compression

members and vertical tie rods between the four-bar top and bottom booms.

Red Army Bridge, Taishun, Zhejiang, China

This woven timber arch 'lounge bridge' was built in 1954. Its overall length is 39m and it spans 33m between 6m-wide stone abutments. The eaves of its pitched roof project out above the side cladding. The bridge is named to commemorate a successful river crossing here by the Red Army in 1937.

Red Bridge, Meore Kesalo, Kvemo Kartli, Georgia

The Red Bridge over the Debeda River carries the Tbilisi-Ganja road between Georgia and Azerbaijan and, although the customs post is in Georgia, the area on both sides of the border crossing is surrounded by minefields. For a long time the bridge was used by pedestrians only and there were many civilian casualties here from mine explosions. The 175m-long stone bridge itself, which is considered to date from the twelfth century, has four pointed arches – a central larger arch with a single smaller one at each side – and, separated slightly, another larger arch.

Red Bridge, Meore Kesalo

Red Bridge, Yerevan, Ararat, Armenia

An earlier structure across the Hrazdan River here was destroyed by an earthquake in 1679. Its replacement, which was built of a red-coloured volcanic stone, was originally about 80m long and had two main pointed arches. These were flanked by two smaller arches located on the river banks. The bridge was restored in the 1830s but is now just an attractive ruin.

Red Gate Aqueduct, Augsburg, Bavaria, Germany

Although it is believed the Romans built an aqueduct to bring water into Augsburg, the basics of the present water system, including three water towers, date from 1416 and a new large hydraulic wheel to raise the water into the towers was built in 1548. In 1777 Johann Christian Singer built the two-level structure that carries a road above the aqueduct across the city's defensive Red Torwall ditch. This bridge, which has separate rows of six segmental brick arches supporting each of the decks, was totally renovated between 2005 and 2010 and became a UNESCO World Heritage Site in 2019.

Red Gate Aqueduct

Red Women's Bridge, Huixian, Henan Province, China

This stone road bridge and aqueduct has a segmental main arch spanning 65m that supports a seventeen-arch arcade 95m long carrying the 3m-wide road deck. An irrigation channel runs along one side of the road. The structure dates from 1971 and is of historical interest because it was built by eighty young peasant women.

Redzinski Bridge, Wroclaw, Lower Silesia, Poland

Built between 2008 and 2011, the cable-stayed Redzinski Bridge was built to carry the six-lane A8 motorway on a bypass away from the city centre of Wroclaw and it crosses 18m above the Oder River via a river island. With an overall length of 1,742m, it is the longest bridge in the country and its main 122m-tall pylon makes it also the country's tallest bridge. There are four spans – the central cable-stayed pair each 256m long and two 50m-long approach spans – and each carriageway is separately supported from a different one of the two upper vertical sections of the H-shaped pylon. The bridge was designed by Jan Biliszczuk.

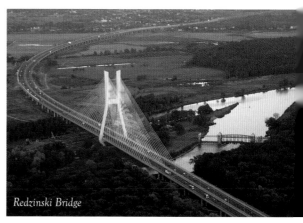

Redzinski Bridge

Regensburg Bridge, Regensburg, Bavaria, Germany

The first recorded bridge across the Danube at Regensburg was a wooden one built by Charlemagne in the late eighth century. The present stone bridge, built between 1135 and 1146 a short distance to the west of the earlier bridge site, is considered to be a masterpiece of medieval bridge building. There were three defensive towers, two of which were destroyed during the Thirty Years' War, and a chapel demolished in 1829. The bridge remained the only river crossing between Ulm and Vienna for centuries and was the only bridge in the town until the 1930s. Its overall length is 309m and it has sixteen arched spans. The old town, including the bridge, is a UNESCO World Heritage Site.

Regensburg Bridge

Rendsburg High Bridge, Rendsburg, Schleswig-Holstein, Germany

Because earlier swing bridges carrying the railway over the Kiel Canal here were causing unacceptable delays to trains when they were opened for water traffic, this dual-purpose cantilever bridge was built in 1913 as part of a 2½km-long viaduct. It consists of a main span 140m long, with 42m clearance below for water traffic, but which also acts as a transporter bridge by having a suspended gondola for carrying road traffic and foot passengers across the canal.

Rendsburg High Bridge

Replot Bridge, Korsholm, Ostrobothnia, Finland

Built between 1994 and 1998, Finland's longest bridge connects the island of Replot to the mainland. The two main 83m-high pylons for the cable-stayed structure are in the shape of stretched lozenges. These straddle the deck and support a vertical extension in which the two semi-fan systems of stays are anchored. The span between these pylons is 250m and the overall length of the 12m-wide bridge is 1045m.

Rewa Bridge, Nausori, Fiji

Opened in 2006 to replace an earlier steel structure, this four-lane concrete beam bridge crosses the Rewa River to link Suva and Nausori. It has seven main 50m-long spans and a smaller 39m-long span at each end.

Rhaetian Railway Bridges, Filisur, Graubünden, Switzerland

The whole of this 384km-long, single-track, narrow-gauge, electrified railway, sometimes called the 'Little Red', has been declared a UNESCO World Heritage site and includes 612 bridges, of which a few are noted below. **Landwasser Viaduct, Filisur** is a 136m-long, 65m-high, six-arched masonry structure with a longest span of 20m. It is built on a 2% gradient and a curve with a radius of 100m. The 110m-long stone **Brusio Spiral Viaduct** has nine semicircular-arched spans, each 10m long, and curves round completely on itself with a radius decreasing from 70m to 50m, passing through one of its own arches as it descends on a 7% gradient. The reinforced concrete **Langwies Viaduct**, built in 1914, is claimed to be the world's first large-scale concrete rail bridge. It is 284m long and has thirteen spans, the longest being 100m. The **Wiesen Viaduct** is a masonry-style structure but built of precast concrete blocks. It is 210m long and 89m high with seven semicircular arches and is well known for Ernst Ludwig Kirchner's paintings of it.

Brusio Spiral Viaduct

Rhine Bridge, Bad Säckingen, Baden-Württemberg, Germany

The first timber bridge over the Upper Rhine on this site was built in 1272 but it and successors were washed away several times, including in 1570. Following this, the bridge was rebuilt on masonry piers but the superstructure was still destroyed in 1633 and 1678. The current structure, locally known as the Holzbrücke, was built around 1700 and the piers were rebuilt in the 1960s. The bridge originally carried vehicles but since 1979 it has been used only as a footbridge. It is a queen post structure with an overall length of 668ft, has nine spans, is 16ft wide and is the longest timber covered bridge in Europe.

Rialto Bridge, Venice, Veneto, Italy

There had long been timber bridges here over Venice's Grand Canal, with collapses recorded in 1444 and 1524. This world-famous single-span segmentally-arched crossing was completed in 1591 to a design by Antonio da Ponte. The wide deck is edged by a footway each side, fronted by attractive corbelled balustrades, inside which are a double row of shops with arched frontages and a further central footway. Above the crown of the bridge each row of shops is interrupted by a grand open-pedimented arch that links the three footways. The bridge has been painted by many artists including Canaletto, his pupil Francesco Guardi, and Turner.

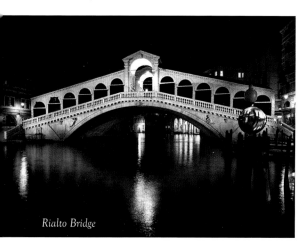
Rialto Bridge

Richmond Bridge, Richmond, Tasmania, Australia

The masonry bridge over the Coal River, built by Major Thomas Bell using convict labour between late 1823 and early 1825, is the oldest bridge in Australia and is on the Australian National Heritage List. It has four main arches and two smaller end arches, all with brick arch rings, and there are stonework spandrels and parapets. The parapet walls are level over the main arches but slope down over the end arches. The bridge has an overall length of 135ft long and is 25ft wide.

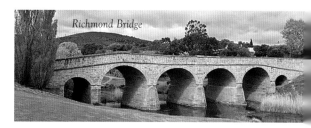
Richmond Bridge

Richmond-San Rafael Bridge, Richmond, California, USA

Together with its approaches, this double-decked bridge across San Francisco Bay has an overall length of 8,851m and is one of the longest bridges in the world. Each deck has three traffic lanes carrying Interstate 580. There are two separate main cantilever structures, each crossing one of the two shipping channels through the Bay. These structures consist of a central frame with a cantilever arm on each side and they are linked by nine trussed girder spans. There are a further sixty-two girder and truss spans. The bridge opened in 1956 and its chief designer was Norman Raab.

Rimac River Bridge, Rimac, Lima, Peru

Also called the **Stone Bridge**, this bridge over the Rimac River was built in 1608 by the architect Juan del Corral to provide a permanent link between Rimac and Lima by replacing an earlier wooden structure. It consists of five main stone arches. The bridge, which is sometimes called the Bridge of Eggs because 10,000 sea bird eggs were included in the mortar, is within the historic centre of Lima that has been a UNESCO World Heritage site since 1988.

Rio Cobre Bridge, Spanish Town, Middlesex, Jamaica

This bridge, erected in 1802, is a roughly one-third size version of the 1796 Wearmouth Bridge in Sunderland, England (see *AEBB*), which was for many years the world's longest cast iron bridge. The Jamaican structure was designed by Thomas Wilson (who had been one of the designers of the original bridge in Sunderland) and manufactured by Walker & Co at their Rotherham works in England. It was exported in 320 pieces weighing a total of 87 tons and erected between abutments broadly similar in appearance to those of the original. The bridge has four ribs spanning 82ft, each consisting of twenty voussoirs roughly 4ft long by 3ft deep with a further half-length voussoir at the crown and its deck is 15ft wide. The bridge featured on a 15c stamp issued by Jamaica in 1972. The World Monuments Fund placed the bridge in its Watch in 1998 but storm water damaged the foundations in 2000, it was further damaged by a hurricane in 2004 and political conflicts then prevented repair work from starting. However, the bridge was fully restored in 2010. The Jamaica National Heritage Trust declared the bridge a National Monument but it failed to become a UNESCO World Heritage Site because of local violence.

Rio Cruces Bridge, Torobayo, Valdivia, Chile

Opened in 1987, this so-called triangular bridge (in elevation its profile is like a low-pitched roof) crosses the Cruces River just before it joins the Valdivia River and connects Teja Island with the mainland. The reinforced concrete structure has thirteen spans with A-shaped intermediate piers.

Rio Cruces Bridge

Rio Goascoran Bridge, El Amatillo, La Unión Department, El Salvador

Replacing a 1942 structure, the bridge over the Goascoran River linking El Salvador and Honduras was completed in 2008 as part of the Pan American Highway that runs through eighteen countries. The 480ft-long structure is at an altitude of 6,500ft in the Sierra Madre mountain range.

Rio Negro Bridge, Manaus, Amazonas, Brazil

The four-lane cable-stayed road bridge over the Negro River (a major tributary of the bridgeless Amazon) at Manaus was opened in 2011 to carry the AM-070 highway linking Manaus with Iranduba. The main part is a two-span structure either side of an unusual central pylon. This has three tiers with, first, a lower section beneath the deck consisting of four vertical corner columns. Above this deck is a middle tier, with four slightly inclined columns, and this is surmounted by a fleche-like upper tier, its top 185m above water level. The two 200m-long side spans are supported by diagonal stays from the two upper stages of this pylon, giving 55m clearance between river level and the underside of the deck. The overall length of the bridge is 3,600m and there are 103 approach spans.

Rio Negro Bridge

Rio Seco Aqueduct, Almuñécar, Andalusia, Spain

The aqueduct bridge over the River Seco is one of three similar structures that are part of a water-carrying trough 8km long built by the Romans in the first century CE to provide water for the town of Almuñécar. With an overall length of 90m, it has two tiers of semicircular arches standing on piers. The lower tier, which has three arches, in turn supports six of the piers for the upper tier of eleven arches. The aqueduct has a maximum height of 17m above the river.

Rio-Antirrio Bridge, Patras, Greece

This reinforced concrete motorway bridge over the Gulf of Corinth, which was opened in 2004, has an overall length of 2,880m and is 27m wide. The cable-stayed structure has three 560m-long main spans and two 286m-long end spans, making it one of the world's longest multi-span cable-stayed bridges. The spans are supported by four pylons reaching to 160m above sea level. Each of these pylons has four inclined legs meeting beneath a vertical column in which the stays are anchored, an arrangement that gives the bridge its distinctive appearance. Designed by Berdj Mikaelian, the bridge is officially called the Charilaos Trikoupis Bridge after a former Prime Minister who first envisaged a link at this site in 1880.

Rio-Antirrio Bridge

Rio-Niteroi Bridge, Rio de Janeiro, Brazil

Built between 1968 and 1974, this 8¼ mile-long box girder bridge carries eight lanes of the BR-101 federal highway across Guanabara Bay to connect the cities of Rio de Janeiro and Niteroi, with two-thirds of the overall length being above water. Most of the structure is of prestressed concrete but the 300m-long and 72m-high central navigation span, together with the flanking span on each side, are of steel. The standard span length on the approach viaducts is 80m and the bridge deck is 27m wide and 4.7m deep.

Ritsurin Garden Bridges, Takamatsu, Kagawa, Japan

The Ritsurin Garden is a traditional Japanese daimyo strolling garden containing lakes and bridges with hills in the background. First laid out in 1625 and opened to the public in 1875, it is one of the most famous historical gardens in Japan and was designated a Special Place of Scenic Beauty (National Treasure Garden) in 1953. Perhaps the best-known view is of **Nanko Bridge**. The timber beam structure has five spans, the intermediate supports being simple frames consisting of two vertical posts and cross beams.

Nanko Bridge

River Neva Bascule Bridges, St Petersburg, Leningrad Oblast, Russia

Also called the Peter the Great Bridge, **Bolsheokhtinsky Bridge** across the Neva River was opened in 1911 after three years of construction and links the Smolny Cathedral area in the centre of the city to the factory area of Okhta on the river's right bank. It is 334m long and 23m wide and consists of two segmental steel arches spanning 136m, between which is a double-bascule section spanning 48m. The downstream ends of the central piers, which house lifting mechanisms for the bascules, are each decorated with a tall masonry lighthouse-type tower.

The original **Liteyny (Foundry) Bridge** was completed in 1879 to replace a temporary floating bridge, fourteen people being killed in two accidents during its construction. It now has a single-leaf bascule that is 55m long, 34m wide and weighs 3,225 tons – a world record when it was installed in 1967.

Troitsky (Trinity) Bridge, which was the third permanent bridge across the Neva, was opened in 1903 to mark the 200th anniversary of the city's founding. It is 24m wide and 582m long and has three short masonry arches by the right bank leading to six steel arched spans, one of which is a single huge bascule 43m long next to the left bank. The entrance to the bridge is flanked by tall obelisks supporting decorative lighting.

Trinity Bridge, St Petersburg (opening bascule span in foreground)

Rochefort-Martrou Transporter Bridge, Rochefort-sur-Mer, Charente-Maritime, France

Built over the River Charente by Ferdinand Arnodin, this was one of five of these distinctive transporter structures of Arnodin's in France.

It was first opened in 1900 but, after being taken out of service in 1967, it became a classified historical monument in 1976 and was recommissioned in 1994 for use by cyclists and foot passengers during the summer. It was re-opened in 2020 after a major re-building. The main beam, under which the original traveller ran that supported the people- and vehicle-carrying gondola, is a 2m-high 176m-long lattice girder spanning 140m between 67m-tall twin lattice pylons on each bank of the river. This beam is supported by diagonal stays near the pylons and, in the centre, by vertical hangers from suspension cables strung between the pylons. The bridge is now being considered for inclusion as a UNESCO World Heritage site.

Rochefort-Martrou Transporter Bridge

Rochers Noirs Viaduct, Lapleau, Corrèze, France

Built between 1911 and 1913 to carry the Yellow Train Railway's services on a single track across the Luzège River, this bridge was designed by Albert Gisclard. The bridge has an overall length of 172m, with a 140m-long main span between two masonry towers, and its deck is 92m above water level. Each side of both half-bridges is supported by a system of six stays from the towers. Where the highest stay from one tower intersects the stays from the opposite one, a vertical hanger from each intersection point supports the bridge deck below. In addition, there is a separate suspension cable between the towers from which vertical hangers also support the radiating stays. The whole arrangement is thus effectively a statically-indeterminate truss. The deck has particularly elegant wrought iron parapet railings. The railway closed in 1959 and, following conversion, it carried road traffic until 1982 and became a footbridge in 1983. It has been listed as an Historic Monument since 2000 but has been closed to pedestrians since 2005.

Rock Island Railroad Bridge, Yakima, Washington, USA

In 1892 the Great Northern Railway built this single-track bridge over the Columbia River. Its total length was 875ft and its longest span, a through truss, was 417ft long. Immediately next to this was a deck truss, producing an unusual combination. In 1925 the bridge was strengthened by building a second pair of trusses outside the original ones, thus allowing services to continue during construction work and producing a 'bridge within a bridge'.

Rock Island Railroad Bridge

Rockville Bridge, Harrisburg, Pennsylvania, USA

The first bridge over the Susquehanna River on this site, completed by the Pennsylvania Railroad in 1849, was a timber Howe truss bridge on stone piers that carried a single railway line. This was replaced in 1877 by a double-track wrought iron latticework bridge on the old piers. The present 3,820ft-long structure, opened in 1902, is the

world's longest stone arch railway bridge. It has forty-eight 70ft-long segmental arches that span between stone piers with pointed cutwaters, the upstream ones raked back slightly. Originally, the bridge's 52ft-wide deck carried four tracks, but there are now just two, thus minimising the risk of a container being blown off a train into the river as had happened in the late 1990s. The bridge is included in the ASCE list of historic bridges.

Roman Aqueducts, Aspendos, Antalya Province, Turkey

The large bridge-like ruins at Aspendos are part of the remains of a water supply system that was built by the Romans around 200CE to carry water to the town from hills about 12 miles to the north. The main aqueduct structure was 510m long and 5.7m wide and, of the original forty-seven arches, twenty-nine are still standing. There was also a shorter two-tier aqueduct. The cross-section of the water channel was about 2ft wide by 3ft deep, its gradient about $2\frac{1}{2}$% and its capacity about $1\frac{1}{4}$ million gallons of water a day. It was not a gravity system; the water channel was sealed so that the hydrostatic pressure at the initial hills would force the water to the top of water towers at intermediate aqueducts, and the whole system has been described as an inverted syphon. Historians believe that it was badly damaged by an earthquake about 150 years after its completion and never rebuilt.

Aspendos Aqueduct

Roman Bridge, Cordoba, Andalusia, Spain

The bridge at Cordoba over the River Guadalquivir is 250m long and has sixteen semicircular arches spanning between huge rounded and domed cutwaters. However, only two of the arches are original from the first century BCE bridge, built on the orders of the emperor Augustus. Most of the structure dates from an eighth-century rebuild and there was a further major restoration in 2006.

Roman Bridge, Frias, Burgos, Spain

Despite its name, this fortified bridge over the River Ebro is not Roman but dates from the twelfth century. Slightly curved in plan, it is 143m long and 3½m wide and has nine arched spans, two of which are noticeably pointed. Between these two arches there is a fourteenth-century pentagonal gateway tower with a short machicolated section above the roadway through the tower. The piers have pointed cutwaters and two are topped by high level flood relief arches.

Roman Bridge, Mérida, Badajoz, Spain

The Roman bridge over the Guadiana River at Mérida is the longest bridge built by the Romans that still survives and dates from 120AD. It is estimated that the original structure had sixty-two spans over a length of 755m but only fifty-seven arches can be seen today, the longest spanning about 11m. The arches are semicircular and many have flood relief channels above the piers,

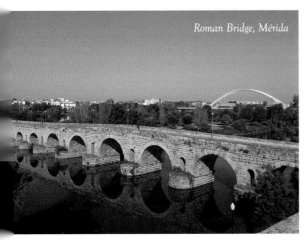

Roman Bridge, Mérida

these having rounded cutwaters on the upstream face. The deck width varies but is generally about 7m. Since 1991 the bridge has been for pedestrians only, with road traffic now using Santiago Calatrava's Lusitania Bridge (q.v.). The Roman bridge is part of the archaeological ensemble at Mérida that is listed as a UNESCO World Heritage site.

Roman Bridge, Mostanica, Niksic Municipality, Montenegro

Mostanica's Roman Bridge is considered to be the oldest bridge in Montenegro, dating from the third century AD. It has five main semicircular arches, each spanning 5.5m, with a smaller and higher arched opening in the upper part of each of the 2m-wide intermediate piers. The bridge was demolished by partisans in 1942 to hinder the Italian army and reconstructed in 1957. The Mostanica River has now dried up.

Roman Bridge, Salamanca, Castile-León, Spain

Only fifteen arches on this bridge over the Tormes River are Roman, probably replacing an earlier timber structure, with the other eleven all dating from 1677. The two parts are separated by an enlarged pier which had once supported a small defensive tower. Since 1954 the Roman bridge has been decorated with a famous statue of a boar. This had first been documented in the thirteenth century and then recovered from the river in 1867. Following the opening of a new crossing in 1973, the bridge is now limited to pedestrian use only. The bridge is shown on the one euro Spanish stamp issued in 2016.

Roman Bridge, Segura, Gipuzkoa, Spain

The Roman bridge across the Elja River at the Portuguese:Spanish border was built in the second century under the Emperor Trajan. It has five semicircular arches: a central one of 10m span flanked on each side by two further arches spanning 7m. These arches have cutwaters on the upstream side only. It is 78m long and nearly 7m wide and its level roadway now connects the EN355 and EX-207 roads.

Roman Bridge, Trier, Rhineland-Palatinate, Germany

It is claimed that the piers for this bridge over the Moselle River date from the second century CE, although the arches were probably built between 1190 and 1490. The total bridge length is 198m and it has eight spans that are generally between 16m and 21m long. The deck is 13m wide. A seventeenth-century print shows three defensive towers on the bridge but nothing of these remains. The bridge is included within the UNESCO World Heritage Site at Trier.

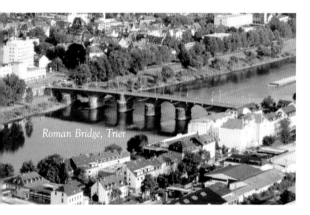

Roman Bridge, Trier

Roman Bridge, Vaison-la-Romaine, Vaucluse, France

This bridge has a single semicircular stone arch spanning 17m directly between the sides of the rocky ravine through which the River Ouvèze flows. It was built in the first century CE and was designated a heritage monument in 1840. Damaged in WWII, it was restored in 1954.

Rombachtal Bridge, Schlitz, Hesse, Germany

Built in 1989, this 986m-long prestressed concrete high-speed railway viaduct has a series of 58m-long column-and-beam approach spans with a central double-length main span that is 116m long. This has an unusual pointed main arch 95m high which was built as two separate half-arches.

Ronzat Viaduct, Gannat, Allier, France

The single-track railway line between Commentry and Gannat crosses the Sioule River on a wrought iron viaduct designed by Gustave Eiffel

and opened in 1869. There are three main lattice girder spans, the longest being 58m long, that span between lattice towers, the tallest of which has eleven tiers. These towers have curved lateral restraints provided by arched braces to the bottom two tiers in a manner later used by Eiffel on his famous Tower.

Roquefavour Aqueduct, Ventabren, Bouches-du-Rhône, France

The Roquefavour Aqueduct across the Arc River, which was built between 1841 and 1847 to carry water from Durance River to Marseilles, is 393m long with a maximum height of 83m and is considered to be the largest stone aqueduct in the world. It has three tiers of arches spanning 16m between piers that are battered out front and back. The structure was designed by the engineer Jean François Mayor de Montricher, who was awarded the Legion of Honour for his work, and has been a National Heritage site since 2002.

Roquefavour Aqueduct

Rosario-Victoria Bridge, Rosario, Anta Fe Province, Argentina

This cable-stayed bridge across the Paraná River was built between 1997 and 2003 as part of a new 60km link between the two cities of its name. The structure is 16m wide, has a main span of 350m, an overall length of 608m and the deck is 56m above the water level. Together with its approach viaducts, the total length of the crossing is 4,098m.

Ross Bridge, Ross, Tasmania, Australia

After the failure of two earlier structures nearby, this very handsome stone bridge was built between 1830 and 1836 to carry a level roadway over the Macquarie River and connect Hobart to Launceston. Designed by local civil engineer John

Ross Bridge

Lee Archer and built by convict labour, this is believed to be Australia's third oldest bridge and is widely considered to be its most beautiful. It has three segmental stone arches, each spanning 9m between piers fronted by cutwaters, and curved stairways at each corner allow pedestrian access to the river banks. The main feature of the bridge is that every one of the thirty-one voussoirs in each arch ring has individually carved decorations. These are either abstract Celtic symbols or represent wheat sheaves, wool bales and the like, and the bridge also has other carvings of people and animals. All the carvings were made directly by, or under the supervision of, convict stonemason Daniel Herbert, who was granted his freedom when the bridge was completed.

Royal Gorge Bridge, Canon City, Colorado, USA

When this suspension bridge, carrying a road 955ft above the Arkansas River, was opened in 1929, it was the highest bridge in the world, taking the record from the Sidi M'Cid Bridge in Algeria (q.v.). The deck is 18ft wide and its total length is 1,260ft with a main span of 880ft. The towers are 150ft high. Since 2015 a new zip line has crossed the gorge here.

Royal Opera House Footbridge, Muscat, Oman

Opened in 2019, this 53m-long stainless steel pedestrian bridge crosses Kharijiyah Street to link the Opera Galleria and the House of Musical Arts. It consists of an air-conditioned half-glazed elliptical tube with decorative internal timber latticework in traditional style.

Rulong (Like a Dragon) Bridge, Quingyuan, Sichuan, China

This covered bridge was built in 1625. Its overall length is 28m, with a clear span of 20m, and its width is 5m. Three sets of siding enclose the main deck passageway, which is covered by three separate pagoda-like roofs, the tallest at the north end being a bell tower and the one in the centre containing altars.

Russky Bridge, Vladivostok, Primorsky Krai, Russia

The Russky bridge was completed in 2012 to mark the opening of the Pacific Economic Cooperation Summit. It crosses the Eastern Bosphorus Strait and links Russky Island to the Vladivostok Peninsula. The bridge has an overall length of 3,100m, which includes eight approach spans, and a main 1,886m-long cable-stayed structure. The central span of this is 1,104m long, making it the longest cable-stayed bridge in the world. The A-shaped pylons are 324m tall and stand on piles founded up to 77m below ground level. The deck, which is 30m-wide and carries four traffic lanes, is an aerodynamic steel box girder. It is supported by 168 Freyssinet parallel-strand cable stays, the lengths of these ranging from 136m to the longest, of seventy-nine strands, which is 580m long and the longest in the world.

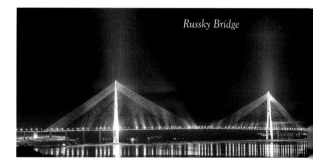

Russky Bridge

Rusumo Bridge, Rusomo, Eastern Province, Rwanda

The single-lane steel arch bridge at Rusomo over the Kagera River, which was designed by Luigi Corradi, was fabricated in Italy and erected between 1971 and 1972. It has an overall length of 100m with a span of 63m. A second structure to allow simultaneous crossing in both directions was opened in 2014.

S

Sacramento River Trail Footbridge, Redding, California, USA

The stressed ribbon bridge here over the Sacramento River, which was designed by Charles Redfield and Jiri Strasky, was opened in 1990. The deck is 11ft wide and spans 418ft. The bridge is supported by 236 cables within the deck which are stressed between anchorages set into the rock face at either end.

Sagasta Bridge, Logrono, La Rioja, Spain

The fourth bridge over the Ebro River at Logrono was designed by Javier Manterola and opened in 2003. It carries twin carriageways, with two separate flanking footways that arch out horizontally from the road, all of which are hung on stays from a single twin-tube steel arch. This arch spans 140m and has a rise of 28m.

Sai Van Bridge, Macau, Guangdong, China

The prestressed concrete cable-stayed double-decked bridge across Praia Grande Bay was opened in 2004. Its overall length is 2,200m with the longest individual span being 180m. The 28m-wide bridge has six road lanes on its upper deck with the lower one enclosed for use during strong typhoons and for a possible future rail link. The appearance of the bridge is distinctive, with each of the two main piers having three 85m-tall legs enclosing tall pointed archways.

St Johns Bridge, Portland, Oregon, USA

St Johns Bridge, designed by David Steinman and completed in 1931, carries the US-30 road over the Williamette River with 205ft clearance between the underside of the deck and river level. With approach spans, the overall length of the bridge is 3,608ft but it has a central suspension span that is 1,207ft long with 430ft-long back spans. The bridge is widely thought to be one of the more beautiful American bridges although some consider it contradictory that a modern structure such as a steel suspension bridge should have a Gothic appearance: the pylon above each main pier has a tall lancet-shaped opening through which the 40ft-wide roadway passes and above this is a further arcade of three pointed arches topped by ornate pinnacles 400ft above river level. The bridge is also unusual technically in that it uses twisted strands of wire, in the way traditional hemp ropes are made up, rather than the usual parallel wire cables.

St Johns Bridge

Sai Van Bridge

St Patrick's Bridge, Cork, County Cork, Ireland

The iconic St Patrick's Bridge was completed in 1861 to replace an earlier bridge from 1789 that had been destroyed by floods in 1853. The 300ft-long and 60ft-wide masonry structure, which has three semi-elliptical arches (the central one spanning 62ft), was designed by the City Engineer Sir John Benson, who was also an

St Patrick's Bridge, Cork

architect. The bridge is most handsome: the piers have wide pilasters and pointed cutwaters, the arches are decorated with carved archivolt rings and keystones, and it is topped off by handsome masonry balustrades and cast-iron lamp standards. Between 2017 and 2019 the bridge was completely refurbished in a €1.2m project.

St Servatius Bridge, Maastricht, Limburg, Netherlands

The Romans built a timber bridge over the Meuse in about 50CE and a successor collapsed in 1275 during a procession, killing some 400 people. The stone replacement bridge, built between 1280 and 1298, is considered to be the oldest bridge in the country and is named after Saint Servatius, who was the first bishop of Maastricht and whose statue now stands on one of the piers. The bridge is 160m long and 12m wide and originally had eight 12m-long arches and a longer timber span at the east end. In the 1930s the **Wilhemina Bridge** was built downstream to relieve the old structure and, at the same time, the two existing easternmost spans were replaced to improve navigation. Where these once were is now crossed by a steel drawbridge built in 1962 that spans 55m. Several arches, which were destroyed by the Germans in 1944, were rebuilt in 1948.

Saint-Clément Aqueduct, Montpellier, Hérault, France

The 880m-long aqueduct at Montpellier, designed by the hydraulic engineer Henri Pitot, was built between 1760 and 1772 to provide the city of Montpellier with water from the St Clément spring 14km away and it currently supplies water to the city's fountains. The structure is mainly

two-storied, with semicircular arch spans varying from 3m to 9m, and the maximum height is 22m. The top tier is slightly narrower and its piers are pierced by archways that provided an upper walkway. The lower stage has 51 arches and in the upper stage there are 182 arches. The aqueduct finishes on a raised platform at the Peyrou Esplanade with a grand three-arch single-storey structure centred on a decorative semi-elliptical arch spanning 18m. This feature was designed by the architect Jean Giral.

St Clement Aqueduct, Montpellier

Samuel Beckett Bridge, Dublin, County Dublin, Ireland

Santiago Calatrava designed the 129m-long cable-stayed opening road bridge over the River Liffey that was completed in 2009. It has a distinctive main pylon 48m high which has an elongated S-shape in the vertical plane and looks a little like a stretched Irish harp. This pylon supports thirty-one stays, anchored along the centre-line of a 27m-wide deck, that are balanced by two much larger stays to the outer edges of a short back span. The bridge opens by rotating horizontally through ninety degrees.

Samuel Beckett Bridge

San Francisco-Oakland Bay Bridge, San Francisco, California, USA

Originally, this crossing consisted of several different structures carrying traffic on two levels. First, there were twin end-to-end suspension bridges each with a 2310ft-long main span and sharing a common anchorage between the two adjacent side spans, then a tunnel and, finally, a 1,400ft-long cantilever structure. The bridge, which was built between 1933 and 1936 by the American Bridge Company to a design by Charles H. Purcell, carried the ten road traffic lanes of Interstate 80 on a 58ft-wide deck with clearance below of 190ft. However, in an earthquake in 1989 part of the upper deck of the cantilever bridge collapsed onto the lower deck and this section was rebuilt as a 258ft-wide single deck, making it the world's widest bridge structure. Later, after the original Bay Bridge East Span had been deemed seismically unsound, it was replaced between 2002 and 2013 by a new structure that became the world's largest self-anchored suspension bridge. This has a single central 525ft-tall tower supporting the two suspension cables over the two asymmetrical spans of 1,263ft and 580ft. Each cable contains 17,399 steel strands. The full bridge is included in the ASCE list of historic bridges.

San Juanico Bridge, Tacloban, Eastern Visayas, Philippines

It is claimed that this bridge was built at the instigation of the President's wife, Imelda Marcos, who came from this area, and it was opened on her birthday in 1973. The S-shaped structure, which is part of the Pan-Philippine Highway, is 2164m long and has forty-three spans. Its longest span of 192m is an arch-shaped truss.

San Juanico Bridge

San Martin Bridge, Toledo, Castile-La Mancha, Spain

An earlier bridge here across the Tagus River was destroyed during the Castilian Civil War of 1368–89 between Henry II of Castile and Peter IV of Aragon. The present San Martin Bridge was completed in 1380 to provide access into the city from the west, entry from the east being over the Alcántara Bridge (q.v.). The bridge, which was designed by Archbishop Pedro Tonorio (whose head is carved on the central keystone), has five pointed arches, the central one spanning 40m, and these support a level deck 27m above river level. There is a large defensive tower at each end. The bridge has been a national monument since 1921 and Toledo is now a UNESCO World Heritage site. Modern times are represented by the zip line next to the bridge.

Sanahin Bridge, Alawerdi, Lori, Armenia

The bridge here over the Debeda River was built between 1192 and 1200 by Queen Vanen in memory of her husband King Abas III. The overall length is 197ft and it has a segmental stone arch, formed of three concentric stone rings, that spans 61ft. The 10ft-wide approach from one side of the river is up more than twenty shallow steps flanked by stepped parapet walls but on the other side the footway leads directly onto the flanks of the adjacent hills. There is a carved lion at the end of the highest parapet section.

Sandö Bridge, Kramfors, Västernorrland County, Sweden

The temporary scaffolding for Sandö Bridge collapsed in 1939 during construction, killing eighteen workmen, and the bridge was not finally completed until 1943. The segmental reinforced concrete arch structure crosses the Angerman River in a single span of 264m with a rise of 42m, and was the largest concrete arch in the world until the 305m-span Gladesville Bridge (q.v.) in Australia was opened in 1964. At the centre of the arch the 10m-wide deck is 42m above water level. Each side of the deck above the arch is supported on slim, circular

Sandö Bridge

spandrel columns but, behind the springing points, the deck stands on tall H-shaped frames with rectangular columns. The overall length of the bridge is 810m. After the main north-south coast road was moved to the new Höga Kusten Bridge in 1997, Sandö Bridge underwent major renovations, not re-opening until 2003.

Sanhao Bridge, Shenyang, Liaoning, China

Designed by Man-Chung Tang, this cable-stayed bridge has a single central pier supporting twin horseshoe-shaped reinforced concrete arches that lean away from each other. On each side is a 100m-long span supported by a harp arrangement of cable stays anchored in the arches, with horizontal cables between them. The bridge crosses the Hwun River and was built between 2006 and 2008.

Santa Fe Suspension Bridge, Santa Fe de la Vera Cruz, Santa Fe, Argentina

The suspension bridge here over the Rio Santa Fe was built between 1924 and 1928 to a design by Albert Gisclard that was first used at La Cassagne Bridge in France (q.v.). Instead of being hung from conventional catenary cables the deck is supported by vertical hangers from connection points where diagonal stays from each main tower meet. The 148m-long central span is flanked by back spans that are 74m long. The structure was badly damaged by floods in 1983 when the river rose nearly 2m. The road now carries Highway 168 and, since 2017, the bridge has been illuminated by a very successful lighting scheme.

Santa Justa Lift, Lisbon, Estremadura, Portugal

Lisbon's famous 45m-tall lift structure, which links the streets in Lisbon's lower historic old centre with Carmo Square in the city's higher part, acts as the main pier for the outer end of the footbridge connecting the upper lift landing to the area round the square. Structurally, the bridge consists of two plate girder spans that are haunched at the central concrete pier. However, these are topped by elegant wrought ironwork panels that form the sides of the enclosed deck, which is covered by a pitched roof. The elevator tower and bridge were designed by Raul Mesnier de Ponsard and opened in 1902.

Sanhao Bridge

Santa Justa Lift

Santa Teresa Bridge, Elche, Valencia, Spain

Work began on the first bridge over the Vinalopó River here in 1705 but it was destroyed by floods in 1751. A new 11m-wide and 77m-long stone bridge, opened in 1756, has two pointed arches. These each span 14m on either side of a 7m-wide central pier with pointed cutwaters on which stand tall triangular towers decorated with pinnacles.

Santa Trinita Bridge, Florence, Tuscany, Italy

Earlier bridges on this site over the Arno River had been swept away in 1259, 1333 and 1357. The replacement for this latter one was built by the architect Bartolomeo Ammannati between 1567 and 1569 and is claimed to be the first bridge in the world built with curved stone arch spans that are neither semicircular nor segmental but are, effectively, semi-elliptical (sometimes called 'basket-handled'). In fact, each half-arch is almost but not quite a quarter-ellipse, these meeting at a very slight point at the mid-span crown with this discontinuity being masked by a carved pendant. The bridge has three of these arches: a central one spanning 32m which is flanked on each side by a 29m-long span. The structure was destroyed in 1944 by the retreating German army and rebuilt in 1958.

São Vicente Bridge, São Vicente, Cacheu Region, Guinea-Bissau

The 730m-long bridge across the Cacheu River, which was completed around 2010, carries a two-lane road between Bissau and the Senegal border. It has seven 85m-long prestressed concrete spans with a gently-arched profile.

Saracens Bridge, Adrano, Sicily, Italy

The main structure over the Simeto River Canyon is a tall pointed stone arch that may date back to Roman times, but the three smaller arches that carry the road over the approaches to one side of the gorge are modern, having been rebuilt after floods destroyed them in 1948. Leading away from the gorge, these are a small pointed arch followed by a large and then a small semicircular arch.

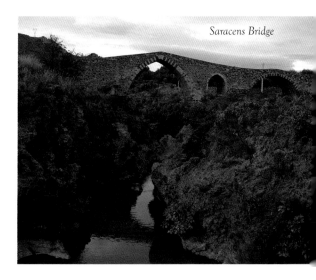

Saracens Bridge

Sart Aqueduct, La Louvière, Hainaut Province, Belgium

The historic engineering works of the Canal du Centre, which were designated a World Heritage Site by UNESCO in 1998, include a number of aqueducts and bridges as well as the well-known hydraulic boat lifts. Built between 1998 and 2002, the 498m-long and 46m-wide Sart Aqueduct became, at 65,000 tons, the heaviest structure ever to be push-launched into position. It is supported by fourteen pairs of concrete columns that are 3m in diameter.

Savage Mills Bridge, Savage, Maryland, USA

The railway bridge at Savage Mills, built by the Baltimore & Ohio Railroad and completed in 1869, is one of America's oldest iron railway

Savage Mills Bridge

bridges still standing and is believed to be the only surviving example of a Bollman truss, a type of structure invented by Wendel Bollman and patented in 1852. In these trusses the deck beams are supported by a fan of metal tie rods radiating out from the top of cast iron columns at the end of each span. The bridge has a mid-river granite pier supporting two equal main spans, each of which is 80ft long, 25ft wide and 25ft tall. In 1966 the bridge was the first Civil Engineering Landmark designated by the American Society of Civil Engineers and became a National Historic Landmark in 2000.

Saxon Switzerland National Park Bastei Bridge, Kurort Rathen, Saxony, Germany

The Bastei natural rock formations in the Saxon Switzerland National Park, which tower above the Elbe River, were linked by a timber footbridge in 1824. This was replaced by the present stone structure in 1851, which is 77m long and has seven semicircular stone arches spanning a 40m-deep ravine. There is a small side-span to the top of a circular masonry tower and the bridge has distinctive masonry parapets.

Saxon Switzerland Bastei Bridge

Scarawalsh Bridge, Scarawalsh, County Wexford, Ireland

Built in 1795 on the line of an old ford, the name of this bridge means 'Walsh's shallow ford' and Scarawalsh Bridge was built by the Oriel brothers of Hampshire to replace an earlier

timber bridge that had been swept away in 1787. Considered to be a particularly fine example of the civil engineering of its time, it crosses the River Slaney on six nearly semicircular stone arches with spans ranging from 22ft to 32ft and an overall length of about 310ft. There is a pleasant vertical profile and the piers have tall pointed cutwaters. The bridge was bypassed downstream by a prestressed concrete structure in 1976 and is now disused.

Schaffhausen Bridge, Schaffhausen, Schaffhausen Canton, Switzerland

The covered wooden bridge over the River Reno in Schaffhausen was designed and built by Hans Ulrich Grubenmann (1709–1783) between 1755 and 1758. Initially, he had designed the structure to have a single enormous span of 111m but the structure as built has two spans of 52m and 59m and consists of polygonal arches within the side walls. Its internal width is 5m and the maximum height under its pitched roof is 5.5m. The bridge was burnt down by retreating French forces in 1799 but was rebuilt. It is decorated with a small cupola on the roof above the intermediate pier.

Schöllenen Gorge Devil's Bridge, Andermatt, Uri, Switzerland

There are a number of bridges across the gorge over the Upper Reuss River, the most famous being the Devil's Bridge.

The first bridge of that name here is thought to have been built in 1306 and this was replaced by a stone structure in 1595. The two structures that are now in use to cross the gorge in this location are a higher-level railway bridge of 1917 and a lower-level road bridge of 1958.

Schwabelweis Bridge, Regensburg, Bavaria, Germany

This road bridge across the Danube was built in 1982. It has a separate, broad, rectangular box-section steel arch on each side of its double carriageway deck. The two arches have different spans and there is no bracing or connection between them.

Schwansbell Bridge, Lünen, North Rhine-Westphalia, Germany

The road bridge over the Datteln-Hamm Canal, which was opened in 1956, is a rare trussed aluminium structure. The deck is 4.5m wide and the bridge spans 44m.

Sciotoville Bridge, Sciotoville, Ohio, USA

The twin-track railway bridge over the River Ohio in Sciotoville, completed in 1916, has two equal 775ft-long spans and was the world's longest continuous truss from then until 1945. The structure has an unusual elevation as the height from deck level to the top of the truss increases over the central pier to a maximum of 129ft. The bridge was designed by Gustav Lindenthal.

Sciotoville Bridge

Seaka Bridge, Seaka, Quthing District, Lesotho

The bridge which carries the A2 road across the Senqu River (Orange River in South Africa), which was re-erected here in 1950, had been manufactured in England and first erected elsewhere in 1882. The steel structure consists of four trussed bowstring steel arches supported by intermediate reinforced concrete piers. In 1974 the bridge featured on Lesotho's 15c postage stamp.

Seán O'Casey Bridge, Dublin, County Dublin, Ireland

The bridge across the River Liffey between Custom House and City Quays, opened in 2005, is a twin

cantilever balanced cable-stayed footbridge which rotates on a central bearing to allow river traffic to pass. The overall length of the bridge is 98m, with each of the arms being 44m long, and it is 4½m wide. The bridge, which was designed by Brian O'Halloran Architects and O'Connor Sutton Cronin Consulting Engineers, won an International Architecture Award in 2007.

Sebara Dildiy (Broken Bridge), Motta, Amhara, Ethiopia

Fasilides, emperor of Ethiopia between 1632 and 1667, is credited with building the Sebara Dildiy over the Blue Nile in about 1660. It had two main arches of stone, one of which was destroyed during WWII by local patriots to hinder the advance of Mussolini's armies into Ethiopia (then called Abyssinia). For many years anyone wishing to cross the gap had to be pulled across by rope but in 2001 American brothers Ken and Forrest Frantz organised construction of a simple trussed girder span. A new wire cable suspension footbridge, funded by the charity Bridges to Prosperity, now crosses nearby.

Sebara Dildiy (Broken Bridge)

Second Penang Bridge, Bayan Lepas, Penang, Malaysia

Carrying the E28 expressway across the Selatan Strait part of the South China Sea, this bridge links Penang Island to the mainland. It was opened in 2014 after six years of construction and, at 24km long, is the longest bridge in Southeast Asia with 17km being over water. Because of its length, the bridge follows a double 'S' curve in plan to help prevent drivers becoming drowsy and includes seismic expansion joints. The 250m-long main span is a cable-stayed box girder structure with a harp layout of stays and provides 30m clearance above the shipping channel. The back-spans on each side are 118m long and the support pylons are 103m high.

Segovia Aqueduct, Segovia, Castile-León, Spain

This Roman aqueduct, the principal symbol of Segovia, was completed in the early second century and was used to supply water to the city until the middle of the nineteenth century, although part of the single-tier section had to be rebuilt following destruction by the Moors in 1072. The 63ft-high two-tiered main section has forty-four double arches and there are a further seventy-nine single arch spans, all built of granite. At its centre, a low upstand wall stretches across two full and two half-arches of the bottom arcade and originally held a memorial bronze inscription, although the letters were lost a long time ago. The aqueduct is part of the UNESCO World Heritage site at Segovia.

Segovia Aqueduct

Seishun Footbridge, Tsumagoi, Gunma, Japan

The Seishun Footbridge, opened in 2006, is an underslung self-anchored suspension bridge, with its deck structure being a U-shaped, prestressed concrete girder. This is supported by the two main suspension cables that are anchored to the ends of the 57m-long box girder, with the supports themselves, between the cables and the deck, being five intermediate inverted V-shaped pylons. A secondary suspension cable at deck level just outside the parapets was used to adjust the sag of the deck.

Séjourné Viaduct, Fontpédrouse, Pyrénées-Orientales, France

The granite railway viaduct here across a deep ravine containing the River Têt was designed by the architect Paul Séjourné and has one of the most distinctive elevations of any bridge structure. The lower part of the two-tier viaduct consists of a tall pointed arch over the ravine with small round-headed arches in the spandrels. This pointed arch supports the central pier of a four-span upper arched structure. On these upper-tier piers stand semicircular arches with a single opening across both spandrels. At either end of these four spans are main abutment piers, founded on the flatter ground above the ravine, which are flanked on each side by standard approach viaducts. The railway was opened in 1910.

Séjourné Viaduct

Sekong Bridge, Sekong, Laos

The road bridge at Sekong over the Mekong River carries National Road 16B connecting Sekong with Dakchueng. Opened in 2018, the extradosed 300m-long stayed structure has three spans that are 80m, 110m and 65m long.

Semey Bridge, Semey, East Kazakhstan, Kazakhstan

The 1,086m-long suspension bridge at Semey carries twelve motorway lanes of the M38 road across the Irtysh River and was built between 1998 and 2001. The 35m-wide steel box girder deck is supported on cables spanning 750m between 91m-high steel pylons.

Senbonmatsu Ohashi Megane Bridge, Osaka, Japan

The deck of the three-span steel box girder bridge across the Yodo River is 36m above water level and, in order to reach ground level, each end of the road descends on an approach viaduct. However, the unusual feature of these viaducts is that, as they descend, they describe two complete superimposed circles. Both levels of deck are supported on six radial frames, each with a top and middle cross beam. The bridge was completed in 1973.

Senegambia Bridge, Farafenni, North Bank Division, Gambia

Construction of this bridge over the Gambia River began in 2015 and was completed in 2019. It has an overall length of 1,900m with a longest span of 100m and its deck is 12m wide. It carries the Trans-Gambia Highway and N4 road and is part of what will become the 2,830-mile-long Trans-African Highway.

Senneville Bridge, Souillac, Savanne District, Mauritius

An early timber bridge over the des Anguilles River, about 3 miles upstream from its mouth at Souillac, was built in the early days after the British had captured the French colony in 1810, but the present structure dates from 1878. Originally built to carry trains, the three-span bridge has two tapering octagonal collared piers supporting plate girders and is 241ft long. The A9 road from Souillac to Britannia now uses the bridge.

Seonimgyo Bridge, Seogwipo, Jeju Volcanic Islands, South Korea

Seonimgyo Bridge is a steel deck arch structure that is 128m long and 4m wide and carries pedestrians 78m over the stream between two parts of the Cheonjeyeon Waterfall. It is also called the **Seven Nymphs Bridge** after the Korean legend of seven beautiful nymphs descending from heaven at night. Between the top and bottom booms of the trusses on either side of the arch there are elaborate depictions of the seven nymphs, each nymph being 7m long and playing a musical instrument. The bridge is lit at night by traditional-style stone lanterns. It opened in 1984 and is now part of the Jeju World Heritage site.

Seonimgyo Bridge

Seonyu Footbridge, Seoul, Gyeonggi Province, South Korea

This 469m-long footbridge across the Han River, which was opened for the 2002 World Cup, links Sunyudo Island to Seoul and was designed by Rudy Ricciotti. The prestressed concrete structure is made of special high-performance concrete with 2 percent of its mass consisting of 0.2mm diameter reinforcing steel fibres that are 13-15mm long. The 120m-long main span is an arched rib, which is 4.3m wide and 1.3m deep, made from six precast sections. At the crown of the arch, pedestrians walk on a 4.3m-wide deck following the arch curve but, from the quarter-span points, the deck is supported on steel beams spanning between the arch rib and vertical and angled columns based on the arch abutments. There are post-and-beam approach spans at each end of the bridge beyond the abutments. Tuned mass dampers were fitted to reduce vibrations.

Seonyu Footbridge

Seri Wawasan Bridge, Putrajaya, Malaysia

This 37m-wide cable-stayed road bridge across Putrajaya Lake, which was opened in 2003, has an overall length of 240m and its 168m-long main span is supported by thirty pairs of cable stays in a semi-fan layout. The 96m-high main pylon is inclined forward over this span and is balanced by twenty-one pairs of back stays together with unusual sail-shaped steel tie-back arches. The bridge is illuminated by coloured light displays at night.

Seri Wawasan Bridge

Severan Bridge, Arsameia, Adiyaman Province, Turkey

The Severan Bridge across the Chabinas River, named after its builder the Roman emperor Septimius Severus, was completed in about 200BCE and is therefore one of world's oldest bridges. It has an overall length of about 120m with a 34m-long nearly-semicircular main arch and is 7m wide. There is also a small flood-relief arch at one end. The parapet on each side is stepped and the bridge is decorated at each end with a pair of Roman columns about 10m high. The structure was renovated in 1951 and again in 1997.

Severan Bridge

Seythenex Bridge, Albertville, Haute-Savoire, France

Considered to be a pioneering structure when it was completed in 1912, this bridge consists of four stone arches, the longest spanning 41m, that are covered in reinforced concrete. It is 125m long and crosses 53m above a narrow valley. The largest of the spandrel spaces between the upper surfaces of these arches and the underside of the deck is filled by twelve spandrel posts with a horizontal mid-height tie, all in reinforced concrete. The bridge now carries cyclists and pedestrians only.

Shadorvan Bridge, Shushtar, Khuzestan Province, Iran

This ruined bridge (also called **Valerian's Bridge** or **Caesar's Bridge**) was built in the third century CE and stands on a roughly 500m-long overflow weir across the Karun River. It carried the

Shadorvan Bridge

Shaharah Bridge

rising at each end, the bridge's elevation looks somewhat like an hourglass on its side. There is an arched stone gateway at one end protecting access into the town.

highway between the ancient city of Ctesiphon and Pasargadae. Originally, the arched bridge had more than forty spans but it partly collapsed in 1885 and only twenty-five arches still stand, these spanning from 6m to 9m between piers with thicknesses of up to 6m. In 2009 the bridge was designated a UNESCO World Heritage Site.

Shah Amanat Bridge, Chittagong, Bangladesh

Completed in 2010, this prestressed concrete box girder bridge carries the country's N1 highway over the Karnaphuli River. With an overall length of 950m, the structure consists of three 200m-long stayed main spans and a 115m-long side span at each end. Apart from four vehicle lanes and two footways, the 24m-wide bridge unusually also has two lanes dedicated to rickshaws and similar local transport. Three of the piers consist of four, short and chunky inclined legs. The crossing was designed by High-Point Rendel.

Shaharah Bridge, Shaharah, Amran Governorate, Yemen

At more than 2500m above sea level, the seventeenth-century arched stone footbridge here, designed by the architect Salah al-Yaman, spans 20m across a 200m-deep gorge between two mountain ranges and connects steeply-stepped footways up the mountainsides at either end. Since only the central part of the 3m-wide deck is level, with stepped approaches down to and up from it, and the parapets follow this by

Shahi Mughal Bridge, Jaunpur, Utter Pradesh, India

Like the Chuha Gujar Bridge at Peshawar in Pakistan (q.v.), this bridge is in the Mughal tradition of having multiple pointed stone arches, usually twelve but in this case fifteen. Completed in about 1567 to a design by the Afghan architect Afzal Ali, the Shahi Bridge crosses the Gomti River and is also sometimes known as the **Akbari Bridge**. The pointed cutwater at the end of each pier is surmounted by a tall and colonnaded, elongated octagonal pavilion with a domed roof. The structure was rebuilt after seven arches collapsed during an earthquake in 1934 and is 654ft long and 26ft wide. A new bridge was built

Shahi Mughal Bridge

nearby in 2006. Rudyard Kipling's poem *Akbar's Bridge* mentions the bridge and makes the point that a bridge is of more use than a mosque.

Sheikh Jaber Causeway, Kuwait City, Kuwait

With its full name of Sheikh Jaber Al-Ahmad Al-Sabah Causeway, the massive US$3 billion construction project to provide road links across Kuwait Bay from Kuwait City north-eastward to Silk City in Subiya, Northern Kuwait (the 36km-long Main Link) and westward to Doha (the 12km-long Doha Link) lasted from 2013 to 2019. The causeway carries two 17m-wide carriageways, each of three lanes, and the structure for the low-level bridges mainly consists of 60m-long precast concrete girders. There are also two artificial islands. The keynote feature, however, is the 177m-span cable-stayed bridge on the Main Link, the 151m-tall curved pylon for which was inspired by the traditional Kuwait sailing boat, that provides 23m vertical clearance at high tide. This pylon consists of a concrete column connected by horizontal and diagonal steel members to a vertical steel truss that takes the place of the normal back stays. The causeway was designed by Systra and built by a consortium led by the Hyundai Engineering and Construction Company.

Sheikh Zayed Bridge, Abu Dhabi, United Arab Emirates

Designed by Hyundai Engineering and Construction Company to look like the undulating sand dunes of the desert, this 842m-long road bridge crosses Maqta Channel to connect Abu Dhabi Island to the mainland. A series of four steel arches spans between concrete piers, the main arch spanning 140m with its apex 63m above sea level, and the 61m-wide reinforced concrete deck carries two four-lane carriageways and footways. The bridge, which was opened in 2010, won the International Road Federation's Global Road Achievement Award.

Shinkyo (Sacred) Bridge, Nikko, Tochigi Prefecture, Japan

The timber bridge across the Daiya River, which was built in 1636 and marks the entrance to the shrine that was established here in 767, is considered to be one of Japan's most beautiful bridges. Passage over the structure was originally restricted to high personages. The bridge is 28m long and 7m wide and consists of arched timber beams standing on two intermediate piers. It was listed as a World Heritage site in 1999.

Shinkyo (Sacred) Bridge

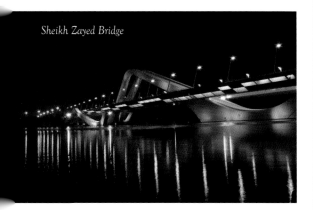

Sheikh Zayed Bridge

Shiosai Bridge, Shizuoka, Honshu, Japan

This unusual prestressed concrete bridge for cyclists and pedestrians, which was completed in 1995, consists of a four-span continuous stressed-ribbon and crosses the estuary of the Kiku River. The two central spans are each 61m long and these are flanked at each end by a 55m-long span. The flat 264m-long and 3m-wide deck is supported off the sagging line of this ribbon by six intermediate I-shaped concrete posts.

Shiosai Bridge

Sidi Rached Viaduct, Constantine, Constantine Province, Algeria

The 107m-high road bridge across the River Rhummel gorge was the tallest bridge in the world when it was built between 1908 and 1912. It is 12m wide and has an overall length of 447m with twenty-seven arches. Its main span is 70m long and has arcaded spandrels containing four

arches on each side. The ground on which it stands is unstable and the bridge was damaged by a landslide in 2008.

Sierre Footbridge, Sierre, Valais, Switzerland

This unusual asymmetric structure over the River Rhône was built between 1997 and 1998 to replace an earlier timber footbridge. It is effectively half a suspension bridge: it has a single V-shaped pylon at one end with each projecting leg of the V supporting a separate suspension cable. Radial hangers from these cables support the outer edges of the 54m-long arched tubular steel and timber deck, the remaining 14m of the total 68m-long span being cantilevered out in concrete from the abutment on the other bank of the river.

Sigiri Bridge, Sigiri, Busia County, Kenya

The three-span Sigiri Bridge over the Nzoia River is a 9m-wide and 100m-long composite deck structure consisting of twelve steel plate girders with their top flanges encased in reinforced concrete. Designed and built by the Chinese Overseas Construction & Engineering Company, the bridge collapsed during construction in 2017 when one end of the 50m-long central span sheared off near one of the river support piers injuring twenty people. The structure was safely completed and opened in 2018.

Sinamalé Bridge, Malé, Northern, Maldives

The country's first inter-island sea bridge connects the Maldives island of Malé to its airport on Mulhulé Island and Hulhumalé. The structure, which has an overall length of 2,100m and a width of 20m, was built between 2015 and 2018. After approach viaducts at each end, there are four main arched spans and the three intermediate piers contain large openings between the spans. At the Malé end of the bridge there is a dramatic Islamic gateway arch supported by stays from flanking minarets. The project was built by the Chinese and has been called the 21st Century Maritime Silk Road.

Singapore River Bridges, Singapore

There are a number of interesting footbridges across the Singapore River including the following. The **Cavenagh Bridge**, built as a road bridge in 1869, is an early type of chain stay structure. The piers are buttressed masonry columns topped by unusual-looking enclosed cappings. The bridge remained in use as a pedestrian bridge after its traffic was transferred to the replacement Anderson road bridge, opened in 1910. The **Alkaff Bridge** of 1999 is a 55m-long bow-shaped truss bridge, but with the arched section beneath the deck level built to replicate the appearance of earlier local boats called tongkangs. The **Helix Bridge**, which was opened in 2010 to complete a pedestrian way around Marina Bay, is 280m long and has a tubular truss in the form of the DNA double helix. This truss has five spans: two 45m-long end spans flanking three 65m-long main spans. The curved **Jubilee Bridge**, built in 2015 to take pedestrians off the adjacent Esplanade road bridge, is a 220m-long, three-span, post-tensioned reinforced concrete box girder which is only 1.5m deep at mid-span.

There have also been a number of road bridges over the Singapore River including the following. The iron **Elgin Bridge**, which replaced an earlier wooden footbridge built in 1824, was built in 1862 and was itself replaced by a concrete arched structure in 1929. This is 46m long and 25m wide. The **Esplanade Bridge** across the mouth

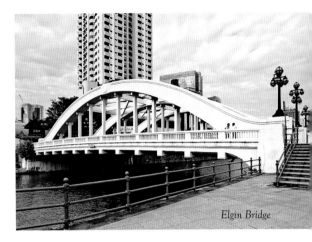

Elgin Bridge

of the Singapore River was opened in 1997 and is 280m long and 46m wide.

Sino-Korean Friendship Bridge, Sinuiji, North Pyongan, North Korea

An earlier bridge across the Amnok River on this site, opened in 1911 and now known as the **Yalu River Broken Bridge**, was destroyed by American bombing in 1951 during the Korean War although the ruins remain. The present Friendship Bridge, which was built by the Japanese Army that occupied Korea between 1937 and 1943, also suffered severe war damage but was repaired and re-opened in 1953 after the end of the Korean War. It is a 3087ft-long combined steel girder and suspension bridge which carries a single rail track and a single carriageway road but no footpaths. The **New Yalu River Bridge**, a cable-stay bridge 9,800ft long, opened in 2020.

Sitra Causeway Bridges, Bahrain

In the 1970s a 2.4km-long causeway was built across Tubli Bay between Bahrain main island of Manama and the smaller island of Sitra and this included two concrete bridges, each carrying twin two-lane carriageways. Between 2006 and 2010 these bridges were replaced by new twin four-lane post-tensioned concrete structures designed by international consulting firm Cowi Almoayyed. The north bridge is 200m long and 58m wide and the south bridge is 400m long and 55m wide.

Helix Bridge

Skeiðarárbrú Bridge, Kirkjubæjarklaistur, Eastern Region, Iceland *

The bridge that was built across one of the outlets from the Skeiðarárbrú glacier to complete the country's Route 1 ring road was opened in 1974. The 904m-long multi-span structure, the longest bridge in the country, consisted of steel beams supported on concrete piers and carried a single carriageway only, with passing places above the piers. The upstream ends of these piers were shaped to force floating ice sheets upwards so they would crack and break. However, the bridge was badly damaged in 1996 by house-sized floating ice boulders, a new bridge was built on an alternative route, and all that now remains of the old structure are bits of twisted girder.

Skjálfandafljót Bridge, Fosshóll, North Iceland, Iceland

The first bridge over the Skjálfandafljót River, a timber structure built in about 1883, was replaced in the 1930s by a simple N-braced truss spanning between stone piers, each of which has a tall internal arch. This bridge was itself superseded in 1972, although it continues in use as a footbridge. The new structure is a beam bridge with two raking ladder-like inclined struts and carries a single-lane road.

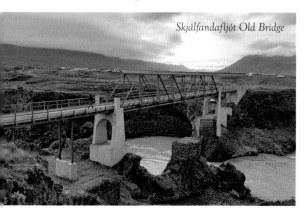

Skjálfandafljót Old Bridge

Skopje Aqueduct, Skopje, North Macedonia

Originally stretching over about 6 miles, the surviving part of the aqueduct is now only about 386m long and consists of fifty-five semicircular main arches with forty-two smaller pointed arches above the piers. The stonework of all the arch rings is very distinctive and the main arches are generally about 5m high with a maximum height of 16m. The aqueduct, which may have been built by the Romans in the first century CE or by the Turks in the sixteenth (although some of the mortar has been carbon-dated to the seventeenth), remained in use until the eighteenth century.

Skopje City Centre Footbridges, Skopje, North Macedonia

Two adjacent footbridges cross the Vardar River in the city centre. The **Eye Footbridge**, built in 2013 leads to the classically-styled Archaeological Museum and is decorated with statues of historical figures. It is believed to be in poor structural condition.

The 83m-long **Art Footbridge** splits into two separate lengths with a 12m-wide focal point link at midspan. It is decorated with twenty-nine one-and-a-half-times life-size statues of artists and musicians, one on the centre link and the others standing on plinths between sections of balustraded parapet, that are the work of Kosta and Viktorija Mangarovski. The bridge structure itself, built in 2014, is a three-ribbed concrete arch with open spandrels.

Skopje City Art Footbridge

Sky Bridge, Vancouver, British Colombia, Canada

Built between 1987 and 1990, this cable-stayed bridge carries three railway lines 45m above the Fraser River: two for the SkyTrain service and a third set of tracks for maintenance use. The

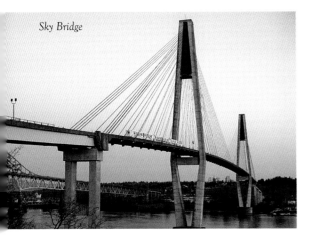

Sky Bridge

supporting pylons are 123m tall, the overall length of the bridge is 616m and its longest span is 340m.

Slauerhoff Flying Drawbridge, Leeuwarden, Friesland, Netherlands

The unusual Slauerhoff bascule tail bridge over the Harlingervaart River was designed by Van Driel Mechatronica and opened in 2000. The concept is simple: a movable deck 15m square is supported on two parallel beams, with the centreline between these beams running along the diagonal of the movable square deck. These beams are then attached to hinges at right angles to the axis. This means that after the square deck section has been raised, it looks as if it were effectively standing on one of its corners. It is claimed that this makes opening and closing the bridge quicker, and the design allows for these operations to be fully automatic.

Slauerhoff Flying Drawbridge

Slaves Bridge, Cuilapa, Santa Rosa, Guatemala

The Spanish conquistadores arrived here in the early sixteenth century but, although they did not complete building the bridge over the Los Eslavos River until 1592, the structure is still one of the oldest in Central America. With eleven arches, it is 117m long and is now bypassed by a modern steel bridge.

Slender West Lake Bridges, Yangzhou, Jiangsu, China

There are several interesting bridges in the Slender West Lake area of Yangzhou. Best known of these is the **Wuting (Five Pavilions) Bridge**, which was built in 1757. This has a semicircular arched bridge spanning between two elongated abutments that extend outward from both faces of the arch. These provide bases for four pagoda-style roofs standing on timber posts at each corner and themselves include a small semicircular arch in each of the projecting faces. The fifth and double-tiered pavilion roof stands over the space above the main water arch and between the other roofs. Because of these five elements this structure was originally called the Lotus Bridge. All five of the timber pavilions were destroyed by fire during the Taiping Rebellion of 1850–64. Nearby is the **Nine-Turn Bridge**, a zigzag bridge with ten deck sections on which the stone balustrade panels are pierced by butterfly-shaped openings.

Five Pavilions Bridge

Smithfield Street Bridge, Pittsburgh, Pennsylvania, USA

The first bridge over the Monongahela River in Pittsburgh was a timber structure built in 1818, but this was destroyed by fire in 1845. It was followed by an eight-span wire rope suspension bridge built by John A. Roebling and by the twin-span steel lenticular truss designed by Gustav Lindenthal that used the piers from Roebling's bridge. Built between 1881 and 1883, this is now Pittsburgh's oldest bridge. It has an overall length of 1,184ft with two 360ft-long spans, was widened in 1889 and again in 1911 and given a new deck in 1995. The bridge has castellated and pinnacled triumphal gateways and is a National Historic Civil Engineering Landmark.

Smithfield Street Bridge

Smolen-Gulf Bridge, Ashtabula, Ohio, USA

This modern timber covered bridge, which replaced an earlier iron bridge, was built in 2008. It crosses 93ft above the Ashtabula River and has four spans between reinforced concrete column-and-beam piers. The bridge, a Pratt truss structure, was designed by John Smolen and is 613ft long with an internal height of 14ft. It has a clear internal width of 30ft and 5ft-wide covered footways each side. It is considered to be the longest covered bridge in the world.

Soca Railway Bridge, Solkan, Gorizia, Slovenia

The Bohinge Railway running between Vienna and Trieste crosses the River Soca by a single-track line on a bridge designed by the Austrians, architect Rudolf Jaussner and engineer Leopold Oerley. This structure has an overall length of 220m, which included a main segmental arch over the river that spanned 85m and carried five secondary arches in each spandrel. This arch was destroyed by retreating Austro-Hungarian forces in 1916 but was rebuilt by Italian railway engineers between 1925 and 1927. In this rebuild the number of secondary arches in each spandrel was reduced to four. The bridge was attacked again during the Second World War and was hit by a bomb but this failed to explode. The 85m-long main arch is considered to be the longest stone railway arch in the world and is now a Slovenian technical monument.

Soca Railway Bridge

Sonjuk Bridge, Kaesong, North Hwanghae, North Korea

This little three-span stone slab clapper bridge, about 7m long and 2½m wide, was built in 1290 and in 1392 was the site of the assassination of statesman Jeong Mong-ju. In about 1700 a descendant built low stone posts and parapet rails around the bridge, to keep it as an unused and sacred memorial. He also built a replacement bridge alongside but since 1780 the whole site has been a national monument and closed to traffic. It is now a UNESCO World Heritage Site.

Souillac Bridge, Souillac, Lot Department, France

The road bridge across the Dordogne River in Souillac, designed by Louis Vicat and built between 1812 and 1824, is considered to be one of the first post-Roman uses of mass concrete in bridges. (However, it should be noted that the binder was different from the later Portland cement binder patented by Joseph Aspdin in 1824 and used in the Garden Bridge at Grenoble – q.v.) The overall length of the bridge is 180m and it has seven semi-elliptical arched spans each 22m long. It is classical in style with rounded cutwaters and a cornice below the parapet.

Source of the Nile Bridge, Njeru, Buikwe District, Uganda

Also known as New Jinja Bridge, this cable-stayed structure across the Victoria Nile River was opened in 2018 to replace the **Nalubaale Bridge** of 1954. Its two pylons are in the shape of inverted Ys and it has an overall length of 525m and width of 23m. Interestingly, it is not supposed to be used by people riding motor bikes or cycles, who are restricted to the old bridge.

South Viaduct, Savannah, Georgia, USA

This bridge in Savannah was completed in 1852 to carry the Georgia Central Railroad over the Savannah & Ogeechee Canal and West Boundary Street, near what is now the Georgia State Railroad Museum, and continued in use until 1971. Designed by the engineer Augustus Schwaab

Central of Georgia South Viaduct (Emily Beck)

(1824–99) to replace an earlier temporary trestle structure, the structure is 30ft wide and has four 48ft-span brick arches, which at the time were believed to be the largest brick arch spans in the country. Interestingly, these are semi-elliptical – a shape of arch not commonly used on American railway viaducts. The bridge is also distinctive because the brickwork is unusually decorative, including such features as corbel tables, pilasters and blind parapet arcades. A similar second bridge, the **North Viaduct**, was completed in 1860 and is now used a footbridge connecting parts of the Savannah College of Art & Design campus.

South Washington Street Bridge, Binghamton, New York, USA

Completed in 1887, this road bridge over the Susquehanna River has three identical lenticular trusses each spanning 161ft. The unusual continuing sinusoidal lines of successive arch and tie make a very attractive picture and the portal bracing frames at each end of the bridge are topped by a plaque commemorating the design and construction firm, the Berlin Iron Bridge Company.

Southern Bridge, Riga, Latvia

The Southern Bridge in Riga across the Daugava River was finished in 2008, although not all access roads were completed until 2013. The prestressed concrete extradosed bridge has five main spans, each 110m long, and two end spans. It is 34m wide and there is 13m clearance above water level. Its main feature is the arrangement of the eight cable stays on each side of the six support pylons, the tops of which are at right angles to the bridge's axis. However, from here the stays splay out to anchorages along the centre line of the deck. The bridge was hugely expensive, costing more than 800 million Euros, and was facetiously nicknamed 'The Golden Bridge'.

Spean Praptos Bridge, Chi Kiaeng, Siem Reap, Cambodia

The 87m-long late-twelfth-century structure across the Chi Kreng River, also known as **Dragon Bridge** or **Kampong Kdei Bridge**, is

Spean Praptos Bridge

Spectacle Bridge

a rare survivor from the Khmer empire, being built during the reign of King Jayavarman VII (1125-1218). It consists of twenty-one narrow corbelled laterite stone arches, the face on one side being battered back, and is 14m high. It was once the longest corbelled arch bridge in the world and remained in use as part of Highway 6 until a new bridge was built a short distance away in 2006. Each edge of the deck, which is 14m wide, is safeguarded by a low, continuous pillar-and-lintel open balustrade, the ends of which are marked by carved figures of a naga – a multi-headed mythical creature. Below each naga is a further carving of a guardian figure with sword. The bridge features on Cambodia's 5,000 riel banknote.

Spectacle Bridge, Lisdoonvarna, County Clare, Ireland

The bridge here was built in 1875 by the County Surveyor John Hill (1812–1894) to carry traffic 46ft above water level where the River Aille runs in a deep and narrow gorge. Effectively, the road is supported on a solid masonry wall about 72ft long which has, at its base, an 18ft-span semicircular stilted-arch opening through which the river passes. Standing on this arch, and directly above it, is a further 18ft diameter circular tunnel through the wall, which serves to reduce the amount of masonry needed for the whole structure. The appearance of the two

arches when seen from the river have led to the name Spectacle Bridge being adopted.

Speicherstadt Island Bridges, Hamburg, Lower Saxony, Germany

The Speicherstadt Harbour in Hamburg, which became a UNESCO World Heritage Site in 2015, includes a number of interesting bridges. **Köhlbrandbrücke**, completed in 1974, has an overall length of nearly 4km long and its main 325m-long cable stay span is supported from 135m-high pylons. The 44m-long steel arched **Brooksbrücke**, which was built in 1887, has a tall statue at each corner. **Poggenmühlen-Brücke** has twin bridges linking an island to each bank. **Ellerntorsbrücke** was built in 1668 and has three semicircular masonry arches. **Lombardsbrücke**, which carries both road and rail links, has long, low, semi-elliptical arch spans. Built in 1884, **Feenteichbrücke** has a brick arch with a span of

8m and decorative columns at each corner that support streetlamps. **Baakenhafenbrücke**, one of the newest bridges, was opened in 2013 and is 165m long. The central 30m-long section can be floated out on a barge if a larger vessel needs to pass through.

Spring Creek Trestle Bridge, Hanover, Montana, USA

One of the most outstanding of the trestle bridges built by American railroad companies is this one that the Burlington Northern Railway constructed over the Big Spring Creek in 1930 and that still stands. Designed by the engineer H.S.Loeffler, it is 1,392ft long and is six tiers high.

Stanley Bridge

Stadlec Chain Bridge

Spring Creek Trestle Bridge

Stadlec Chain Bridge, Stadlec, South Bohemia, Czech Republic

Originally built in 1848 to provide a road crossing over the Vltava River at Podolsko, this chain bridge was dismantled in 1960 into 2,000 stone blocks and 1,100 pieces of wrought and cast iron which were placed into storage. Eventually the bridge was re-erected 18km from its original site to cross the Luznice River in Stadlec, and was recommissioned in 1975. Its total length is 157m and its timber deck is 6m wide. The twin bar chains on each side of the deck, mounted one above the other, are supported by classical-style stone towers, each with a single semicircular arch for the traffic to pass through. A new timber deck was installed in 2007 and the structure is now believed to be the only preserved chain bridge in the country. It is on the Czech List of National Cultural Monuments and has also appeared on the Kc8 stamp issued in 1999.

Stanley Bridge, Alexandria, Alexandria Governorate, Egypt

This bridge was built in the 1990s across Stanley Bay to improve the corniche. It has six spans and its overall length is 400m. The bridge is decorated with tall, three-storey towers at each corner which have semicircular arches at pavement level, pointed arches above and are topped by open colonnaded pavilions.

Stanley Bridge

Stari Most, Mostar, Herzsegovina-Neretva Bosnia-Herzegovina

The Ottoman architect Mimar Hajrudin was commissioned by Sultan Suleyman to replace an existing but unsatisfactory timber bridge over the River Nerevta in Mostar. His first attempt failed but in 1566 he successfully completed the new structure. Following a battle in 1652, in which the Ottomans defeated the Venetians, defensive towers were built just off each end of the bridge. In 1993 it was deliberately destroyed by shellfire during the Serbian-Muslim War and rebuilt in 2004. The bridge (its name meaning Old Bridge in English) has a roughly segmental stone arch spanning 100ft with its crown, from which boys dive or jump into the river to entertain tourists, being 66ft above the water level.

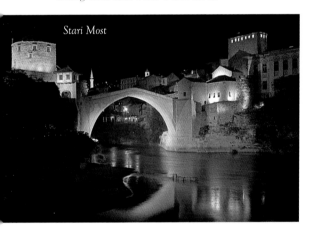

Stari Most

Starrucca Viaduct, Lanesboro, Pennsylvania, USA

When it was completed by the New York & Erie Railroad in 1848 this was the largest stone railway viaduct in the world with a total length of 1,040ft and a maximum height of 100ft. It was also one of the first in which the pier foundations included concrete. There are seventeen nearly-semicircular spans of 50ft and there are voids within the spandrels to reduce the dead loading on the arches. The viaduct was built with a 26ft-wide deck to carry a single broad-gauge track but replacement twin tracks were installed in 1886. The structure was listed as a National Historic Civil Engineering Landmark in 1973.

Starrucca Viaduct

State Castle Moat Bridge, Ceský Krumlov, South Bohemia, Czech Republic

The castle in this town overlooking the Vltava River was founded in the thirteenth century and a timber bridge over the western side of the moat was first built in the fifteenth century. The present structure (also called **Cloak Bridge** in English) was reconstructed in the period 1764–1777. It consists of three tiers of tall semicircular masonry arches which support an open deck, with three further enclosed passageways above. The bridge is 29m long and 6.4m wide and the main deck is 39m above the moat. The castle is now a UNESCO World Heritage site.

State Castle Moat Bridge

Station Bridge, Beersheba, Northern Negev, Israel

This 210m-long enclosed footbridge above railway tracks provides a link between Beersheba

Station and a new Advanced Technologies Park. It has two spans of 110m and 70m between V-shaped piers. Each span is supported by a lenticular steel truss on either side of the deck, with the plane of these trusses tilted inwards, and in elevation the bridge looks like a pair of eyes. The structure, which was completed in 2016, was designed by Rokach Ashkenazi Engineers with Bar Onan Architects and won the prize for the best long-span footbridge at the Footbridge 2017 Conference in Berlin.

Stauffacher Bridge, Zurich, Switzerland

The structural engineering design of this 40m-span, mass concrete, three-hinged segmentally arched road bridge across the River Sihl, completed in 1899, was by Robert Maillart. Maillart had just joined the city's public works department and this was his first bridge design. The city authorities, however, insisted that the bridge should be suitably ornamented by their architect Gustav Gull. They were probably not disappointed: the structure was completely clad in masonry, the parapets were supported on corbels, there was a grand escutcheon over the central hinge and there were four tall corner pylons capped by seated lions and decorated with elaborate lamps on each pylon face.

St-Léger Viaduct, Saint Chamas, Bouches-du-Rhône, France

The stone railway viaduct here crosses the Touloubre River and was built in 1848 to a design by the engineer Gustave Desplaces. The 31m-high

St-Léger Viaduct

structure is 385m long and contains seventy-four arches, the longest spanning 6m, and is unusual for the distinctive arrangement of these arches. These are semicircular and span between alternate piers, this interlocking producing a lens-shaped aperture above each pier. The piers themselves are two-staged, the tall lower stage terminating in an impost from which the arch rings spring. A narrow steel footway also spans between the bottom stages of the main piers.

Stone Bridge, Skopje, Greater Skopje, North Macedonia

The Stone Bridge across the Vardar River, which is also known as **Dusan Bridge** after a king of Serbia, was built between 1451 and 1469 on earlier Roman foundations. Originally it was 214m long and 6.3m wide and had thirteen semicircular arches spanning between pointed cutwaters. However, since the river has been narrowed within embankments, only four of these now cross water, the others providing walkways for pedestrians between terraces and gardens lying behind the embankment walls.

Stone Bridge, Zamora, Castile-León, Spain

It is believed that a Roman bridge over the Douro River in Zamora was destroyed in the tenth century and that a new structure was in place by the end of the thirteenth century. This had twenty-two pointed arches and was protected by two defensive towers flanking the entrance. Later, these towers were demolished and a triumphal archway across the road was built topped by an elegant Flamenco spire. However, during major reconstruction between 1905 and 1907, this was removed and the

Stone Bridge

bridge was rebuilt with an overall length of 280m containing fifteen pointed arches and including high-level flood relief arches over the piers. The old battlements on the parapets were also removed. When the new prestressed concrete Poets Bridge was opened upstream in 2013, the old structure was restricted to pedestrians only.

Storseisundet Bridge, Kristiansund, Møre og Romsdal, Norway

The bridge here across the Storseisund was constructed between 1983 and 1989 as part of a project to develop tourism in the archipelago of islands near Kristiansund by building an 8km-long connecting road. This bridge, which is 260m long, has a main arch spanning 130m with propped-cantilever half-arched back spans. The maximum clearance beneath the arch is 23m. Because of its curved and steep approach, the bridge has gained a reputation as being scary to drive over.

Strimka Lan Bridge, Kamianets-Podolski, Khmelnytskyi, Ukraine

The road bridge across the River Smotrych here was opened in 1973 and has been a national architectural monument since 2008. With an overall length of 379m, it is a beam bridge supported by a linked pair of inclined struts at each end, the span between the points where the struts meet the underside of the deck being 174m.

Sulphite Railroad Bridge, Franklin, New Hampshire, USA

The Sulphite covered bridge over the Winnipesaukee River was built in about 1897 for the Boston & Maine Railroad. The 234ft-long structure consists of three Pratt truss spans, the central one being 180ft long, and the bridge is distinctive for having the rail track run on top of the structure rather than through it, the barn-like enclosure being to protect the trusses from the weather. The railway closed in 1973 and the bridge is believed to be the world's only surviving covered bridge of this type.

Sultan Omar Ali Saifuddien Pedestrian Bridge, Bandar Seri Begawan, Brunei-Muara, Brunei

The S-shaped curved footbridge here over the River Kedayan connects the city centre to the Seri shopping centre. The interesting feature about this bridge is the crescent moon shape of the steel sculpture in the middle of the bridge.

Sundial Bridge, Redding, California, USA

This unusual bridge, designed by Santiago Calatrava and opened in 2004, crosses the Sacramento River and carries a footway and cycleway to link the north and south areas of the Turtle Bay Exploration Park. The cable-stayed structure has a single 217ft-high mast, which is angled at 42° from the horizontal (its weight meaning that backstays are not required) and stands on a tripod base. The mast also acts as the gnomon for a giant sundial. The translucent deck, which is supported on a triangular truss beneath it, is 700ft long and 23ft wide and is illuminated from below at night.

Sundial Bridge

Sungai Kebun Bridge, Bandar Seri Begawan, Brunei-Muara, Brunei

The single-pylon cable-stayed bridge, which carries a dual carriageway road across the Brunei River, was built between 2014 and 2017. At the time of its completion its 622m overall length was claimed to be the world's second longest bridge of this type. The main span is 300m long and the pylon

Sungai Kebun Bridge

is 157m high. At its top is an Islamic dome that is nearly 9m in diameter and weighs 9½ tonnes.

Suramadu National Bridge, Surabaya, East Java, Indonesia

Opened in 2009, this bridge across the Teluk Lamong Channel links Surabaya, Indonesia's second city and once the main city of the Dutch East Indies, to the island of Madura. With an overall length of 5,438m it is the longest bridge in Indonesia. The main part of the crossing over the shipping channel has a cable-stayed central span 434m long with 192m-long side spans and these are supported by piers 146m high.

Suransuns Bridge, Thusis, Graubünden, Switzerland

The stress-ribbon footbridge across the Hinterrhein River in the Viamala Gorge spans 40m. The 1.1m-wide deck is made of granite slabs, which are each 60mm thick and stretch 2,150mm along the bridge length, that are laid over prestressed flat steel bars. Slim steel handrails protect the sides of the deck. The bridge, which was opened in 1999, was designed by Jürg Conzett.

Suspension Bridge, La Ferme, Rodrigues Island, Mauritius

This wire rope suspension bridge spans 350m across the Cascade Pistache River ravine and its deck is only about 1ft wide. At mid-span there is a platform suspended below the walkway which is used as the starting point for a pendulum jump into the river.

Svilengrad Bridge, Svilengrad, Haskova Province, Bulgaria

Svilengrad Bridge was the first building designed by Mimar Sinan, the famous Ottoman architect, and was built between 1528 and 1529 to become a major link between Europe and Asia. With an overall length of 295m, it crosses the Maritsa River in twenty-one stone arches, the biggest being 18m long, and is 6m wide. About halfway along the bridge on one side a large stone memorial stands on an enlarged pier and extends high above the parapet wall. In 1766 some of the arches were brought down by a flood and rebuilding was not completed until 1809. Further damage was caused in 1912 when the Ottoman army tried, but failed, to demolish the bridge during the First Balkan War.

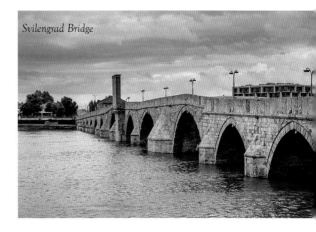

Svilengrad Bridge

Svinesund Bridge, Svinesund, Västra Götaland County, Sweden

Svinesund Bridge, completed in 2005, crosses the Iddefjord to connect Norway and Sweden. It has a reinforced concrete two-hinged hollow half-through arch which spans 247m and has a rise of 30m. On each side of this arch, and suspended from it on cable hangers, are twin-cell steel box girders that carry the four-lane E6 route 92m above water level. This bridge was designed and built by Bilfinger Berger.

Old Svinesund Bridge, an earlier structure about 800m away which now carries minor traffic, was built between 1939 and 1946, construction work being delayed when, in 1942, lightning set off explosives designed to deny use of the bridge to an enemy. The bridge has an overall length of 420m with semicircular stone arch approach spans flanking a main span. This 165m-long main span is a reinforced concrete arch with the road deck supported above it on spandrel posts.

Swandbach Bridge, Hinterfultigen, Bern, Switzerland

Designed by Robert Maillart and considered to be his masterpiece, this deck-stiffened reinforced concrete arch has been described as 'a work of art in modern engineering'. Its 150mm-thick deck, which carries a 4.9m-wide road, is curved horizontally and supported by eight 160mm-thick spandrel cross walls standing on a slender polygonal arch that spans 37m. This supporting arch, which is only 200mm thick, is curved on plan along its inside edge but its outside edge follows a straight line across the steeply-sided valley. This results in its width

varying from 4.2m to 6m, thereby helping to resist centrifugal forces generated by the road traffic. The structure, which was completed in 1933, is listed by the Swiss authorities as a heritage structure of national importance.

Swiss Bay Footbridge, Onsov, South Moravia, Czech Republic

The interesting Swiss Bay suspension footbridge, designed by civil engineer Jiri Strasky to cross the eastern end of Vranov Lake and replace a ferry service, was completed in 1993. The twin suspension cables span 312m between common suspension points at the top of each of the inverted V-shaped pylons and, at the bottom of their catenary arc, are on either side of the deck. The back-stay length is 30m. The stress ribbon deck, which is both slightly arched and slightly waisted, consists of precast concrete segments that are only 0.42m thick. This gives the bridge a span:depth ratio of 630:1, making it one of the most slender in the world. The bridge also carries gas and water pipes.

Sydney Harbour Bridge, Sydney, New South Wales, Australia

The iconic 'coathanger' bridge across Sydney's famous harbour is a spandrel-braced, two-pin, through-arch steel structure with a main span of 1,650ft. Designed by the engineer J.J.C. Bradfield and built by the UK firm of Dorman, Long & Co, it was opened in 1932. Its overall length is 3,770ft and it is still the world's tallest steel arch bridge, being 440ft from water level to the highest point of the structure. Its 160ft-wide

Sydney Harbour Bridge

Swandbach Bridge

deck carries eight traffic lanes and two rail lines and the overall weight of the steelwork is 52,800 tonnes. The main span is bookended by the masonry towers above each abutment. The Bridge Climb Sydney, up 1,390 steps on the eastern side of the bridge to a climax at 134m above water level, was opened in 1998.

Symingyi Viaduct, Jingpeng Pass, Inner Mongolia, China

The Symingyi viaduct on the Jitong Railway's single-track line has fifteen spans, each of which is roughly 30m long. The reinforced concrete structure has round-ended piers which flare out beneath the pier heads to support deep beams with shallow web stiffeners.

Synod Hall Footbridge, Dublin, County Dublin, Ireland

This footbridge, completed in about 1875 by the architect George Street as part of a major ecclesiastical area in Dublin, crosses Winetavern Street to connect the western end of the medieval Christ Church Cathedral to Synod Hall. Since 1993 the Hall has been the site of a permanent heritage centre about Viking and medieval Dublin called Dublinia. The enclosed bridge, which carries an internal stairway on each side, has a single segmental four-ribbed stone arch with four pointed windows above each spandrel and a single tall pedimented window above the arch crown.

Széchenyi Chain Bridge, Budapest, Central Hungary, Hungary

The idea of a fixed link across the Danube between the twin cities of Buda and Pest was promoted by the Hungarian statesman Count István Széchenyi (1791–1860). The bridge was then designed by William Tierney Clark (1783–1852), who had previously designed the first Hammersmith Bridge (see *AEBB*), and construction was supervised by the Scotsman Adam Clark (1811–1866). Built between 1839 and 1849, it is a chain suspension bridge with a main span of 202m between impressive masonry towers like formal gateways and has an overall length of 376m. The structure was badly damaged in 1944 and rebuilt between 1947 and 1949.

Széchenyi Chain Bridge

Tachogang Lhakang Bridge, Paro, Paro Province, Bhutan

The iron chain suspension bridge here over the Paro River is claimed to have been built in about 1420 by Thangtong Gyalpo (1385–1464), nearly 300 years before the UK's Wynch Bridge in 1741 (see *AEBB*). The bridge was swept away by floods in 1969 but a replacement modern suspension footbridge was quickly constructed. In 2005 the old bridge was fully restored using some of the original chains. These, which are decorated with colourful prayer flags, are strung between three-storey white-painted towers with low-pitched roofs like Swiss chalets.

Tachogang Lhakang Bridge

Tacoma Narrows Bridges, Seattle, Washington State, USA

The first suspension bridge across part of Puget Sound on this site, with its 2,800ft-long main span being the third longest in the world at the time, was built between 1938 and 1940. It immediately began to show aerodynamic instability even in low winds, with alternate sides of the deck moving up and down while simultaneously twisting along its axis, and was nicknamed Galloping Gertie. It collapsed in 40mph winds just over four months after it was opened and its failure – now known to have been caused by self-exciting aeroelastic

flutter – remains one of the best-known bridge collapses ever. A replacement structure, opened in 1950, is 40ft longer and 21ft wider than the original. The crossing capacity was greatly increased when an additional bridge was opened in 2007. The original bridge is included in the ASCE list of historic bridges.

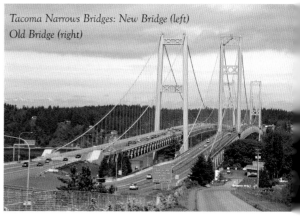

Tacoma Narrows Bridges: New Bridge (left) Old Bridge (right)

Takutu River Bridge, Lethem, Upper Takutu-Upper Essequibo Region, Guyana

Opened in 2009, this bridge across the Takutu River links Lethem in Guyana to Bonfim in Brazil. It is a three-span beam structure with curved soffits. The support pier on each river bank consists of twin-slab concrete columns. Interestingly, the bridge marks the only land border in the Americas between countries with left- and right-hand-drive rule-of-the-road and the route in one direction crosses the other on a flyover bridge just inside Guyana territory.

Taman Negara Canopy Walkway, Kuala Tahan, Pahang, Malaysia

This treetop walkway in a Malaysian national park connects nine tree platforms and is on average about 40m above ground level. The

Taman Negara Canopy Walkway

Tancarville Bridge

footway is ½m wide but the sides up to waist level are formed of expanded mesh. Altogether, the wire rope suspension bridges total 530m in length and are considered to make up the world's longest canopy walkway. Although originally built for research purposes, it was opened to the public in 1993 but was closed for part of 2019 following damage by falling trees.

Tancarville Bridge, Le Havre, Seine-Maritime, France

When this suspension bridge across the River Seine was opened in 1959 its 608m-long main span was the longest in Europe. The overall length was 1420m, the deck width 13m and the total height of the piers was 123m. Between 1996 and 1999 all the cables, which had corroded, had to be replaced.

Tappan Zee Bridge, New York, New York, USA

The first bridge here across the Hudson River, opened in 1955, was a three-span cantilever bridge. This had a 1,212ft-long main span and, with approach spans, an overall length of 16,013ft, but was demolished in 2019. The current bridge consists of two identical but independent dual-span cable-stayed structures which were built between 2013 and 2018. The 419ft-tall masts, which lean outwards slightly on each side of the deck, support a 1,200ft-long main span, but the overall length of the bridge is 16,368ft, or about 3.1 miles. The bridge decks are 96ft and 87ft wide, carrying between them eight lanes of traffic and a reversible bus lane. A one-off consortium of major engineering organisations called Tappan Zee Constructors was set up to undertake all the design and construction work.

Tarn Bridge, Quézac, Lozère, France

The crossing here over the River Tarn was completed in 1450 to enable pilgrims to visit Quézac's Black Madonna sanctuary, which dates from the twelfth century. The bridge has six

Tarn Bridge

nearly semicircular arches, some of which were rebuilt in the seventeenth/eighteenth centuries. There are tall pointed cutwaters on the four central piers. A decorative arch stands on the parapet on one side of the bridge.

Tarragona Aqueduct, Tarragona, Catalonia, Spain

Also known as the **Devil's Bridge**, this 249m-long traditional-looking Roman aqueduct was built in the first century BC. It has two tiers of arches, with eleven arches between tapered piers on the lower level and twenty-five arches between square piers above. All the arches span 6m and spring from heavy impost courses. The aqueduct is part of the UNESCO World Heritage City of Tarragona.

Tauber Bridge, Rothenburg ob der Tauber, Bavaria, Germany

It is believed that this medieval walled town's bridge, with a second bridge standing on the deck of a lower one, was built around 1330 and given defensive fortifications during a rebuild following earthquake damage in 1356. The upper level was built after floods had isolated the original crossing. The upper arches collapsed in 1790 and were rebuilt the following year. The approach to the eastern end was widened in 1925 and in 1945 the whole structure was blown up by the retreating German army. Rebuilding did

not start for a further ten years and the bridge was re-opened in 1956. The lower bridge has six segmental stone arches, the longest spanning 13m, with high-level flood relief openings near the top of some of the piers, and the width of the upper deck is 5m.

Te Rewa Rewa Bridge, Taranaki, New Zealand

This unusual footbridge, completed in 2010, is supported by a 70m-span tubular steel arch that crosses diagonally over the bridge deck, supporting it on nineteen fabricated curved hangers connected to just one side of the deck. This means that the curvature of the hangers increases as their points of attachment to the arch become progressively further off the vertical. With its stark white paint finish, the design is said to be modelled on the rib cage of a whale. The bridge provides a perfect frame for the view of the snow-clad Mount Taranaki seen looking along the bridge length.

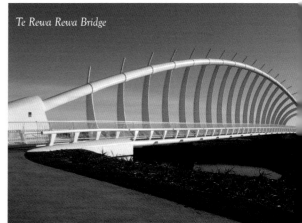

Te Rewa Rewa Bridge

Tauber Bridge

Temburong Bridge, Bandar Seri Bagawan, Brunei-Muara, Brunei

This 30km-long sea crossing over the Brunei River estuary provides a road link between Bandar Seri Begawan, the capital of Brunei, and Temburong, an area that had previously been completely separated from the capital by the sea and part of Malaysia. The bridge, completed in 2020, consists of prefabricated box girder viaducts and two cable-stayed main spans.

TGV Bridges, Ventabren, Bouches-du-Rhône, France

There are two interesting TGV bridges in Ventabren. The 308m-long **Arc River Bridge** carries the high-speed railway line across the river on seven 44m-long spans. Each span has a pair of inverted arches which meet at midspan but lean outwards so that the tie member runs along the underside of the deck about one third of the deck width in from its edge. The ends of the arches connect to a steel framework at the tops of the six intermediate T-headed round columns. The **A8 Motorway Bridge**, which is the longest viaduct on the Mediterranean TGV line, is a prestressed concrete viaduct that is 1,730m long and has a 100m-long main span across the motorway. Its two beams, each weighing 3,600 tonnes, were built parallel to the motorway on each side of it and were then rotated into their final positions.

Third Mainland Bridge, Lagos, South West, Nigeria

The road bridge across the Lagos Lagoon connecting Lagos Island to the mainland was opened in 1990. At 11.8km long it was the longest bridge in Africa until Cairo's 6th October Bridge (q.v.) was opened in 1996. Most of the spans are 45m long and consist of slightly arched concrete beams spanning between four-column piers. The deck is 33m wide with four traffic lanes in each direction.

Three Countries Bridge, Huningue, Haut-Rhin, France

The name of this bridge records that it crosses the Rhine between Huningue in France and Weil am Rhein in Germany and is within 200m of Switzerland. It is also claimed to be the world's longest single-span bridge restricted for use by cyclists and pedestrians only. The main structure is a very flat 230m-span steel arch with a span:rise ratio of 11.5:1 from which is suspended a 5m-wide deck. Designed by Leonhardt, Andrä & Partner, it opened in 2007 and in 2008 it won the German Bridge Construction Prize awarded by the German Federal Chamber of Engineers and the German Association of Consulting Engineers. The first bridge on this site was destroyed by French troops in 1797 and a later pontoon bridge here had the same fate inflicted on it in 1944 by Allied forces.

Three Countries Bridge

Three Sisters Bridges, Pittsburgh, Pennsylvania, USA

These three nearly-identical bridges, which carry Pittsburgh's Sixth, Seventh and Ninth streets across the Allegheny River, were built between 1924 and 1928 and were America's first self-anchored suspension bridges. All three bridges have unusually large eyebars and are painted a distinctive yellow colour. The **Sixth Street Bridge**, also named the Clemente Bridge, was called the Most Beautiful Steel Bridge of 1928 by the American Institute of Steel Construction. It is the fourth to be built on the site, following a wooden covered bridge of 1820, a Roebling suspension bridge of 1880 and a bowstring truss bridge of 1892 designed by Theodore Cooper. The overall length of the bridges is between 939ft and 1000ft, the central span is between 372ft and 419ft and each carries four traffic lanes.

Sixth Street Bridge

Three-Arched Bridge, Venice, Veneto, Italy

Built to replace an earlier sixteenth-century timber structure and with a distinctive brick and stone elevation, this is the last remaining three-arch bridge in Venice. Originally called **St Job's Bridge**, it crosses the Cannaregio Canal and is considered to be one of Venice's main bridges. It was built in 1688 to a design by the architect Andrea Tirali (1657–1737) and has a large decorative shallow niche above each of the two mid-canal piers. The bridge, which was painted between 1765 and 1770 by Francesco Guardi (1712–1793) as 'The Three-Arched Bridge at Cannaregio', was restored in 1794.

Three-Arched Bridge

Ti Peligre Bridge, Cange, Centre, Haiti

The 200ft-long suspension bridge across Haiti's Thomonde River was built by Bridges to Prosperity (B2P), a charitable organisation, founded by Professor Bryan Cloyd, of civil engineering students from Virginia Tech, USA. Work started in November 2009 but, following an earthquake in 2010, completion was delayed until March 2011.

Tiberius Bridge, Rimini, Italy

Also known as the **Bridge of Augustus**, the Roman emperor who had directed that the project should be started, the Tiberius Bridge was completed by his successor in 20CE to carry the Via Aemilia over the Marecchia River. Its overall length between abutments is 63m and its five semicircular arches average about 8m in

Tiberius Bridge

span. The piers, which have an average width of about 4½m, are faced with decorative and pedimented niches.

Tilikum Crossing, Portland, Oregon, USA

The cable-stayed bridge over the Willamette River, built between 2011 and 2015, is unusual is that it is restricted to light rail, buses, cyclists and pedestrians, with cars and lorries being prohibited. It has five spans, the main one being 780ft long and its overall length is 1,720ft. The normal deck width is 60ft, although one of the features of the bridge is the enlarged pedestrian belvederes around the piers, and another feature is the LED flood-lighting system. The structure was designed by T.Y. Lin International.

Timber Bridges (Spreuer and Chapel), Lucerne, Switzerland

For centuries there were three world-famous timber covered bridges across the River Reuss in Lucerne, but the High Bridge was demolished in the nineteenth century. The oldest of the original structures, **Spreuer Bridge**, was built in the thirteenth century to connect mid-river mills

Chapel Bridge

to the right bank and in 1408 was extended to the left bank and at angle. It was rebuilt after being destroyed by floods in 1566. The bridge contains forty-five paintings in a Dance of Death sequence. Work first started in 1333 on the **Chapel Bridge**, a covered wooden footbridge now 672ft long, which is the world's oldest surviving bridge of its type. It now links St Peter's Chapel to the south bank. A distinctive feature is the water tower, which is linked to the main structure by a separate short bridge. Chapel Bridge was badly damaged by fire in 1993 but re-opened in 1994.

Titlis Suspension Bridge, Engelberg, Obwalden, Switzerland

The Titlis Cliff Walk in the Swiss Alps includes this 98m-long suspension bridge, the 1m-wide deck of which carries walkers 500m above ground level and 3040m above sea level. The structure, claimed to be the highest suspension bridge in Europe, opened in 2012.

Tjörn Bridges, Tjörn, Bohuslän, Sweden

The first bridge here, the 532m-long Almö Bridge, was completed in 1960 to connect Tjörn Island to the mainland. It consisted of two 3.8m-diameter arched steel tubes spanning 290m, the longest arch span in the world at the time. On 18 January 1980 the structure collapsed after it was hit by a ship, resulting in the deaths of eight people in vehicles that fell into the sea before the bridge could be closed.

Tjörn Bridges

The second Tjörn Bridge was opened in 1981 to reconnect Tjörn Island and the mainland. This cable-stayed structure has an overall length of 664m, with a 386m-long main span between 106m-high H-shaped piers, and its 15m-wide steel box-girder deck carries Highway 106 at a height of 46m above water level.

Tokyo-Haneda Monorail Bridge, Kawasaki, Kanagawa, Japan

There are not many bridges in the world that only carry monorail tracks. In Japan the twin-track Tokyo Monorail Haneda Airport Line runs above the Keihin Canal on a simple concrete trestle bridge near the Kamome-bashi footbridge. It was opened in 1964.

Tomoegawa Bridge, Chichibu, Saitama, Japan

The bridge crossing the Arakawa River consists of twin steel segmental arches spanning 172m that are connected together along their length by simple struts, forming a structure sometimes known as a ladder arch. The edges of the 8m-wide deck are supported from the arches by a network arch system of criss-crossing diagonal stays. The bridge was opened in 1994.

Tonegawa River Bridge, Katori, Chiba, Japan

Built in 1972 across the Tone River, the Tonegawa Bridge carries the E51 Tohoku expressway road. The eight-span steel bridge has twin box girders, with each carriageway running inside one. The box girder sides consist of Warren girders and the roof of X-braces.

Tonneins Bridge, Tonneins, Lot-et-Garonne, France

This arched road bridge across the Garonne River was designed by Eugene Freyssinet and completed in 1922. It has five spans of 46m and is distinctive for what must be among the largest-ever circular openings in the spandrels above the piers.

Toupin Viaduct, Saint-Brieuc, Côtes-d'Armor, France

This unusual structure, which was designed by the engineer Louis Harel de la Noë, was

Toupin Viaduct

Traversina Footbridge

completed in 1904 to carry a railway 44m above
the Gouédic River. The approaches on each side
have two and four masonry arches but the main
part of the structure has seven tall 15m-long
spans. Each of the six intermediate piers consists
of lower storeys of brick columns supporting
further tiers of reinforced concrete columns
and beams. Above these are rolled I-section
steel parabolic arch ribs, which span 15m with
spandrel columns supporting the deck above,
with latticework ties at springing level. The
reinforced concrete deck is 44m above water
level. The bridge was later converted to carry
road traffic and restored in 2013 and has been a
registered historic monument since 2014.

Traversina Footbridge, St Moritz, Graubünden, Switzerland

An earlier bridge here was destroyed by a
rock fall in 1999. This 56m-span replacement
footbridge across the 70m-deep Traversina
River gorge was designed by Jürg Conzett and
built in 2005. Although a suspension bridge, it
is unusual because, while both ends of the two
suspension cables are anchored into opposite
faces of the gorge at the same level, the two ends
of the deck are hung at different distances below
the suspension cables, becoming a stairway at
the higher end. Moreover, these hangers are
not simple vertical members but are arranged so

that they, the cables and the deck edge railing
together effectively form a three-dimensional
truss, lenticular in plan, thus greatly reducing the
tendency of the deck to sway in use.

Treponti, Comacchio, Ferrara, Italy

The town of Comacchio is built on thirteen
different small islands and the unusual Treponti
bridge was built in 1638 to a design by the
architect Luca Danese in order to provide a

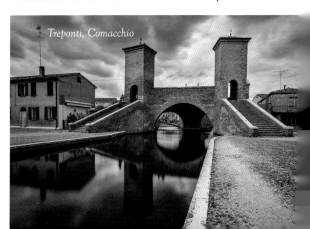

Treponti, Comacchio

footway link between the city centre, the main Pallotta Canal and four smaller waterways. The brick bridge structure consists of three main arches and five stairways (three at the front and two behind, with each one passing over one of the waterways). There are also two guard towers with low pyramidal roofs.

Trift Bridge, Gadmen, Bern, Switzerland

The first Trift Bridge, built in 2004, was a simple 170ft-span suspension footbridge 300ft above part of Lake Triftsee and considered to be the longest such structure in the Alps. Following the retreat of the glacier as a result of global warming, the footbridge was replaced in 2009 by another one at a higher level.

Trim Bridge, Trim, County Meath, Ireland

Ireland's oldest unaltered bridge, probably built in about 1393, is Trim Bridge over the River Boyne. It has four pointed arches, each spanning 15ft to 16ft between 8ft-wide piers, with pointed cutwaters at each pier end. The deck is about 17ft wide.

Trimiklini Double Bridge, Trimiklini, Limassol District, Cyprus

The bridge over the Kouris River at Trimiklini was first built in 1901 with a single segmental stone arch. However, after the introduction of motor transport, it was felt that the gradients down to and up from the bridge were too steep and a second similar arch was built above the first in 1917. It is believed to be the only double bridge on the island.

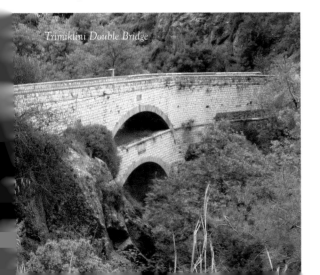

Trimiklini Double Bridge

Triplets Bridges, La Paz, Pedro Domingo Murillo, Bolivia

The Triplets Bridges project, opened in 2010, consists of three similar three-span road bridges over the gorges of the Choqueyapu River. The largest structure, the **Kantutani Bridge**, has a main span of 114m. The bridges are of prestressed concrete, but the extradosed bridge design was chosen to keep the pylon heights down, the highest being only 40m above deck level.

Triplets Bridges

Trumpf Footbridge, Ditzingen, Baden-Württemberg, Germany

This unusual structure, designed by Schlaich Bergermann, and opened in 2018, crosses a busy street to connect two separate parts of the Trumpf Group headquarters. Pedestrians using the slightly-arched, 28m-long footway must walk along a 2.2m-wide specially coated strip on 2cm-thick stainless steel plates. The corners of these plates have been twisted downward to form the four triangular support points for the whole structure, thus making the steel plate membrane doubly-curved. These supports are about 20m apart along the bridge axis and 10m apart at right angles to the axis. The sheets have had several thousand slots and holes cut into them by Trumpf lasers, giving them the appearance of a net, while along the footway are much smaller drilled holes that are capped with glass plugs. On either side of the footway are glass-panelled parapet walls.

Tsaritsyno Park Bridges, Moscow, Russia

The estate that now contains Tsaritsyno Park was bought by Catherine the Great in 1775 but the buildings, which include the large main Bolshoy Palace, were not finally completed until 2007 and they and the park have been open to the public since then. The 1,000-acre park has several interesting bridges, some of which are described below. The two oldest were built by the architect Vasily Bazhenov (1738–99). The first of these is the **Figurny Bridge**, built in 1776, which has a single main arch over the entrance driveway that runs along a small ravine. This arch spans between drum towers, which have tall pointed windows above that are topped by battlements, and supports a colonnaded walkway connecting the upper and middle ponds. Bazenhov's **Gothic Bridge** at the entrance to the park was built during the period 1778–84. The central section, with its three pointed arches, is framed by tall, slim columns with a decorative patera above each of the outer ones and has a deep cornice on which small circular bi-coloured finials stand. There are further arches in the approaches on each side. The **Grotesque Bridge**, which was built by Ivan Yegotov in 1805 to replace an earlier bridge, has a single segmental stone arch and the whole structure is decorated with rusticated stone dressings. The bridge was restored in 2007. The **West Arch Footbridge**, which was also restored in 2007, is an elegant iron structure with a gently arched deck supported on three main arches and a half-arch at each end.

Tsarskoye Selo Park Bridges, Saint Petersburg, Leningrad Oblast, Russia

Tsaskoye Selo is now a UNESCO World Heritage site that includes the parks at both the Alexander and Catherine Palaces. In the park at the Catherine Palace, Catherine the Great built a Chinese village which was approached by three bridges. The cross-shaped **Krestovy Bridge**, dating from 1779, consists of a pair of half-arches on either side of a stream, each half-arch supporting a staircase with all four meeting under a central pagoda over midstream. The **Dragon Bridge** of 1785 has four winged dragons of zinc and the **Chinese Bridge**, built in the same year, is a segmental stone arch over another stream. It is decorated with imitation coral branches and pink granite vases.

At the nearby Alexander Palace the **Marble Bridge** is a copy of the Palladian Bridge that was built at Wilton House in England in 1737 (see *AEBB*) and described later by the writer James Lees-Milne as 'one of the most beautiful buildings in all England'. It consists of a five-bay Ionic colonnade linking two single-bay arched and pedimented end pavilions and was built in 1776. The **Chinese Bridge** is an openwork iron bridge with filigree-style metal railings and decorative little towered gateways across the roadway above each abutment and the **White Bridge** is another iron structure with a low segmental arch crossing a stream.

Figurny Bridge

Marble Bridge

Tsing Ma Bridge, Hong Kong, China

Construction of this bridge was part of a huge project to build a new Hong Kong airport off the island of Lantau, Hong Kong's biggest island, and to link this, together with the Tsing Yi and Ma Wan islands, to the mainland. When it was completed in 1997 the Tsing Ma suspension bridge was the second longest in the world, with a total length of 2,160m and a main span of 1,377m. The concrete pylons are 206m tall and each contains 65,000 tons of steel and concrete. The 41m-wide structure, which was designed by the British engineers Mott MacDonald, is double-decked, with twin three-lane highways above two further road lanes and a double-track railway link to the airport.

Tsitsikamma Suspension Bridge, Knysna, Western Cape, South Africa

The Tsitsikammer National Park has three separate suspension footbridges but the main one spans 77m and carries walkers 7m above water level at the point where the Storms River meets the Indian Ocean. The structure was originally built in 1969 but was later rebuilt to provide greater stability.

Tunca Bridge, Erdine, East Thrace, Turkey

The bridge at Erdine across the Tunca River was built between 1606 and 1615 and has ten pointed stone arches. The central pier has a tall pointed arch through it and there are also similar arches on either side of the adjacent piers. The overall length of the bridge is 136m and it is 7m wide. There is a small tower called the Kitabe Pavilion on the upstream face in the middle of the bridge and the upstream ends of the piers are protected by cutwaters with sloping tops. The structure was restored in 2008 and, since 2015, has been illuminated by LEDs.

Tunkhannock Viaduct, Nicholson, Pennsylvania, USA

The reinforced concrete structure across the Tunkhannock Creek was designed by the Delaware, Lackawanna & Western Railroad's engineer Abraham Burton Cohen and built between 1912 and 1915. It has ten main deck-arch spans, each of which has twin 180ft-long ribs supporting an arcade of eleven further arches. Overall, the bridge is 2,375ft long and the maximum height of the deck above the creek is 240ft. A further interesting statistic about the bridge is that it used 167,000 tons of concrete, nearly half of this below ground: for one pier foundation the excavation was 138ft deep. During building work a Blondin cableway was used to carry construction materials along the bridge from the abutments. The bridge is now a National Historic Civil Engineering Landmark.

Tunca Bridge

Tunkhannock Viaduct

U Bein Bridge, Amarapura, Mandalay, Myanmar

This 1200m-long timber footbridge across the shallow Taungthaman Lake, named after the local mayor who built it between 1849 and 1851, has 482 spans. There are also four wooden rest and shelter pavilions along its length resulting in there being a total of 1,089 supporting posts, although some are now concrete replacements. The bridge, which is approached at each end by a short brick structure, is famous for being the oldest and longest teak bridge in the world.

Ucanha Bridge, Tarouca, Viseu, Portugal

The Ucanha Bridge across the Varosa River is claimed to be Portugal's only remaining fortified bridge. The Romans had probably built an earlier crossing here and there is documentation showing that in 1465 the abbot of Salzedas commissioned construction of this bridge and authorised tolls, although these ended in 1527. It has a main pointed stone arch flanked on each side by two progressively smaller arches and has been called 'the most beautiful bridge in Portugal'. A bridge cross dated 1865 stands on one parapet. On the right bank the road passes onto the bridge through a semicircular-arched tunnel at the bottom of the three-storied 20m-high defensive tower.

Umshiang Double-Decker Root Bridge, Cherrapungi, Meghalaya State, India

The Meghalaya area of north-east India, considered to be one of the wettest places on earth, is where the so-called living bridges are most common. Lieutenant Henry Yule, in a journal published in 1844, was the first to record how they were grown from the Indian rubber tree *Ficus elastica*. The Umshiang Root Bridge, which has been grown and trained by local Khasi tribespeople, has a secondary root system, growing from high up its main trunk, that has been guided across the river to the opposite bank where it has established itself in the soil. After about a dozen years the tree was soundly rooted on both sides of the river, was more than 15m long and 1½m wide, and could be crossed on foot. Tree bridges like this can span up to 100ft and be over a century old but there are no accurate records about their age.

Umshiang Double-Decker Root Bridge

Unification Bridge, Rome, Italy

The Unification Bridge across the River Tiber was opened on 11 May 1911 to mark the fiftieth anniversary of the proclamation of Rome as the capital city of the Italian state in memory of the Italian Risorgimento. Designed by Giovanni Antonio Porcheddu, who was the sole Italian franchisee of the Hennebique system, the bridge was one of the first reinforced concrete structures in Italy. It has a single segmental arch spanning 100m and is 21m wide.

Unity Bridge, Mtambaswala, Mtwara Region, Tanzania

The Unity Bridge across the Ruvuma River, built between 2005 and 2010, links Tanzania and Mozambique and helps to shorten the length of the Cape to Cairo Road. The structure is a prestressed concrete box girder, which is 720m long and has eighteen spans, that carries two traffic lanes on its 14m-wide deck. It was designed by Norconsult and built by the China Geo-Engineering Corporation.

Ushibuka Haiya Bridge, Kumamoto, Japan

Designed by the architect-engineer Renzo Piano and Ove Arup & Partners, the Ushibuka Haiya Bridge was opened in 1997 to link Gezu Island to Shimoshima Island. Curved in plan and arched in elevation, it has an overall length of 883m with a 150m-long main span and reaches

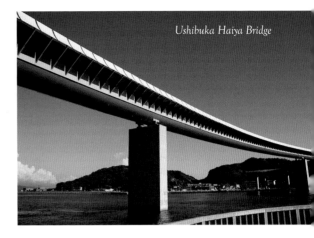

Ushibuka Haiya Bridge

a maximum height of 19m above the sea. The bridge is distinctive for the upward-curving wind deflectors along the upper half of the continuous 5m-deep box girders. In 1998 the Japan Society of Engineers awarded the bridge the Tanaka Prize.

V

Valens Aqueduct, Istanbul, Marmara Region, Turkey

This aqueduct was first built in the second century CE and rebuilt in the fourth century, being completed in 368. Where the water-carrying channel within the city was well above the ground the structure was originally 971m long and consisted of seventy-three semicircular arches of which fifty-five were two-tiered, generally with a span of 4m and a height of up to 29m.

Valentré Bridge, Cahors, Lot, France

This famous Gothic-looking fortified bridge was built over a period of about seventy years and completed in 1378. There are three machicolated towers, one at each end of the bridge and one in the middle, each of which is about 40m high above water level. Originally, there was also a separate outer barbican at each end but only the one at the city end remains. The bridge has six main spans across the River Lot and a further span at each end leads from the river bank onto the tower. The arches span 16½m between piers with pointed cutwaters and, overall, the bridge is 138m long and 5m wide. It was greatly restored in 1880 and, since 1998, has been a UNESCO World Heritage site. It is also on one of the Pilgrims' Routes to Santiago de Compostela.

Vallikraavi Footbridge, Pärnu, Pärnu County, Estonia

The bridge for pedestrians and cyclists that crosses the moat here consists of a glued and laminated timber beam in the shape of a very flat arch, the ends of which are connected together by a steel tie rod. The bridge, which is 13m long and 3.4m wide, was built in 2010 and won the country's national award for being the outstanding glulam structure of the year.

Vang Bridge, Vang, Gorno-Badakhshan Province, Tajikistan

This suspension bridge across the Panj River, one of a series of bridges funded by the Aga Khan Foundation, was opened in 2011. It is 216m long and has three-tiered steelwork pylons.

Vanne Aqueduct, Pont-sur-Yonne, Bourgogne-Franche-Comté, France

This structure, claimed to be the first mass (unreinforced) concrete bridge, was built between 1870 and 1873. With an overall length of 1495m, it has 162 spans the longest being 40m. Part of the structure has two levels of semicircular arches, the lower level ones with pilasters between them that continue upwards as columns framing a tall arch through each pier for a footway, while the upper level arches support the deep aqueduct trough. There is a decorative circular recessed feature above these piers. In other parts of the

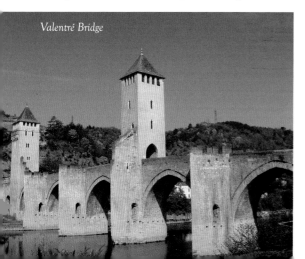

Valentré Bridge

Vanne Aqueduct

aqueduct, where the arch spans are greater and there is no footway within the structure, the main arches are segmental with tall secondary arches bridging the spandrel area above the piers.

Vanvitelli Aqueduct, Caserta, Italy

The aqueduct lying within the park of the royal palace at Caserta was designed by Luigi Vanvitelli for Charles of Bourbon and built between 1753 and 1762. The main three-tiered masonry section crosses the Valle di Maddaloni and has a maximum height of 183ft. From top to bottom, there are forty-three, twenty-nine and nineteen round-headed arches in each of the tiers and its overall length of 1,736ft made it the longest bridge structure in Europe at the time it was built. The piers of the middle tier arches are pierced by small openings to provide a walkway and one of the lower tier arches has an additional decorative surround. The structure is now a UNESCO World Heritage site.

Vanvitelli Aqueduct

Varda Viaduct, Karaisali, Adana Province, Turkey

Completed in 1916, this viaduct carries a single-track railway line on a sharp curve 98m above the Çaki Deresi canyon. Approached by four semicircular spans at each end, there are three main segmental arches and the spandrels above each of the two tall central piers are filled by an arcade of five semicircular arches. The structure was designed by Philipp Holzmann.

Vasco da Gama Bridge, Lisbon, Estremadura, Portugal

This cable-stayed bridge was opened for the 1998 Lisbon World Exposition and to relieve pressure on the 25 April Bridge (q.v.) and commemorates the fifth centenary of explorer da Gama's return from India in 1498. The overall length of the crossing over the Tagus River is more than 10 miles but the main cable-stay span is 420m long.

Vasco da Gama Bridge

Vatuwaqa Bridge, Suva, Rewa Province, Fiji

This new 86m-long prestressed concrete road bridge with four 20m-long spans was built by the Chinese between 2015 and 2017 to replace an earlier straight crossing that also had four spans. The new bridge is curved on plan in order that it could be built away from the previous structure but make maximum use of the existing approaches.

Verrazzano Narrows Bridge, New York, USA

Giovanni da Verrazzano was the Italian explorer whose ship, in 1524, was the first from Europe to enter New York harbour. The suspension bridge named after him that now spans the harbour entrance was opened in 1964, and its later lower deck in 1969. The total length of the bridge is 13,700ft and its main span is 4,260ft long, a world record until Britain's Humber Bridge (see *AEBB*) opened in 1981. The suspension towers

Verrazzano Narrows Bridge

viaduct was the largest metal arch in France. Its total length is 410m with a main span of 220m and, where it crosses the Viaur River, it has a maximum height of 117m. Structurally, it consists of nearly-symmetrical double cantilevers connected by a pin joint where these meet at mid-span. The viaduct, which was designed by Paul-Joseph Bodon, is now a listed historic monument.

Viaur Viaduct

are 693ft high, each anchorage contains 700,000 tons of steel and concrete, and the bridge is 103ft wide, with seven traffic lanes on the upper deck and six on the deck below. The bridge was designed by Othmar Ammann.

Vessy Bridge, Geneva, Switzerland

The road bridge over the Arve River in Geneva was designed by Robert Maillart and completed in 1937. It is a 10m-wide three-pinned, three-ribbed reinforced concrete arch structure with a span of 56m and an overall length of 79m. Each end of the deck is supported above the relatively flat arch (its span:rise ratio is 12:1) on three narrow X-shaped cross walls. The concrete, which had deteriorated badly, underwent major restoration treatment in 1992.

Vézère Bridge, Les Eyzies-de-Tayac-Sireuil, Dordogne, France

The original crossing here over the River Vézère was a simple stone bridge with three 20m-span segmental arches containing decorative oculi above the piers. In 1999 the structure was widened by the addition of a cantilevered footbridge on one side designed by the architect Alain Spielmann. A new 6.5m-wide concrete slab was placed over the existing arches within which were embedded projecting steel brackets to support a sinusoidal laminated timber deck with a metal railing each side.

Viaur Viaduct, Tanus, Tarn, France

When completed in 1902 as part of the Carmaux-Rodez railway line, this steel, single-track railway

Victoria & Alfred Harbour Bridge, Capetown, Western Cape, South Africa

The footbridge in Capetown Harbour crosses the channel between the Victoria and Alfred Basins. Built in 1997, the first structure had a 2m-wide deck and was a single-masted cable-stayed girder, the mast and back-stays of which were mounted on a turntable. This could swing open like a gate on a gatepost so vessels could use the channel. In 2019 the bridge was replaced by a sleeker swinging structure, also cable-stayed, but in which the back-stays are anchored to a balance beam that stands up like a raking mast. The 42m-long deck is 4m wide and can open in less than a minute.

Victoria Falls Bridge, Victoria Falls, Matabeleland North, Zimbabwe

Bridging the Second Gorge just downriver from the famous waterfalls on the Zambezi River between Zambia and Zimbabwe is this parabolic steel arch bridge. Built in 1905 as part of Cecil Rhodes's grand (but still unfulfilled) scheme for

Victoria Falls Bridge

with large rounded cutwaters which continue through to parapet level to provide recesses for pedestrians off the 4½m-wide deck. The three original defensive towers were demolished in 1777 and the bridge was widened in 1828.

Vihantasalmi Bridge, Mäntiharju, Åland, Finland

Built in 1999, this bridge was the outcome of a design competition won by the Finnish engineering firm Rantakokko for the replacement of an earlier steel bridge. The five-span structure, which is 168m long and 14m wide, includes three 42m-long A-shaped glulam through trusses, with the end spans being beams made from a timber-concrete composite. The structure is considered to be the longest timber main road bridge in the world.

Vihantasalmi Bridge

a Cape to Cairo railway, it was designed by Sir Douglas Fox & Partners and prefabricated in England by the Cleveland Bridge & Engineering Company before its 1500 tons of steelwork were shipped out for erection. Spanning 513ft at a height of 420ft above the rushing torrent, it carries a road, single-track railway and footpath. There is now a bungy jump from the bridge. The bridge is included in the ASCE list of historic bridges.

Victory (Haghtanak) Bridge, Yerevan, Ararat, Armenia

This bridge in the Armenian capital, which carries a dual carriageway road 34m above the Hrazdan River, was built in 1945 to mark victory over Germany. The structure is 25m wide and its overall length of 200m includes three main segmental masonry arches flanked at each end by two smaller semicircular arches. Each of the open spandrels of the main arches contains three tall semicircular-arched openings.

Vienne Bridge, Confolens, Charente, France

The stone arched bridge over the Vienne River was probably built in the fourteenth century. It is 127m long and has ten arches. The four arches at the town end are all pointed, with the largest having a span of just over 12m, while the remaining six arches are all segmental. The upstream ends of the piers are protected

Vimy Memorial Bridge, Ottawa, Ontario, Canada

Shortly after its completion in 2014, the bridge across the Rideau River was named the Memorial Bridge after the First World War battle at Vimy Ridge in April 1917. More than ten thousand Canadians were killed or wounded

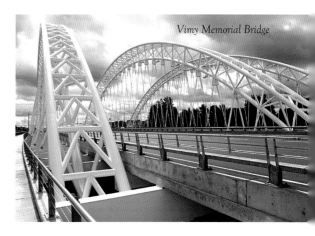

Vimy Memorial Bridge

in the fighting in which four Canadian divisions captured the ridge from the Germans, a battle now considered to be the country's most important in its national self-awareness. The bridge has three parallel tubular steel double arches spanning 125m from which the deck is supported on inclined hangers. This 41m-wide deck carries eight vehicle lanes, two cycle ways and two footpaths. In 2015 the structure was awarded the Gustav Lindenthal Medal.

Vine Bridge, Koulé, Nzérékoré Prefecture, Guinea

The vine bridge over the Loffa River at Koulé, which is 56m long, consists of more than 3,200 specially selected lianas and is renewed every year.

Vistula Bridges, Dirschau, Eastern Pomerania, Poland

The first of the two railway bridges over the Vistula River at Dirschau was built between 1850 and 1857. Designed by Johann Carl Wilhelm Lentze, the 837m-long lattice girder structure has six spans, each 131m long. The wrought iron girders were nearly 12m deep and 6m apart. Although there was only a single railway track, there was also an external footway. The German army specified that, because of its strategic importance, the bridge should be properly defendable and, at each end of the bridge, it was therefore protected by a barbican in front of twin, tall rectangular towers. Smaller cylindrical towers were also built on either side at each intermediate pier. These defences were designed by the architect Friedrich August Stüler in medieval style. Between 1888 and 1891 a second parallel bridge was built 40m downstream to carry a double-track railway, with the older bridge being converted for road use. The new structure consisted of wrought iron lenticular trusses. Parts of both bridges were blown up by the Polish army on 1 September 1939 to try to prevent the German invasion of Poland. After the war both bridges were repaired and further reconstruction of the bridges and towers started in 2012. The older bridge was added to the ASCE List of Historic Engineering Milestones in 2004.

Vitim Bridge, Kuandinsky, Siberia, Russia

The old railway bridge across the Vitim River, which was probably built in the 1880s, is just 6ft wide and is 1870ft long with ten spans. The deck is 50 ft above water level. The plate girder structure is supported by triple-column piers and, with the railway tracks having been lifted after a new five-span trussed girder bridge was built a short distance upstream, the deck now consists only of timber planks, some missing, which can easily become slippery. Moreover, there are no protective side railings and just simple timber baulks mark the edge of the deck. It is therefore considered to be especially exciting and dangerous to drive over the old structure and it is part of the Baikal-Amur-Mainline (BAM) Road used by thrill-seekers.

Vittoria Emanuele II Bridge, Rome, Italy

This road bridge over the River Tiber was designed by the architect Ennio de Rossi in 1886 but building work did not start until 1905 and the bridge was opened in 1911. The white marble structure, which has three segmental arch stone spans, is 108m long and 20m wide. It is exuberantly decorated with large sculptured groups above each pier and tall corner pylons with bronze Winged Victories.

Vizcaya Transporter Bridge, Portugalete, Biscay, Spain

The transporter bridge linking Portugalete and Getxo across the Neruion River was the first such crossing to be built in the world. Designed by Ferdinand Arnodin, who had patented the concept, and the Basque architect Alberto de Palacio, it was opened in 1893. In 1937, during the Spanish Civil War, it was partly demolished by removal of the stayed girder between the supporting pylons. It was re-opened in 1941 with the girder now supported by suspension cables. The main wrought iron structure consists of double 61m-high latticework pylons at each end, between which steel suspension cables span to support a 164m-long deck truss 40m above water level. This deck supported a low-level travelling gondola that carried six vehicles as well as foot

Vizcaya Transporter Bridge

passengers between the low river banks. A major programme of improvements in 1995 included the installation of access lifts and a suspended footway for pedestrians as well as of a new plastic and aluminium gondola that can carry 200 passengers and six cars. The bridge has been a UNESCO World Heritage site since 2006.

Vlaardingse Vaart Bridge, Vlaardingen, South Holland, Netherlands

Spanning 45m across the Vlaardingse Vaart Canal and opened in 2009 is this unusual bridge for pedestrians and cyclists. It consists of a rectangular-section longitudinal tube, with the four sides made up from 400 small-diameter tubes welded together in overlapping Warren girders. As the tube crosses the canal the cross section rotates through nearly ninety degrees so that the top and bottom faces at one end become the two sides at the other. The slightly-arched bridge deck is suspended within the tube. The bridge was designed by West 8 Architects and ABT Engineers.

Vytautas the Great Bridge, Kaunas, Kaunas County, Lithuania

Earlier bridges across the Nemunas River in Kaunas included a pontoon bridge first built in 1794 and the German army built another pontoon bridge in 1915. These crossings were generally known as the **Aleksotas Bridge** before 1930 and again between 1940 and 2008. A new stone structure, completed in 1930 and named Vytautas the Great Bridge after an earlier grand duke of Lithuania, was destroyed by bombing in WWII, rebuilt in 1948 and reconstructed again in 2005. This bridge is 16m wide with an overall length of 256m consisting of six steel plate girder spans. The central one of these is a shorter lifting span to allow large vessels to pass and is flanked by tall piers clad in stone.

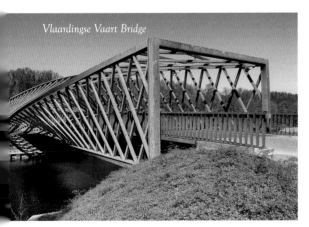
Vlaardingse Vaart Bridge

Vytautas the Great Bridge

Wadi El Kuf Bridge, Al Bayda, Cyrenaica, Libya

Designed by Riccardo Morandi and built between 1965 and 1972, this bridge crosses the Kouf Valley at a height of 172m, the second highest bridge in Africa (after Bloukrans Bridge – q.v.). The structure, which has an overall length of 477m, consists of two A-shaped piers with prestressed concrete stays supporting a 282m-long main span. Interestingly, the 13m-wide bridge deck is not supported by the main A-frame columns but by separate V-shaped piers standing inside the A-frames. The bridge was closed in 2017 after potential fractures were identified in the structure, but was quickly re-opened for light traffic.

Wadi Laban Bridge, Riyadh, Riyadh Province, Saudi Arabia

The Wadi Laban Bridge is a cable-stayed (semi-fan) road bridge carrying six traffic lanes across a dry river bed. The bridge has an overall length of 763m with a 405m-long main span. The two reinforced concrete pylons are 176m high. The deck, 85m above the wadi, is made from precast concrete sections prestressed together and its

Wadi Laban Bridge

Danyal Saeed Photograph

width of 36m made it the widest in the world of this type at the time the bridge was completed in 1997.

Wan'an (Eternal Peace) Bridge, Pinghan, Fujian, China

It is claimed the first bridge on this site was built before 1279, but many of the earlier crossings were destroyed in local conflicts. Rebuildings have been recorded in 1742, 1845, 1932 and 1954. This latest structure has six woven timber arched spans, the longest being 15m, an overall length of 98m and width of 5m. The bridge deck is approached by steps at each end and the structure includes a large temple with elaborate wood carvings.

Wanxian Bridge, Chongqing, Sichuan, China

When it was completed in 1997, this road bridge spanning 425m over the Yangtze River became the world's longest reinforced concrete arch, taking the record from Krk Bridge (q.v.). The overall length of the structure is 865m and it is 24m wide.

Webb Bridge, Melbourne, Victoria, Australia

Part of this 145m-long structure was originally built in 1986 as the **Webb Dock Rail Bridge** over the Yarra River to connect the dock to the city railway system, but the line was closed in 1992. In 2003 a new bridge carrying a footway and cycle way was opened. The earlier structure, which was given new enclosing hoops, now leads to a new 80m-long curving ramp extension also with hoops, but these feature a random covering network of connecting pieces. The landmark design, which was by the architectural firm Denton Corker Marshall and artist Robert Owen, was given the Australian Joseph Reed Award for Urban Design.

West End Bridge, Pittsburgh, Pennsylvania, USA

Sited one mile downstream from where the Allegheny and Monongahela Rivers join to form the Ohio River, this steel bowstring arch bridge was designed by the Allegheny County Department of Public Works. It was built between 1930 and 1932 and carries the 41ft-wide roadway of US-19. The arch has a main span of 780ft and, with approaches, the overall length of the bridge is 1891ft. Its deck is 66ft above water level, making it the city's highest bridge.

West Lake Bridges, Hangzhou, Zhejiang, China

The West Lake Cultural Landscape, just south of the Yangtze River at Hangzhou, is a World Heritage Site with a number of bridges. The **Pavilion Bridge** is a small three-span stone bridge topped by a pagoda with a two-tiered roof, the **Broken Bridge** built in 1922, is a small, three-span, semicircular stone arch approached by later spans formed from stone slabs, and there is also a hump-backed bridge consisting of three semicircular stone arches.

Weygand Bridge, Sassandra, Gbôklé, Ivory Coast

Completed in either 1932 or 1947 (the available sources differ), this reinforced concrete bridge has a segmentally-arched main span over the Sassandra River with a through deck suspended on hangers from the arch. There is a curved approach of eight subsidiary spans, each span consisting of the outward-leaning leg from the V-shaped pier on each side linked together by a shallow arch. The gap at the top of the piers between the outward-sloping legs is filled by another shallow segmental arch. The result of this arrangement is that the elevation of each pier looks like the plan view of a slice of cake.

Whirlpool Rapids Bridges, Niagara Falls, New York State, USA

The first bridge across the narrowest point of the Niagara Gorge downstream from the Falls, which stood from 1848 to 1855, was the **Niagara**

Falls Suspension Bridge*. This carriage and footbridge was designed by Charles Ellett and consisted of four 1,190ft-long cables on each side of the 8ft-wide timber deck spanning 625ft between 80ft-high timber towers. The structure was followed by the double-decked rail-over-road combined suspension and stay **Niagara Falls Railway Suspension Bridge***, designed by John A. Roebling, which was built between 1852 and 1855. This bridge had a span of 625ft and was supported by four cables spanning between four separate masonry towers with a separate pair for each deck, thus creating the bridge's distinctive double-sag appearance. As well as the vertical-plane stays between the tower tops and the stiffening trusses, there were also a large number of horizontal stays between theses trusses and the edges of the gorge. Initially, the stiffening trusses were of timber but these were changed to wrought iron in 1880 and the supporting masonry towers were similarly updated in 1886. (Interestingly, because of a lack of standardisation in early American railways, the single-track line was built with four rather than two separate rails to enable vehicles of three different track gauges to use the bridge.) In 1896 work started on a totally new replacement structure to carry heavier trains and the **Niagara Railway Arch Bridge**, designed by Leffert L. Buck was opened in 1898. This 780ft-long steel structure has a

Niagara Railway Arch Bridge

two-pinned main span that is 550ft long. Like its predecessor it is double-decked, with a 26ft-wide roadway and two 11ft-wide footways below a 32ft-wide railway deck. The bridge is included in the ASCE list of historic bridges. There has also been a series of other bridges nearer the Falls, the latest of these since 1941 being the Rainbow Bridge (q.v.).

White Nile (Omdurman) Bridge, Khartoum, Khartoum State, Sudan

The road bridge here, which crosses the White Nile to link Khartoum and Omdurman, was built between 1924 and 1926 by Dorman Long & Co. With an overall length of 613m it has eight steel truss spans supported by seven intermediate river piers each consisting of a pair of round columns.

Omdurman Bridge

White River Concrete Arch Bridge, Cotter, Arkansas, USA

The bridge here, also known as Cotter Bridge and R.M. Ruthven Bridge, was designed by James Barney Marsh, the design being patented as the Marsh Rainbow Arch, and this was the largest of the type to be built. Its overall length is 1805ft and it has five main 216ft-span parabolic concrete arches that are 70ft high. The 24ft-wide deck carries road US62B across White River. Interestingly, it was constructed using a Blondin cableway to get materials onto the structure. The bridge, which was opened in 1930, was added to the ASCE List of Historic Landmark Bridges in 1986.

Whitworth Aqueduct, Abbeyshrule, County Longford, Ireland

The Royal Canal crosses the River Inny on this 165ft-long and 35ft-wide aqueduct. There are five segmental arches each spanning 20ft and the abutments, piers and spandrel walls are all wider at their bases than at parapet level.

Wignacourt Aqueduct, Valetta, South Eastern, Malta

This aqueduct was built between 1610 and 1615 by the Order of Saint John to bring water from springs in Dingli and Rabat to about 30,000 people in the new capital city of Valetta. The system was designed by Padre Giacomo and, after his death, was completed by Bontadino de Bontadini, an hydraulics expert from Bologna. The above-ground parts of the aqueduct consist of 361 arches over a length of 9 miles. Most of the arches are semicircular but, on sections where the water channel is lower, they are segmental. The aqueduct includes three inspection towers and several fountains.

Wignacourt Aqueduct

Winding Patapat Viaduct, Pagadpud, Luzon, Philippines

The viaduct that runs along the edge of Pasaleng Bay is 1,300m long and 31m above sea level. Opened in 1986, the simple column and beam structure was built to deal with the problem of landslides frequently carrying away the road.

Wörlitz Garden Bridges, Dessau-Wörlitz, Saxony-Anhalt, Germany

The landscaped park and gardens at Wörlitz were the first to have been constructed in Germany and were praised by Goethe. The **White Bridge**, a timber arched structure, was based on the first Kew Bridge (see *AEBB)* and the **Iron Bridge** was a quarter-scale replica of the world's first iron bridge in Coalbrookdale (see *AEBB).* The whole Garden Kingdom, with its thirty-two bridges, has been a UNESCO World Heritage Site since 2002.

Wye Bridge, Keddie, California, USA

The Western Pacific Railroad's viaduct over the Feather River Canyon at Keddie, which was completed in 1931, is unusual as it carries two single-track railway lines that meet at one end of two separate, curved, plate girder structures to form a Y-shape in plan. Immediately after the junction the line disappears into a tunnel.

Wye Bridge

Xidong Bridge, Sixi, Taishun, China

This woven timber arch bridge across Dongxi Creek was first built in 1570 and rebuilt in 1745 and again in 1827. The span over the river between the stone abutments is 26m but the overall length of the main 5m-wide covered passageway is 42m. In the centre of this is a raised second floor, containing an altar bay, which has a separate pagoda-like roof structure.

Xihoumen Bridge, Zhoushan, Zhejiang, China

This suspension bridge across Hangzhou Bay, opened in 2009, has a main span of 1,650m between support piers that are 211m high. This is the second longest suspension span in the world after Japan's Akashi Kaikyo Bridge (q.v.).

Xu Pu Bridge, Shanghai, China

This cable-stayed bridge carries eight lanes of traffic over the Huangpu River to connect the Xuhui and Pudong districts of the city. Built in 1997, it has a 590m-long main span, and is 36m wide. The two main A-shaped pylons are 590m tall.

Yadanabon Bridge, Sagaing, Mandalay, Myanmar

Built in 2008 to replace the previous 1934 Ava Bridge over the Irrawaddy River, this steel bridge has three main fixed arches, each spanning 224m, with trussed approach spans on each side. The overall length is 1,711m, making it the longest bridge in the country. The deck, which is 15m wide, carries a four-lane motorway and two footways.

Yangpu Bridge, Shanghai, China

Opened in 1993, this cable-stayed bridge is larger than Shanghai's earlier Nanpu Bridge (q.v.) of 1991. Its overall length is 8,354m, of which the main span is 602m, and the 30m-wide deck carries six lanes of traffic. The bridge's inverted Y-shaped pylons are 223 tall, needed to provide 48m clearance for shipping on the Huangpu River.

Yangtze River Bridge, Nanking, Jiangsu, China

Completed in 1969, this major bridge is a double-decked structure carrying a four-lane road above a twin-tracked railway. The 1,574m-long main structure has ten spans, the longest of which is 160m, and stretches between tall masonry abutment towers. These contain viewing platforms for pedestrians. A major difficulty in the bridge's construction was that the river is up to 70m deep from water level to the bedrock on which the piers were founded.

Yangtze River Bridge, Wuhan, Hubei, China

This bridge's double-decked main structure consists of nine steel girder spans, each 128m long, and carries an 18m-wide road above two railway tracks. There are footways on both sides of each deck. The six-storey abutment tower at each end includes viewing galleries and is topped with two pavilions. Leading up to these towers

Yangtze River Bridge

are a further 514m of approach spans. The bridge was completed in 1957.

Yavuz Sultan Selim Bridge, Istanbul, Marmara Region, Turkey

Previously called the **Third Bosphorus Bridge**, this steel road-rail crossing near the Black Sea entrance to the Bosphorus Strait was designed by Michel Virlogeux and built between 2013 and 2016. It is a hybrid cable-stay and suspension structure with an overall length of 2164m and a main span that is 1,408m long. The deck girders are 5.5m deep. The structure is the third tallest bridge in the world, its two inverted V-shaped pylons reaching a height of 330m. The bridge deck, at 59m wide, makes it also one of the world's widest bridges. The crossing failed to generate projected income and it became necessary to refinance the US$2.7 billion cost. The bridge's name honours the memory of Sultan Selim I who was the father of Suleyman the Magnificent, sponsor of the famous Ottoman architect Mimar Sinan whose bridges included the Mehmed-Pasha Sokolovic Bridge (q.v.). The Yavuz Sultan Selim Bridge received the WAN Best Bridge Award in 2017 and the International Road Federation's Global Road Achievement Award in 2019.

Yayabo Bridge, Sancti Spiritus, Sancti Spiritus Province, Cuba

Although the town was founded in 1514, the bridge here across the Yayabo River was not completed until 1831 and it seems odd this should be one of the country's oldest bridges. Its total length is 85m and it has five semicircular stone arches, the longest having a span of 10m. The entrance to the bridge at one end is flanked by a pair of tall, polished stone columns. The bridge was added to the National Heritage of Cuba in 1995.

Yellow Bridge, Nusa Lembongan, Bali, Indonesia

The thirty-year-old bridge between the islands of Nusa Lembongan and Nusa Ceningan is well-known to visitors to Bali. The deck, which is about 100m long and only 1.5m wide, is hung from cables spanning between cross-braced steel pylons standing on T-shaped concrete piers. It collapsed in 2016 with the deaths of eight people and more than thirty others injured. In 2017 new cables were installed between the original steel pylons and a new deck was hung.

Yibna Bridge, Yavne, Central District, Israel

The bridge across the Nahal Sorek River, completed in about 1274, was one of a series of bridges that Sultan Baybars built in Palestine. Baybars was the Mamluk sultan who inflicted

the first major defeat on the Mongol army in 1260. The 11m-wide and 48m-long structure has three pointed arches, the central one being significantly the largest. These arches span between piers with pointed upstream cutwaters and pilasters downstream.

Yiheyuan Summer Palace Bridges, Beijing, China

The imperial gardens in Beijing contain more than thirty bridges that were originally built between 1751 and 1764 to link islands in the man-made Kunming Lake and to provide causeways around the lake edge. They include the Yudai and Shiqigong Bridges, possibly the two best-known bridges in the whole of China. The **Yudai (Jade Belt) Bridge**, also known as the Camel's Back Bridge, is a marble footbridge with a slightly pointed high arch that was needed to provide clearance below for the emperor's dragon boat. The **Shiqigong (Seventeen Arches) Bridge**, also of marble, is the largest in the gardens and provides the only connection to Nanhu Island. It is 150m long and 8m wide and is decorated with 544 different carved marble lions on the parapets. Both bridges were rebuilt following severe damage in both the Second Opium War of the 1850s and the Boxer Rebellion of 1900. The Summer Palace is a World Heritage site.

Seventeen-Arch Bridge

Yibna Bridge

Yongle Bridge, Tianjin, Hebei, China

The Yongle Bridge (previously called the **Chihai Bridge**) is a six-lane three-span viaduct across the Hai River. It is significant, however, because also spanning the river directly above the bridge

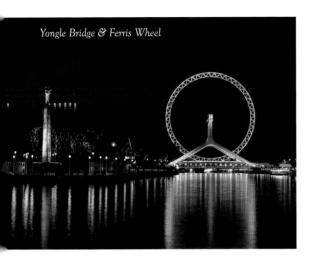

Yongle Bridge & Ferris Wheel

is a 120m-tall Ferris wheel (the third biggest in the world). This is called the Tientsin Eye and has forty-eight capsules each holding up to eight passengers. The wheel's axle is supported on inverted Y-shaped frames, the lower ends being founded on abutments on the river banks.

Yongqing (Eternal Celebration) Bridge, Taishun, Zhejiang, China

Built in 1797, this two-span bridge uses timbers cantilevered out from its masonry abutments and mid-river pier to support the deck. This

has an enclosed passageway 33m long with, in the centre, timber steps leading up to a loft which contains shrines and has a pagoda-style roof.

Yu Yuan Bazaar / Garden Bridge, Shanghai, China

One of Shanghai's major tourist attractions is the Yuyuan Garden (Garden of Quiet Joy) with its teahouse approached across a lotus pond by the ten-legged zigzag timber footbridge. The first bridge here was built in the period 1559–1577 but the present structure dates from major restoration work on the garden completed in 1961.

Yuwen (Nourish Culture) Bridge, Taishun, Zhejiang, China

Built in 1839, this bridge has a semicircular stone arch spanning 7.6m supporting a wooden passageway which is 23m long and 4m wide.

Z

Zarasas Bridge, Ditkūnai, Utena County, Lithuania

The skywalk bridge here, which was opened in 2011, follows a nearly circular route about 100m long over part of the shoreline of Lake Zarasas. The bridge deck spans between twin steel edge joists, which are supported 17m above water level by seven X-shaped tubular steel piers.

Zayanderud Bridges, Isfahan, Iran

There are eleven historic combined bridge-weirs over the River Zayanderud in Isfahan that provide crossings and help to regulate the river's flow. **Pol-e-Khaju Bridge**, perhaps the best known of these, was built by Shah Abbas II in about 1650 on the foundations of an earlier stone bridge from the late fourteenth century. It is 133m long by 12m wide, has twenty-three pointed arches and the downstream face of the weir is stepped. The longer brick and stone **Siosepol Bridge** (also known as **Allah Verdi Khan Bridge**) is 298m long by 15m wide with thirty-three pointed arches and was built by Shah Abbas I between 1599 and 1602 after making the

Siosepol (Allah Verdi Khan) Bridge

city his new capital. Both these bridge-weirs have double-decked brick and stone superstructures, each consisting of arcades of pointed arches. On the Pol-e-Khaju the lower ones stand on the top of the weir and carry enclosed footpaths and an upper road, with further arches above to support the roof over these passageways whereas, on the Siosopol, an enclosed footway on each side flanks an open roadway. Another feature of these two bridges is an ornate pavilion about half way along. Older still is the **Shahrestan Bridge**, the stone foundations of which have been roughly dated to the fifth century, but which now has thirteen eleventh-century pointed brick arches spanning between boat-like stone piers. This bridge is 108m long and 5m wide. There is a secondary pointed arch above eight of these piers and a pavilion at one end of the structure.

Zeeland Bridge, Nieuwerkerk, South Holland, Netherlands

Carrying the N256, together with a cycle route along its western side, the Zeeland Bridge across the Eastern Scheldt estuary is the country's longest bridge at 5022m and has fifty-one spans. Most of these are 95m-long but one has two lifting bascule leaves. The bridge was built between 1963 and 1965 and is about 5 miles east of the Eastern Scheldt Bridge (q.v.).

Zezelj Bridge, Novi Sad, Vojvodina, Serbia

The Yugoslav engineer Barnko Zezelj designed the first road-rail bridge here over the River Danube, which was built of prestressed concrete tied arches in 1961. It was destroyed in 1999 during the NATO peacekeeping campaign – aimed at encouraging Yugoslavia to withdraw from their ethnic cleansing campaign against Albanians in Kosovo – and a similar-looking structure was eventually rebuilt in 2018. Now

New Zezelj Bridge

carrying a double-track railway on a deck that also ties the arch ends together, the steel ladder arch has two segmental arch spans – of 219m and 177m – and a total length of 474m.

Zhangjiajie Glass Bridge, Zhangiajie, Hunan Province, China

The Zhangjiajie glass-decked suspension bridge opened in 2016 and, when built, was the world's tallest and longest glass bridge. Its 14m-wide deck, which consists of ninety-nine glass panels each 60cm thick, spans 430m across a canyon, is 300m above ground level and can carry 800 visitors. The two suspension cables are made up of nineteen wire strands, each containing ninety-one wires of 5.1mm diameter, and the planes of hangers between cable and deck incline outwards. The bridge was designed by the engineering firm China Railway Major Bridge Reconnaissance & Design Institute with architect Professor Haim Dotan. Shortly after it was opened the bridge had to be immediately closed for urgent strengthening when visitor numbers were ten times what had been expected.

Zhaozhou Bridge, Zhaozhao, Hebei, China

This stone bridge over the Xiao River, built between 581 and 605 by Li Chun, is also known as **Anji Bridge**, meaning 'Safe Crossing' and was one of the five great ancient bridges of China. It is considered to be the earliest arch bridge in the world in which the roadway is not laid directly over the arch. Instead, the separate deck structure is above the main arch,

leaving openings in the spandrels that thus reduce the load the arch has to carry. These smaller arched openings also allow flood waters to pass through therefore reducing sideways pressure on the arch during times of high river flow. There are twenty-eight parallel arch rings spanning 37m, with the individual voussoirs of each ring connected together by iron dovetails, and the keystone on the outer ring is decorated with a carved animal head. The sides of the deck are protected by stone parapets which include inset small stone columns projecting above the parapet line. The bridge was added to the ASCE List of Historic Civil Engineering Landmark Bridges in 1989.

Zhivopisny Bridge, Moscow, Russia

This cable-stayed bridge across the Moskva River is most unusual as, instead of crossing more or less at right angles to the river, it is S-shaped in plan crossing the water at an oblique angle. Near the river banks the bridge deck structure stands on circular piers but in midstream it is supported by seventy-eight stays that are hung from a 105m-high parabolic steel arch that crosses above both the river and the bridge deck. A drum-shaped enclosure, which is suspended below the crown of the arch, was planned to house a restaurant, but that part of the project has never been completed. The bridge itself was opened in 2007.

Zhivopisny Bridge

Zubizuri Bridge, Bilbao, Biscay, Spain

The 'White Bridge' (the English meaning of its Basque name) is a tied arch footbridge, also known as the **Campo Volantin Bridge**, that crosses the Nervion River in Bilbao. It was designed by the architect:engineer Santiago Calatrava and opened in 1997. The bridge is 75m long and has a slim parabolic steel arch that spans 75m with a rise of 15m and supports the deck on thirty-nine stay cables. The arch, the plane of which is tilted away from the vertical towards the centre of radius for the curved deck, has hangers supporting both ends of under-deck cross beams. There have been disputes between Calatrava and the city council regarding the impracticality of some design features such as glass bricks set into the footway that became slippery when wet.

Zuoz Bridge, Zuoz, Graubünden, Switzerland

Robert Maillart's reinforced concrete bridge carries a road across the Inn River on a slight curve and was completed in 1901. The three-hinged, twin-celled hollow box arch bridge has a main span 38m long. Unfortunately, the bridge elevation has been slightly spoilt by attached pipework.

Zuwulufu Bridge, Kandewe, Rumphi District, Malawi

It is believed that the first natural fibre suspension bridge here, linking Kandewe to villages on the north side of the Rukuru River, was built in about 1904 but, because of its low height above the river, it was soon swept away by flood waters. Later rebuilds have kept the lowest part at least 5m clear of normal river levels. The suspension ropes are made from palm fronds and the U-shaped deck from woven bamboo, the whole structure reportedly looking like a giant elongated basket. In recent times the natural suspension ropes have been supplemented by wire cables and the bridge had become a tourist attraction.

Bridge Lists

All bridges can be classified, on the basis of how their structure works, as being in one of five main types: arch, beam, cantilever, stayed or suspension bridge. However, there are a number of possible other listings, based on special characteristics of bridges' design, usage, history etc., which can help to identify interesting bridges that might otherwise remain largely unknown, and a few of these are noted below. This is not meant to be an exhaustive set of references but more a guide to examples of bridges with unusual or interesting features (bridges no longer existing marked *). The headings to these seventy-six lists are as follows:

Aluminium Bridges
Ancient Bridges
Aqueducts
ASCE List of Historic Civil Engineering Landmark Bridges
Bailey Bridges
Border Bridges
Bridge Collapses and Failures in Use
Bridge Failures During Construction
Bridge Fountains
Bridges as Destinations
Bridges Built for Olympics, World Cups & World Fairs
Bridges Damaged or Destroyed by Earthquakes
Bridges Damaged or Destroyed by Fire
Bridges Damaged or Destroyed by Floods and Ice
Bridges Damaged or Destroyed in Civil Unrest and Non-World Wars
Bridges Damaged or Destroyed in World Wars I and II
Bridges Damaged or Destroyed by High Winds
Bridges Featured in Paintings by Famous Artists
Bridges Featured on Banknotes, Coins and Postage Stamps

Bridges on Cycling and Walking Trails
Bridges Stranded by Changes in the Course of Rivers
Bridges with Decorative Towers
Castle Bridges
Chain Suspension Bridges
Chapel Bridges
Charitable Bridge Building
Copy Bridges
Corbelled Arch Bridges
Covered Bridges
Crib Bridges
Devil's Bridges
Double-Decked Bridges
Early Concrete Bridges
Extended Construction Times of Bridges
Floating and Pontoon Bridges
Folly Bridges
Gisclard Suspension Bridges
Fortified Bridges
Glass Bridges
Glulam Bridges
Highest Bridges
Hotels with Bridges
Inhabited Bridges
Longest Bridges and Spans
Lounge, Pagoda, Pavilion, Teahouse and Temple Bridges
Medieval Bridges
Memorial Bridges
Modern Enclosed Bridges
Mono-Cable Suspension Bridges
Monorail Bridges
National Heritage / Monument Bridges
Natural Fibre Bridges
Observation Deck Bridges
Opening Bridges
Packhorse Bridges

Pilgrim Routes over Bridges
Pipeline Bridges
Pontoon Bridges
Prize-Winning Bridges
Record-Breaking Bridges
Re-Located Bridges
Ribbon Bridges
Sea Bridges
Self-Anchored Suspension Bridges
Stone Slab Bridges
Timber Bridges

Transporter Bridges
Trestle Bridges
Triumphal Archways and Gateways on Bridges
UNESCO World Heritage Bridges
Unusual Bridge Features
Unusual Truss Designs
Widest Bridges
World Monuments Fund Bridges
World's Best Bridges
Zigzag Bridges

ANCIENT BRIDGES

This entry, which covers bridges with entries in this book that were built before 1000CE, is separated into **BCE Bridges** (for non-Roman structures built before the beginning of the Common Era), **Roman Bridges** and **First Millennium Bridges** for other bridges built between 1CE and 1000CE. (For bridges built between 1000 and 1500CE, see the separate main entry **Medieval Bridges**.)

BCE Bridges

Bridges, all or parts of which are claimed to date from before the common era (BCE) began, include:

Anlan Suspension Bridge, Dujiangyan City, Sichuan, China [third century BCE]
Caravan Bridge, Izmir, Smyrna, Turkey [850BCE]
Girsu Bridge, Tello, Iraq [c. 2000 BCE]
Kazarma Bridge, Arkadiko, Peloponnese, Greece [1300–1200BCE]
Pons Fabricius, Rome, Italy [62 BCE]
Ponte Salario, Rome, Italy [first century BCE]
Severan Bridge, Arsameia, Adiyaman Province, Turkey

Roman Bridges

Bridges built by the Romans throughout the parts of the world they colonised, of which a major part remains, include:

Adana Roman Bridge, Adana, Anatolia, Turkey
Adži-Paša's Bridge, Podgorica, Montenegro
Ancient Roman Bridges, Rome, Italy

Bridge of Apollodorus, Drobeta-Turnu, Romania *
Bridge of Augustus, Narni, Terni, Italy
Burgo Bridge, Pontevedra, Galicia, Spain
El-Kantara Bridge, Constantine, Constanine Province, Algeria
High Level Aqueducts, Caesarea, Israel
Kirkgoz Kemeri Bridge, Limyra, Antalya Province, Turkey
Kızılçullu Aqueducts, Smyrna, Izmir Province, Turkey
Pompey's Bridge, Mtskheta-Mtianeti, Georgia
Pons Aelius, Rome, Italy
Pons Aemilius, Rome, Italy
Pont du Gard, Nîmes, Bouches-du-Rhône, France
Pons Fabricius, Rome, Italy
Pont Flavien, Saint-Chamas, Bouches-du-Rhône, France
Pont Julien, Bonnieux, Vaucluse, France
Ponte Milvio, Rome, Italy
Rio Seco Aqueduct, Almuñécar, Andalusia, Spain
Roman Aqueducts, Aspendos, Antalya, Turkey
Roman Bridge, Cordoba, Spain
Roman Bridge, Mérida, Badajoz, Spain
Roman Bridge, Mostanica, Montenegro
Roman Bridge, Salamanca, Castile-León, Spain
Roman Bridge, Segura, Gipuzkoa, Spain
Roman Bridge, Trier, Rhineland-Palatinate, Germany
Roman Bridge, Vaison-la-Romaine, Vaucluse, France
Segovia Aqueduct, Segovia, Castile-León, Spain
Shadorvan Bridge, Shushtar, Khuzestan Province, Iran
Tarragona Aqueduct, Tarragona, Catalonia, Spain
Tiberius Bridge, Rimini, Italy

Valens Aqueduct, Istanbul, Turkey

First Millennium Bridges
Non-Roman bridges and aqueducts that were built before 1000CE include:

Ain Diwar Bridge, Ain Diwar, Al Hasakah Governorate, Syria
Alcántara Bridge, Alcántara, Cáceres, Spain
Ponte dei Saraceni, Adrano, Catania, Italy
Skopje Aqueduct, Skopje, North Macedonia
Zhaozhou Bridge, Zhaozhao, Hebei, China

Five Great Ancient Bridges of China
China has a magnificent history of bridge-building and five of its historic structures are usually referred to as the Five Great Ancient Bridges. These are:

Anping (Peace) Bridge, Fujian, Hebei, China (built 1138–51)
Guangji / Xiangzi (Great Charity) Bridge, Chaozhou, Guangdong, China (built 1170)
Loyang (Ten Thousand Peace) Bridge, Chuanchow, Fujian, China (built 1053–59)
Marco Polo (Lugou) Bridge, Beijing, China (built 1189–1192)
Zhaozhou Bridge, Zhaozhao, Hebei, China (built 581–605)

AQUEDUCTS

The first aqueducts were built in ancient times as a means of bringing water to major cities and, later, aqueducts were needed for commerce by enabling load-carrying vessels to travel through areas where there were no natural waterways. Entries for aqueducts include:

Águas Livres Aqueduct, Lisbon, Estremadura, Portugal
Aguila Aqueduct, Nerja, Andalucia, Spain
Alloz Aqueduct, Alloz, Navarre, Spain
Amoreira Aqueduct, Elvas, Alentejo, Portugal
Bar Aqueduct, Stari Bar, Montenegro
Briare Aqueduct, Briare, France
Cabin John Aqueduct, Cabin John, Maryland, USA
Carioca Aqueduct, Rio de Janeiro, Brazil
Carpentras Aqueduct, Vaucluse, Vaucluse, France
Delaware Aqueduct, Lackawaxen, Pennsylvania, USA
Dongguan City Aqueduct Bridge, Guandong, Guangzhou, China
Duck Creek Aqueduct, Metamora, Indiana, USA
Grand Maître Aqueduct, Sens, Yonne, France
Håverud Aqueduct, Håverud, Västra Götaland County, Sweden
High Level Aqueducts, Caesarea, Israel
Jundushan Aqueduct, Yanqing, Beijing, China
Kızılçullu Aqueducts, Smyrna, Izmir Province, Turkey
Magdeburg Water Bridge, Magdeburg, Saxony-Anhalt, Germany
Orb Aqueduct, Beziers, Hérault, France
Padre Templeque Aqueduct, Hidalgo, Eastern, Mexico
Pont du Gard, Nîmes, Bouches-du-Rhône, France
Ponte delle Torri Aqueduct, Spoleto, Perugia, Italy
Querétaro Aqueduct, Santiago de Querétaro, Querétaro, Mexico
Red Gate Aqueduct, Augsburg, Bavaria, Germany
Red Women's Bridge, Huixian, Henan Province, China
Rio Seco Aqueduct, Almuñécar, Andalusia, Spain
Roman Aqueducts, Aspendos, Antalya, Turkey
Roquefavour Aqueduct, Ventabren, Bouches-du-Rhône, France
Saint-Clément Aqueduct, Montpellier, Hérault, France
Sart Aqueduct, La Louvière, Hainaut Province, Belgium
Segovia Aqueduct, Segovia, Castile-León, Spain
Skopje Aqueduct, Skopje, North Macedonia
Tarragona Aqueduct, Tarragona, Catalonia, Spain
Valens Aqueduct, Istanbul, Turkey
Vanne Aqueduct, Pont-sur-Yonne, Bourgogne-Franche-Comté, France
Vanvitelli Aqueduct, Caserta, Italy
Whitworth Aqueduct, Abbeyshrule, County Longford, Ireland
Wignacourt Aqueduct, Valetta, South Eastern, Malta

ASCE LIST OF HISTORIC CIVIL ENGINEERING LANDMARK BRIDGES

The American Society of Civil Engineers started listing landmark structures in 1966, the first being the Bollman Truss structure of Savage Mills Bridge in Maryland, USA, and the list now includes more than 265 structures around the world. The following forty-one of these listed bridges (and their date of listing) appear in this book:

Alvord Lake Bridge, San Francisco, California, USA [1969]

Blenheim Covered Bridge, Blenheim, New York, USA [1983]

Brooklyn Bridge, New York, New York, USA [1972]

Cabin John Aqueduct, Cabin John, Maryland, USA [1972]

Carrollton Viaduct, Baltimore, Maryland, USA [1982]

Choate Bridge, Ipswich, Massachusetts, USA [2008)

Cornish-Windsor Covered Bridge, Cornish, New Hampshire, USA [1970)

Cortland Street Bridge, Chicago, Illinois, USA [1981)

Delaware Aqueduct, Lackawaxen, Pennsylvania, USA [1972)

Duck Creek Aqueduct, Metamora, Indiana, USA [1992]

Eads Bridge, St Louis, Missouri, USA [1971]

Fink Deck Truss Bridge, Lynchburg, Virginia, USA [1979]

Forth Railway Bridge, South Queensferry, City of Edinburgh, Scotland [1985]

Frankford Avenue Bridge, Philadelphia, Pennsylvania, USA [1970]

Golden Gate Bridge, San Francisco, California, USA [1984]

John A. Roebling Suspension Bridge, Covington, Kentucky, USA [1982]

Lacey V. Murrow Memorial Bridge, Seattle, Washington, USA [2008]

Lake Pontchartrain Crossing, New Orleans, Louisiana, USA [2013]

Kinzua Viaduct, Westline, Pennsylvania, USA [1982]

Mackinac Bridge, Mackinaw City, Michigan, USA [2010]

Malleco Viaduct, Collipulli, Auraucania Region, Chile [1994]

Manhattan Bridge, New York, New York, USA [2009]

Maria Pia Bridge, Porto, Portugal [1990]

Menai Suspension Bridge, Menai Bridge, Isle of Anglesey, Wales [2002]

Niagara Suspension Bridge, Niagara Falls, New York State, USA [1992]

Poughkeepsie Bridge, New York, New York, USA [2009]

Quebec Bridge, Quebec, Canada [1987]

Querétaro Aqueduct, Santiago de Querétaro, Querétaro, Mexico [1995]

Rockville Bridge, Harrisburg, Pennsylvania, USA [1979]

San Francisco-Oakland Bay Bridge, San Francisco, California, USA [1986]

Savage Mills Bridge, Savage, Maryland, USA [1966]

Smithfield Street Bridge, Pittsburgh, Pennsylvania, USA (1975]

Starrucca Viaduct, Lanesboro, Pennsylvania, USA [1973)

Sydney Harbour Bridge, Sydney, New South Wales, Australia [1988]

Tacoma Narrows Bridge, Seattle, Washington State, USA [2013]

Tunkhannock Viaduct, Nicholson, Pennsylvania, USA [1975]

Victoria Falls Bridge, Victoria Falls, Matabeleland North, Zimbabwe [1995]

Vistula Bridges, Dirschau, Eastern Pomerania, Poland [2004]

Whirlpool Rapids Bridges, Niagara Falls, New York State, USA [1992]

White River Concrete Arch Bridge, Cotter, Arkansas, USA [1986]

Zhaozhao Bridge, Zhaozhao, Hebai, China [1989]

BAILEY BRIDGES

Bailey Bridges, a particular type of temporary military bridge first developed in the UK in 1941 by Sir Donald Bailey (1901–1985), are now often used in civil emergencies or to supplement existing inadequate structures. True to the initial principles of the Bailey design – robust modular design, fully interchangeable standard components, speed and simplicity of erection – companies such as Mabey Bridge (an Acrow Group Company) have gone on to develop a wide range of proprietary modular steel bridging solutions for permanent, temporary, military and emergency use, which have been installed in over 150 countries worldwide. The entries in this book for original Bailey bridges and later Mabey versions include:

Banab River Bridge, Madang, Papua New Guinea

Coppename Bridge, Jenny, Coronie District, Suriname

Cuscatlan Bridge, San Miguel, San Miguel Department, El Salvador

Demerara Harbour Bridge, Georgetown, Guyana

Groot Aub Bridge, Groot Aub, Hardap Region, Namibia

Juba Nile (Freedom) Bridge, Juba, Central Equatoria, South Sudan

Loto Samasoni Bridge, Apia, Samoa

Medjez-el-Bab Bridge, Béja, Tunisia

Nakabuta Bridge, Sigatoka, Fiji

BORDER BRIDGES

Often a boundary between two countries runs down the centreline of a separating river, a bridge over that river providing a link between the countries. This can therefore be classed as a border bridge and, depending on the relations between the countries, might be equipped with immigration and customs checkpoints at each end. Occasionally, however, such bridges can be at (or very near) tripoints or quadripoints where three or four countries meet. Some of the border bridges in this book include:

Al Salam Peace Bridge, El Qantara, Ismaelia, Egypt. This road bridge across the Suez Canal links Africa and Asia.

Bad Säckingen Covered Bridge, Bad Säckingen, Baden-Württemberg, Germany. The first Rhine bridge here was built in 1272 to link Germany and Switzerland.

Bevera Viaduct, Varese, Italy. The railway carried by this bridge links Italy and Switzerland.

Bosphorus Bridge, Istanbul, Marmara Region, Turkey. Opened in 1973, the bridge here was the first fixed link between the continents of Asia and Europe.

Bridge of No Return, Kaesong, North Hwanghae, North Korea. The footbridge across the Military Demarcation Line between North Korea and South Korea was where prisoners-of-war could be exchanged but they could not later change their minds.

Eiffel Bridge, Ungheni, Moldova. This border bridge carries a railway line between Romania and Moldova.

Friendship Bridge, Ciudad del Este, Alto Paraná Department, Paraguay. This bridge crosses the Parana River to connect Paraguay and Brazil.

Glienicke Bridge, Potsdam, Brandenberg, Germany. This 'Bridge of Spies' near Berlin was where agents captured during the Cold War were exchanged.

International Bridge, Valenca, Viana do Castelo, Portugal. The Valenca Bridge over the River Minho unites Portugal and Spain.

Kazungula Bridge, Kazungula, Southern Province, Zambia. The bridge here, which crosses between Zambia and Botswana at a quadripoint, is curved horizontally to avoid crossing over the territories of either Zimbabwe or Namibia.

Platjan Bridge, Mathathane, Central District, Botswana. The Limpopo River bridge here links Botswana and South Africa.

Red Bridge, Meore Kesalo, Kvemo Kartli, Georgia. The Tbilisi-Ganja road crosses between Georgia and Azerbaijan here and the area on both sides of the border crossing is surrounded by minefields.

Takutu River Bridge, Lethem, Upper Takutu Region, Guyana. This bridge marks the only land border in the Americas between countries with left- and right-hand-drive rule-of-the-road.

Three Countries Bridge, Huningue, Haut-Rhin, France. The bridge over the Rhine at Huningue provides a connection between France and Germany and passes within 200m of Switzerland.

BRIDGE COLLAPSES AND FAILURES DURING CONSTRUCTION

Bridges are sometimes particularly susceptible to collapse while they are being built. This can be as a result of either insufficient technical attention being given to identifying unusual temporary loads and stresses that can arise during construction and ensuring that the incomplete structure can withstand these, or of the workforce on site not properly following the correct procedures that have been laid down for the building process. Examples of bridge failures during construction include:

College Footbridge, Lyon, Rhône, France [1844 (8 killed)]

Liteyny (Foundry) Bridge, St Petersburg, Leningrad Oblast, Russia [two accidents 1875 and 1879 (14 killed)]

Quebec Bridge, Quebec, Canada [1907 (75 deaths), 1916 (13 killed)]

Queen Juliana Bridge, Willemstad, Curacao, Venezuela [1967 (15 killed)]

Sandö Bridge, Kramfors, Västernorrland County, Sweden [1939 (18 killed)]

Sigiri Bridge, Sigiri, Busia County, Kenya [2017 (20 people injured)]

BRIDGE COLLAPSES AND FAILURES IN USE

Many bridges have collapsed or failed in some way when in use over the years, whether as a result of inadequate design or construction or from being overwhelmed by the forces of nature. Excluding bridge failures during construction (q.v.), those damaged or destroyed in civil unrest and war (q.v.) and those destroyed by fire, by tornado, by floods and ice (q.v.), bridge collapses and failures in use have included:

Ashtabula Bridge, Ashtabula, Ohio, USA [probably fatigue in materials, 1876 (92 killed)]

Avignon Bridge, Avignon, France [floods, 1603, 1605 and 1680]

Banab River Bridge, Madang, Papua New Guinea [collapse of over-loaded corroded structure, 2018]

Caracas-La Guaira Highway Viaducts [earthquake, 1967]

Colonial Bridge of Tequixtepec, Huajuapan de Leon, Oaxaca, Mexico [date of partial collapse unknown]

Covered Bridge, San Martino, Pavia, Italy [collapsed after cliff fall, 1947]

Elhova Footbridge, Elhovo, Yambol Province, Bulgaria [overloaded by people, 1996 (9 killed)]

El Ferdan Railway Bridge, Ismailia, Province of Upper Egypt, Egypt [ship strike, 1947]

El-Kantara Bridge, Constantine, Algeria [part of cast iron arch collapsed, 1951]

Eurymedon Bridge, Aspendos, Antalya Province, Turkey [reason and date unknown]

Gen Rafael Urdaneta Bridge, Maracaibo, Zulia, Venezuela [ship strike, 1964 (7 killed)]

Huc Bridge, Hanoi, Vietnam [fire, 1877; overloaded by people, 1952]

Hunter Street Bridge, Peterborough, Ontario, Canada [reason unknown, 1875]

International Railroad Bridge, Sault Ste Marie, Ontario, Canada [overloaded, 1941]

Jacques Gabriel Bridge, Blois, Loir-et-Cher, France [reason unknown, 1716]

Japan-Palau Friendship Bridge, Koror, Palau [creep, 1996 (2 killed)]

Kutai Kartanegara Bridge, Tenggarong, Indonesia [2011, failure of a hanger during maintenance (60 killed or injured)]

Menai Bridge, Isle of Anglesey, Wales [inadequately-stiffened deck collapsed, 1839]

Old Bridge (Karl Theodor Bridge), Heidelberg, Baden-Württemberg, Germany [ice floe, 1284]

Polcevera Viaduct, Genoa, Lombardy, Italy [corrosion, 2018 (43 killed)]

Pont des Arts, Paris, Ile-de-France, France [hit by barge 1979]

Puente Nuevo, Ronda, Spain [collapsed 1741 (50 killed)]

River Arachthos Bridge, Plaka, Epirus, Greece [flash floods 1863 and 2015]

Shadorvan Bridge, Shushtar, Khuzestan Province, Iran [1885]

Shahi Mughal Bridge, Jaunpur, Utter Pradesh, India [earthquake, 1934]

St Servatius Bridge, Maastricht, Netherlands [overloaded during a procession, 1275 (c. 400 killed)]

Tacoma Narrows Bridge, Seattle, Washington State, USA [aerodynamic instability, 1940]

Tauber Bridge, Rothenburg ob der Tauber, Bavaria, Germany [reason unknown, 1790]

Tjörn Bridges, Tjörn, Bohuslän, Sweden [ship strike, 1980 (8 killed)]

Traversina Footbridge, St Moritz, Graubünden, Switzerland [rock fall, 1999]

Yellow Bridge, Nusa Lembongan, Bali, Indonesia [collapsed, 2016 (8 killed)]

BRIDGE FOUNTAINS

Although aqueducts carried water into cities, some of this then being used in public fountains, only a very few bridge structures themselves have had such decorative additions. Bridges which support fountains include:

Banpo-Jaamsu Bridge, Seoul, Gyeonggi Province, South Korea

Pont des Belles Fontaines, Juvisy-sur-Orge, Paris, France

Wignacourt Aqueduct, Valeta, South Eastern, Malta

BRIDGES AS DESTINATIONS

Normally, the purpose of a bridge is to provide a *connection* between two places. Occasionally, however, the bridge's principal purpose is itself to become the *destination*. Bridges which people visit to take advantage of the facilities they offer include:

Bridge Climbs
The following two bridges with entries in this book have widened their appeal to visitors by establishing proper climbs up them with permanently fitted safety ropes. These are:

Auckland Harbour Bridge, Auckland, New Zealand
Sydney Harbour Bridge, Sydney, New South Wales, Australia

Bridge Temples
Japanese Covered Bridge, Hoi An, Quang Nam, Vietnam

Bridge Walks
Some bridges, often built specifically to attract tourists, offer various kinds of walk, sometimes attached to the side of a cliff. These include:

Bridge of Immortals, Tangkouzhen, Anhui Province, China

Bridges of Eden, Port Vila, Shefa Province, Vanuatu

Capilano Suspension Bridges, Vancouver, British Columbia, Canada

East Taiheng Glasswalk, Zhangjiajie, Hebei Province, China

Greenisland Railway Viaducts, Antrim, County Antrim, Northern Ireland: Newtownabbey Way

Inca Bridge, Machu Pichu, Cuzco, Peru

Kingdom Centre Skywalk, Riyadh Region, Riyadh, Saudi Arabia

Kinzua Viaduct, Westline, Pennsylvania, USA

Langkawi Sky Bridge, Langkawi Island, Malaysia

Mile into the Wild Walkway, Keenesburg, Colorado, USA

Pinnacle Sky Bridges, Singapore

Queen Mary's Bridge, Füssen, Bavaria, Germany: built to provide views of Neuschwanstein Castle

Taman Negara Canopy Walkway, Kuala Tahan, Malaysia

San Martin Bridge, Toledo, Castile-La Mancha, Spain

Grand Canyon Skywalk, Peach Springs, Arizona, USA

Titlis Suspension Bridge, Engelberg, Obwalden, Switzerland

Zhangjiajie Glass Bridge, Hunan, China

Bungy-Jumping & Zip-Wire-Running Bridges
Because, by their very nature, bridges often cross deep chasms they have become popular sites for bunjee-jumping and zip-wire running, particularly at older structures that have been superseded by new crossings and there is therefore no conflict of interest between the joy-riders and other users of the bridge. Bridges in this book where there is or has been bunjee-jumping, pendulum swinging or zip-wire running include the following:

Arenal Hanging Bridges, La Fortuna, Costa Rica

Asparuhov Bridge, Varna, Bulgaria

Auckland Harbour Bridge, Auckland, New Zealand

Bloukrans Bridge, Knysna, Western Cape, South Africa

Bridges of Eden, Port Vila, Shefa Province, Vanuatu

Chiche Bridges, Quito, Pichincha, Ecuador

Europa Bridge, Innsbruck, Switzerland

Gouritz River Bridges, Albertinia, Western Cape, South Africa

Kawarau Bridge, Queenstown, Otago, New Zealand

Maslenica D8 Bridge, Zadar, Zadar County, Croatia

Mosta Bridge, Mosta, Northern, Malta

Ojuela Bridge, Mapimi, Durango, Mexico

Parkovy Bridge, Kiev, Kievshchyna, Ukraine

Ponte dei Salti, Lavertezzo, Ticino, Switzerland

Royal Gorge Bridge, Canon City, Colorado, USA

San Martin Bridge, Toledo, Castile-La Mancha, Spain

Suspension Bridge, La Ferme, Rodrigues Island, Mauritius

Victoria Falls Bridge, Victoria Falls, Matabeleland North, Zimbabwe

Bridges for Dining, Entertainment and Shopping

Unusual structures which people visit in order to use the entertainment and service facilities that are provided on the structure include:

Banpo-Jaamsu Bridge, Seoul, Gyeonggi Province, South Korea [world's largest bridge fountain]

Bogdan Khmelnitsky Pedestrian Bridge, Moscow, Moscow Oblast, Russia [an enclosed space houses souvenir booths]

Dragon Bridge, Da Nang, Hai Chau, Vietnam [the dragon's head breathes fire and water]

Galata Bridge, Istanbul, Marmara Region, Turkey [the earlier floating bridge is now a floating restaurant only]

Murinsel Bridge, Graz, Bundesland, Austria [two bridges provide access to a floating island with amphitheatre]

Putra Bridge, Putrajaya, Malaysia [the bridge has observation galleries and dining areas]

Yongle Bridge, Tianjin, Hebei, China [a 120m-tall Ferris wheel spans the river directly above the bridge]

BRIDGES BUILT FOR OLYMPICS, WORLD CUPS & WORLD FAIRS

Major international functions such as Expos, Olympics and World Cups can generate huge crowds and to deal with these, and also to take the opportunity of upgrading essential infrastructure as well as providing a lasting 'legacy', they are often associated with spectacular new bridges within the host city. In addition, smaller 'feature' bridges have also sometimes been built as part of the exhibition site itself. The major structures have included:

Alexandre III Bridge, Paris, France [Universal International Exhibition 1900]

Alamillo Bridge, Seville, Spain [Expo 1992]

Bac de Roda-Felipe II Bridge, Barcelona, Catalonia, Spain [Olympics 1992]

Barqueta Bridge, Seville, Andalusia, Spain [Expo 1992]

Bridge Pavilion, Zaragoza, Aragon, Spain [Expo 2008]

Favela da Rocinha Footbridge, Gavea, Rio de Janeiro, Brazil [World Cup 2014]

Liberty Bridge, Budapest, Central Hungary, Hungary [World Exhibition 1896]

Russky Bridge, Vladivostok, Primorsky Krai, Russia [Pacific Economic Cooperation Summit 2012]

Seonyu Footbridge, Seoul, Gyeonggi Province, South Korea [World Cup 2002]

Vasco da Gama Bridge, Lisbon, Estremadura, Portugal [World Exposition 1992]

BRIDGES DAMAGED OR DESTROYED BY EARTHQUAKES AND LANDSLIDES

Earthquakes usually occur on geological fault lines and, if there is relative movement of the ground between different points of support for a structure such as a bridge, partial or full collapse can occur. Alternatively, instead of undergoing physical movement, firm ground can be liquified by the shaking of an earthquake, again leading to partial or full collapse of structures above. Bridges with entries in this book that have suffered earthquake damage include:

Allenby (King Hussein) Bridge, Jericho, Jordan [1927]

Bar Aqueduct, Stari Bar, Montenegro [1979]

Bridge of Augustus, Narni, Italy [1855, 2000]

Bridge of Remembrance, Christchurch, New Zealand [2011]

Caracas-La Guaira Highway Viaducts, Caracas, Greater Caracas, Venezuela [1967]

Castle Bridge, Kamyanets Podilsky, Khelmnytskyi, Ukraine [1986]

Longteng Bridge, Sanyi Township, Miaoli County, Taiwan [1935, 1999]

Malleco Railway Viaduct, Collipulli, Chile [2010]

Ponte delle Torri Aqueduct, Spoleto, Perugia, Italy [2016]

Red Bridge, Yerevan, Ararat, Armenia [1679]

Roman Aqueducts, Aspendos, Antalya, Turkey [c. 350]

San Francisco-Oakland Bay Bridge, San Francisco, California, USA [1989]

Shahi Mughal Bridge, Jaunpur, Utter Pradesh, India [1934]

Sidi Rached Viaduct, Constantine, Constantine Province, Algeria [2008]

Tauber Bridge, Rothenburg ob der Tauber, Bavaria, Germany [1356]

Ti Peligre Bridge, Cange, Centre, Haiti [2010]

Traversina Footbridge, St Moritz, Graubünden, Switzerland [1999]

BRIDGES DAMAGED OR DESTROYED BY FIRE

Early timber bridges were always susceptible to fire, but it is interesting that the following list covers fire damage to bridges over eight and a half centuries:

Chapel Bridge, Lucerne, Switzerland [1993]

Covered Bridge, Lovech, Lovech Province, Bulgaria [1925]

Galata Bridge, Istanbul, Marmara Region, Turkey [1992]

Huc Bridge, Hanoi, Red River Delta, Vietnam [1877]

London Bridge, London, England [1136]

Menai Bridge, Menai Bridge, Isle of Anglesey, Wales [1970]

Ponte degli Alpini, Bassano del Grappa, Vicenza, Italy [1511, 1813]

Poughkeepsie Railroad Bridge, Poughkeepsie, New York, USA [1987]

Schaffhausen Bridge, Schaffhausen, Schaffhausen Canton, Switzerland [1799]

Slender West Lake Wuting (Five Pavilions) Bridge, Yangzhou, Jiangsu, China [1850 and 1864]

Smithfield Street Bridge, Pittsburgh, Pennsylvania, USA [1845]

BRIDGES DAMAGED OR DESTROYED BY FLOODS AND ICE

It is unsurprising that so many bridges have been destroyed by floods over the years (some more than once – such as Avignon Bridge and Santa Trinita Bridge in Florence). In the examples below, the damage was by floods unless ice is specifically mentioned.

Anshun (Peaceful and Favourable) Bridge, Chengdu, Sichuan, China [1947]

Avignon Bridge, Avignon, Vaucluse, France [1603, 1605 and 1680]

Besalu Bridge, Besalu, Girona, Spain [1315]

Blenheim Covered Bridge, Blenheim, New York, USA [2011]

Boone Viaduct, Boone, Iowa, USA [1881]

Bridge of Shadman Malik, Samarkand, Samaquand, Uzbekistan [pre-1502]

Chains Bridge, Fornoli, Lucca, Italy [1836]

Charles Bridge, Prague, Bohemia, Czech Republic [1342]

Chengyang Yongji (Wind and Rain) Bridge, Ma'an, Sanjian, Guangxi, China [1983]

Danube Bridge, Linz, Bundesland, Austria [2016]

Father Matthew Bridge, Dublin, County Dublin, Ireland [1385]

Guangji / Xiangzi (Great Charity) Bridge, Chaozhou, Guangdong, China [1174]

Hartland Covered Bridge, Saint-John, New Brunswick, Canada [ice 1920]

Jihong (Rainbow in the Clear Sky) Bridge, Yongpin, Yunan, China [1986]

Kintai Bridge, Iwakuni, Yamaguchi Prefecture, Japan [1950]

Latin Bridge, Sarajevo, Bosnia and Herzegovina [1795]

Lucan Bridge, Dublin, County Dublin, Ireland [c. 1730, 1786]

Mağlova Aqueduct, Istanbul, Marmara Region, Turkey [1563]

Marco Polo (Lugou) Bridge, Beijing, China [c. 1698]

Moyola Park Bridge, Castledawson, Ulster, Northern Ireland [1929]

Old Bridge (Karl Theodor Bridge), Heidelberg, Baden-Württemberg, Germany [ice floe 1284, c. 1565]

Old Bridge, Pisek, South Bohemia, Czech Republic [1768, 2002]

Old Suspension Bridge, Mallemort, Bouches-du-Rhône, France [1872]

Ponte degli Alpini, Bassano del Grappa, Vicenza, Italy [1409, 1567, 1748]

Ponte dei Saraceni, Adrano, Catania, Italy [1948]

Rainbow Bridge, Niagara Falls, Ontario, Canada [ice 1938]

Rhine Bridge, Bad Säckingen, Baden-Württemberg, Germany [1570]

Rio Cobre Bridge, Spanish Town, Middlesex, Jamaica [2000]

River Arachthos Bridge, Plaka, Greece [2015]

Salzach Bridge, Oberndorf, Salzburg, Austria [1899]

Santa Fe Suspension Bridge, Santa Fe de la Vera Cruz, Santa Fe, Argentina [1983]

Santa Teresa Bridge, Elche, Valencia, Spain [1751]

Santa Trinita Bridge, Florence, Tuscany, Italy [1259, 1333, 1357]

Saracens Bridge, Adrano, Sicily, Italy [1948]

Skeiðarárbrú Bridge, Kirkjubæjarklaistur, Eastern Region, Iceland * [ice 1996]

Spreuer Bridge, Lucerne, Switzerland [1566]

St Patrick's Bridge, Cork, County Cork, Ireland [1853]

Svilengrad Bridge, Svilengrad, Haskova Province, Bulgaria [1766]

Tauber Bridge, Rothenburg ob der Tauber, Bavaria, Germany [1356]

Tachogang Lhakang Bridge, Paro, Paro Province, Bhutan [1969]

Zuwulufu Bridge, Kandewe, Rumphi District, Malawi [c .1905]

BRIDGES DAMAGED OR DESTROYED BY HIGH WINDS

Many **Bridges Damaged or Destroyed by Floods** have suffered these disasters as a result of hurricanes and tornados bringing with them torrential rainfall. However, the winds themselves have also directly damaged and destroyed some bridges:

Belize City Swing Bridge, Belize City, Belize [damaged in 1931, 1961 and 1998]

Choluteca Bridges, Choluteca, Choluteca Province, Honduras [damaged 1998]

Kinzua Viaduct, Westline, Pennsylvania, USA [eleven piers and twenty-three spans were blown down, 2003]

Rio Cobre Bridge, Spanish Town, Middlesex, Jamaica [damaged 2004]

Tacoma Narrows Bridge, Seattle, Washington State, USA [shook itself to pieces in 40mph winds, 1940]

BRIDGES DAMAGED OR DESTROYED IN CIVIL UNREST AND NON-WORLD WARS

Because of bridges' fundamental importance in providing easy access over unfordable rivers, they immediately become vital targets in warfare in order to deny use of them to an enemy. Bridges with entries in this book that have suffered during conflicts are noted in this and the following entry:

Águila Aqueduct, Nerja, Andalusia, Spain [Spanish Civil War, 1937]

Alcántara Bridge, Alcántara, Cáceres, Spain [Peninsular Wars, 1809]

Allenby (King Hussein) Bridge, Jericho, Jordan [Six Days War, 1967]

Argenteuil Bridges, Argenteuil, Val-d'Oise, France [Franco-Prussian War, 1870]

Augartenbrücke, Vienna, Austria [Napoleonic Wars, 1809]

Avignon Bridge, Avignon, Vaucluse, France [Albigensian Crusade, 1209–1229]

Besalu Bridge, Besalu, Girona, Spain [Spanish Civil War, 1939]

Bridge of Boyacà, Tunja, Boyacà, Colombia [Battle of Boyacà, 1819]

C.H. Mitchell Bridge, Port Edward, Kwa-Zulu-Natal, South Africa [terrorist bombs, 2002]

Cunene River Bridge, Xangongo, Cunene Province, Angola [Angolan Civil War, 1975–2002]

Cuscatlan Bridge, San Miguel, San Miguel Department, El Salvador [guerrillas, 1983]

Darnytskyi Railway Bridge, Kiev, Kievshchyna, Ukraine [Ukrainian War of Independence, 1920]

Deir ez-Zor Suspension Bridge, Deir ez-Zor, Deir ez-Zor Province, Syria [Syrian Civil War, 2013]

Devil's Bridge, Martorell, Barcelona, Spain [Spanish Civil War, 1937]

Drava Footbridge, Osijek, Osijek-Barania, Croatia [Croatian War of Independence, 1991–1995]

El Ferdan Railway Bridge, Ismailia, Province of Upper Egypt, Egypt [Six-Day War, 1967]

Gabriel Tucker Bridge, Monrovia, Montserrado, Liberia [Liberian Civil War, 2006]

Hardinge Bridge, Paksey, Pabna District, Bangladesh [Bangladesh Wars, 1971–1972]

Jacques Gabriel Bridge, Blois, Loir-et-Cher, France [French Revolution, 1793; Franco-Prussian War, 1870]

Juba Nile (Freedom) Bridge, Juba, Central Equatoria, South Sudan [Sudan Civil War, 1983–2005]

London Bridge, London, England [Danish attack, 1014]

Long Biên Bridge, Hanoi, Red River Delta, Vietnam [Vietnam War, 1967 and 1972]

Loyang (Ten Thousand Peace) Bridge, Chuanchow, Fujian, China [Sino-Japanese War, 1937]

Maameltein Bridge, Maameltein, Tyre, Lebanon [Israeli-Lebanon War, 2006]

Maslenica Bridges, Zadar, Zadar County, Croatia [Croatian War of Independence, 1991]

Metro Bridge, Kiev, Kievshchyna, Ukraine [Polish-Soviet War, 1920]

Mudeirej Bridge, Sawfar, Mount Lebanon, Lebanon [Israeli-Lebanon War, 2006]

Old Bridge (Karl Theodor Bridge), Heidelberg, Baden-Württemberg, Germany [War of the Grand Alliance, 1689]

Old Stone Bridge, Regensburg, Bavaria, Germany [Thirty Years War, 1633; Napoleonic Wars, 1809]

Pont Vieux, Orthez, Pyrénées-Atlantiques, France [Peninsular Wars, 1814]

Ponte Novu, Castello-di-Rostino, Corsica, France [Franco-Corsican War, 1769]

Ponte Salario, Rome, Italy [Gothic War, 535–554; Napoleonic Wars, 1798; Italian Wars of Independence, 1867]

Regensburg Bridge, Regensburg, Bavaria, Germany [Thirty Years' War, 1633]

San Martin Bridge, Toledo, Castile-La Mancha, Spain [Castilian Civil War, 1368–1369]

Schaffhausen Bridge, Schaffhausen, Schaffhausen Canton, Switzerland [Napoleonic Wars, 1799]

Segovia Aqueduct, Segovia, Castile-León, Spain [Moors, 1072]

Sino-Korean Friendship Bridge, Sinuiji, North Pyongan, North Korea [Korean War, 1951]

Slender West Lake Five Pavilions Bridge, Yangzhou, Jiangsu, China [Taiping Rebellion 1850–1864]

Stari Most, Mostar, Herzegovina-Neretva Bosnia-Herzegovina [Serbian-Muslim War, 1993]

Svilengrad Bridge, Svilengrad, Haskova Province, Bulgaria [First Balkan War, 1912]

Three Countries Bridge, Huningue, Haut-Rhin, France [Napoleonic Wars, 1797]

Vizcaya Transporter Bridge, Portugalete, Biscay, Spain [Spanish Civil War, 1937]

Wan'an (Eternal Peace) Bridge, Pinghan, Fujian, China [local conflicts between 1742 and 1954]

Yiheyuan Summer Palace Bridges, Beijing, China [Second Opium War, 1850s; Boxer Rebellion, 1900]

Zezelj Bridge, Novi Sad, Vojvodina, Serbia [Kosovo War, 1999]

BRIDGES DAMAGED OR DESTROYED IN WORLD WARS I AND II

Although bridges have always been destroyed during wars, the losses in the two world wars were far greater and included the bridges listed below that have entries in this book:

Alf-Bullay Bridge, Bullay, Rhineland-Palatinate, Germany [1945]

Aioi Bridge, Hiroshima, Japan [1945]

Anghel Saligny Bridge, Cernavoda, Constanta, Romania [1916]

Augustus Bridge, Dresden, Saxony, Germany [1945]

Bad Kreutznach Bridge, Mainz, Rhineland-Palatinate, Germany [1945]

Betsiboka Bridge, Maevatanana, Betsiboka, Madagascar [1942]

Bizerte Bridge, Tunis, Tunisia [1944]

Borovnica Railway Viaduct, Ljubljana, Province of Ljubljana, Slovenia * [1941, 1944]

Chains Bridge, Fornoli, Lucca, Italy [1945]

Chaumont Viaduct, Chaumont, Haute Marne, France [1944]

Cize-Bolozon Viaduct, Daranche, Tarn, France [1944]

Covered Bridge, San Martino, Pavia, Italy [1945]

Darnytskyi Railway Bridge, Kiev, Kievshchyna, Ukraine [1943]

Djurdjevica Tara Bridge, Zabljak, Northern Region, Montenegro [1942]

Elisabeth Bridge, Budapest, Central Hungary, Hungary [1945]

Enz Viaduct, Bietigheimer, Baden-Württemberg, Germany [1945]

Glienicke Bridge, Potsdam, Brandenburg, Germany [1945]

Gorica Bridge, Berat, Berat County, Albania [1918]

Harbour Footbridge, Zadar, Zadar County, Croatia [1944]

Hohenzollern Bridge, Cologne, North Rhine-Westphalia, Germany [1945]

Jacques Gabriel Bridge, Blois, Loir-et-Cher, France [1940, 1944]

Japoma Bridge, Edea, Littoral Province, Cameroon [1914]

Kuldiga Bridge, Kurzeme, Courland, Latvia [1915]

Langeais Bridge, Langeais, Indre-et-Loire, France [1940]

Langlois Bridge, Arles, Bouches-du-Rhône, France [1944]

Liberty Bridge, Budapest, Central Hungary, Hungary [1945]

Lipcani-Rădăuți Bridge, Lipcani, Bessarabia, Moldova [1941, 1944]

Ma Kham Bridge, Kanchanaburi, Thailand [1945]

Medjez-el-Bab Bridge, Béja, Tunisia [1942]

Mehmed-Pasha Sokolovic Bridge, Visegrad, Republika Srpka, Bosnia and Herzegovina [WWI, WWII]

Metro Bridge, Kiev, Kievshchyna, Ukraine [1941, 1943]

Montjean-sur-Loire Bridge, Montjean-sur-Loire, Maine-et-Loire, France [1940]

Moresnet Viaduct, Plombières, Liège, Belgium [1940, 1944]

Muskauer Park Bridges, Bad Muskau, Saxony, Germany [1945]

Nogat Bridge, Malbork, Zulawy Region, Poland [1945]

Old Bridge (Karl Theodor Bridge), Heidelberg, Baden-Württemberg, Germany [1945]

Old Bridge, Sospel, Alpes-Maritimes, France [1944]

Old Main Bridge, Würzburg, Bavaria, Germany [1945]

Old Stone Bridge, Regensburg, Bavaria, Germany [1945]

Old Suspension Bridge, Mallemort, Bouches-du-Rhône, France [1940]

Plougastel Bridge, Brest, Finistère, France [1944]

Pont des Arts, Paris, Ile-de-France, France [1918, 1944]

Ponte degli Alpini, Bassano del Grappa, Vicenza, Italy [1945]

Pont Galliéni, Lyon, Rhône, France [1944]

Ponte Scaligero, Verona, Italy [1945]

Roman Bridge, Mostanica, Niksic Municipality, Montenegro [1942]

Roman Bridge, Vaison-la-Romaine, Vaucluse, France [WWII]

Santa Trinita Bridge, Florence, Tuscany, Italy [1944]

Sebara Dildiy (Broken Bridge), Motta, Amhara, Ethiopia [WWII]

Soca Railway Bridge, Solkan, Gorizia, Slovenia [1916]

St Servatius Bridge, Maastricht, Limburg, Netherlands [1944]

Svinesund Bridge, Svinesund, Västra Götaland County, Sweden [1942]

Széchenyi Chain Bridge, Budapest, Central Hungary, Hungary [1944]

Tauber Bridge, Rothenburg ob der Tauber, Bavaria, Germany [1945]

Three Countries Bridge, Huningue, Haut-Rhin, France [1944]

Vistula Bridges, Dirschau, Eastern Pomerania, Poland [1939]

Vytautas the Great Bridge, Kaunas, Kaunas County, Lithuania [WWII]

BRIDGES FEATURED IN PAINTINGS BY FAMOUS ARTISTS

Artists have always loved painting bridges – a subject worthy of a book on its own. Many of the structures described in this book have been depicted by well-known artists, including the following very limited selection of some of the best-known bridges and artists:

Argenteuil Bridges, Argenteuil, Val-d'Oise, France [Caillebotte, Monet, Renoir, Sisley, Vlamink]

Augustus Bridge, Dresden, Saxony, Germany [Bellotto, Canaletto, Kokoscha]

Brooklyn Bridge, New York, USA [de Glehn, Hassam, Maze, O'Keefe]

Bridge of Sighs, Venice, Veneto, Italy [many, including Monet, Turner]

Chain Bridge, Nuremberg, Bavaria, Germany [Dürer]

Langlois Bridge, Arles, Bouches-du-Rhône, France [van Goch]

Lokke Bridge, Sandvika, Greater Oslo, Norway [Monet]

London Bridge, London, England [Canaletto, de Jong, Derain, Monet, Turner]

Lucano Bridge, Villanova, Padua, Italy [della Valle, Piranesi]

Medjez-el-Bab Bridge, Béja, Tunisia [Carr]

Old Bridge (Karl Theodor Bridge), Heidelberg, Baden-Württemberg, Germany [Turner]

Pont des Arts, Paris, Ile-de-France, France [many, including Lépine, Marquet, Renoir, Pissarro]

Pont du Carrousel, Paris, Ile-de-France, France [Monet, Pissarro, Renoir, van Goch]

Pont Neuf, Paris, Ile-de-France, France [many, including Brangwyn, Marquet, Pissaro, Signac, Turner]

Pont Royal, Paris, Ile-de-France, France [Girtin, Pissarro]

Ponte delle Torri Aqueduct, Spoleto, Perugia, Italy [Turner]

Ponte Salario, Rome, Italy [Robert]

Rhaetian Railway Bridges (Wiesen Viaduct), Filisur, Graubünden, Switzerland [Kirchner]

Three-Arched Bridge, Venice, Veneto, Italy [Guardi]

Santa Trinita Bridge, Florence, Tuscany, Italy [Turner]

BRIDGES ON CYCLING AND WALKING TRAILS

Some early bridges, such as Valentré Bridge, were built for long-distance pilgrimage routes (see **Pilgrim routes over bridges**), especially to Santiago de Compostela. Later bridges carrying long-distance cycling or walking routes include:

Alf-Bullay Bridge, Bullay, Rhineland-Palatinate, Germany [Kanonenbahn]

Cross Bayou Bridge, Shreveport, Louisiana, USA [Fred Marquis Pinellas Trail]

Dungeness River Bridge, Sequim, Washington, USA [Olympic Discovery Trail]

Grandfey Viaduct, Fribourg, Fribourg Canton, Switzerland

Greenisland Railway Viaducts, Antrim, County Antrim, Northern Ireland [Newtownabbey Way, Sustrans Route 93]

International Bridge, Valenca, Viana do Castelo, Portugal [Santiago de Compostela Pilgrims' Route]

Kakum Canopy Walk, Wawase, Ghana [Kakum Canopy Walk]

Kalte Rinne Viaduct, Breitenstein, Lower Austria, Austria [Semmering Railway Walk]

Kawarau Bridge, Queenstown, Otago, New Zealand [Queenstown Trail]

Route de Grandfey [Grandfey Viaduct, Fribourg, Fribourg Canton, Switzerland

Sacramento River Trail Footbridge, Redding, California, USA [Sacramento River Trail]

BRIDGES STRANDED BY CHANGES IN THE COURSE OF RIVERS

Tropical storms (called hurricanes and cyclones in different parts of the world) are caused by powerful low-pressure weather systems leading to extremely heavy rainfall which can result in river banks being suddenly overwhelmed and, in the most extreme cases, the course of the river being changed. Three bridges in this book have been left isolated in such circumstances:

Ain Diwar Bridge, Ain Diwar, Al Hasakah Governorate, Syria [Tigris River; date not known but later than twelfth century]

Choluteca Bridges, Choluteca, Choluteca Province, Honduras [Choluteca River; 1998]

Lima Bridge, Ponte de Lima, Norte, Portugal [Lima River; between Roman times and 1125]

BRIDGES WITH DECORATIVE TOWERS

Although medieval bridges were often built with fortified towers to help defend the structure and prevent enemy troops from crossing (see **Fortified Bridges**), modern bridges can include towers that are purely decorative. Such bridges with entries in this book include:

Nanjing Yangtze River Bridge, Nanjing, Jiangsu, China

Oberbaum Bridge, Berlin, Brandenburg, Germany

Peacock Island Palace Bridge, Potsdam, Brandenburg, Germany

Stanley Bridge, Alexandria, Alexandria Governorate, Egypt

Tsaritsyno Park Figurny Bridge, Moscow, Russia

CASTLE BRIDGES

Bridges that were part of integrated defensive systems, such as at medieval castles, include the following:

Ashford Castle Bridge, Clonbur, County Mayo, Ireland

Castle Bridge, Kamyanets Podilsky, Khelmnytskyi, Ukraine

Chateau Moat Bridge, Fère-en-Tardenois, Aisne, France

Chateau Nové Mesto and Metuji, Krcin, Eastern Bohemia, Czech Republic

Châzelet Castle Moat Bridge, Châzelet, Centre-Val de Loire, France

Citadel Bridge, Aleppo, Syria

Pernstejn Castle Footbridges, Nedvedice, South Moravia, Czech Republic

Ponte Scaligero, Verona, Italy

State Castle Moat Bridge, Ceský Krumlov, South Bohemia, Czech Republic

CHAIN SUSPENSION BRIDGES

It is believed the first traditional chain link suspension bridges appeared in China between the seventh and tenth centuries CE. The first chain bridge in the UK was Wynch Bridge (see *AEBB*), built in 1741, and the more convenient flat eyebar chains were invented by Captain Samuel Brown in the UK and first used for the Brighton Chain Pier (see *AEBB*) in 1817. Chain bridges with entries in this book include:

Anlan Suspension Bridge, Dujiangyan City, Sichuan, China

Augartenbrücke, Vienna, Austria

Cavenagh Bridge, Singapore

Chain Bridge, Newburyport, Massachusetts, USA

Chain Bridge, Nuremberg, Bavaria, Germany

Chains Bridge, Fornoli, Italy

Coulouvrenière Bridge, Geneva, Switzerland

Elisabeth Bridge, Budapest, Central Hungary, Hungary

Hercilio Luz Bridge, Florianopolis, Santa Catarina, Brazil

Jihong (Rainbow in the Clear Sky) Bridge, Yongpin, Yunan, China

Kawarau Bridge, Queenstown, New Zealand
Luding Bridge, Garze, Sichuan, China
Luzancy Bridge, Luzancy, France
Menai Straits Bridges, Menai Bridge, Isle of Anglesey, Wales
Metro Bridge, Kiev, Ukraine
Moyola Park Bridge, Castledawson, Ulster, Northern Ireland
Stadlec Chain Bridge, Stadlec, Czech Republic
Széchenyi Chain Bridge, Budapest, Hungary
Tachogang Lhakang Bridge, Paro, Bhutan

CHAPEL BRIDGES

Many medieval river bridges included chapels or similar buildings, often built on a lengthened mid-span pier, where priests said masses for the souls of the bridge's benefactors. Bridge chapels that are noted in this book include:

Avignon Bridge, Avignon, Vaucluse, France
Chapel Bridge, Lucerne, Switzerland
Covered Bridge, San Martino, Pavia, Italy
Devil's Bridge, Martorell, Barcelona, Spain
London Bridge, London, England
Merchants' Bridge, Erfurt, Thuringia, Germany
Montauban Bridge, Montauban, Tarn-et-Garonne, France
Old Stone Bridge, Regensburg, Bavaria, Germany
Regensburg Bridge, Regensburg, Bavaria, Germany

CHARITABLE BRIDGE BUILDING

Even though building bridges is an especially expensive type of construction work, many older ones were built by the church to benefit the community rather than for strictly commercial reasons. Modern structures built by charitable organisations are particularly worth recording and this book includes two built by the charity Bridges for Prosperity:

La Marca Suspension Bridge, Villa Azurduy, Chuquisaca, Bolivia
Ti Peligre Bridge, Cange, Centre, Haiti

COPY BRIDGES

Bridges with entries in this book that have been built as a deliberate copy of an earlier structure built elsewhere include:

Rio Cobre Bridge, Spanish Town, Middlesex, Jamaica [copy of Wearmouth Bridge in Sunderland, England (see *AEBB*)]
Tsarskoye Selo Marble Bridge, Saint Petersburg, Northwestern, Russia [copy of Palladian Bridge, Wilton, England (see *AEBB*)]
Wörlitz Garden Bridges, Dessau-Wörlitz, Saxony-Anhalt, Germany [copies of Kew and Coalbrookdale Bridges in England (see *AEBB*)]

CORBELLED ARCH BRIDGES

Most stone arch bridges are built by successively laying specially-shaped stones called voussoirs on a temporary curved support called centring. When the last stone (the keystone) has been laid the arch becomes self-supporting and the centring can be removed. Corbelled arches (sometimes called triangular arches) are laid by progressively cantilevering a series of successive horizontal courses beyond the one below, securing it with a further stone above and across the new joint. When the gap between the separate cantilevered arms has been closed further continuous courses are laid above to strengthen the structure. Examples of corbelled arch bridges in this book include:

Kazarma Bridge, Arkadiko, Peloponnese, Greece
Spean Praptos Bridge, Chi Kiaeng, Siem Reap, Cambodia

COVERED BRIDGES

Covered timber bridges were built so that the roof and sidings (which could relatively easily be replaced if necessary) would protect the timber trusses from the weather and thus reduce the risk of structural deterioration. (See also Lounge,

Pavilion, Pagoda and Teahouse Bridges.) Examples of covered bridges in this book include:

Amstel Park Footbridge, Amsterdam, North Holland, Netherlands

Bad Säckingen Covered Bridge, Bad Säckingen, Baden-Württemberg, Germany

Batuan Bridge, Sanjiang, Guangxi, China

Beijian Bridge, Taishun, Zhejiang, China

Blenheim Covered Bridge, Blenheim, New York, USA

Bogoda Wooden Bridge, Badulla, Uva, Sri Lanka

Bridgeport Bridge, Bridgeport, California, USA

Burr Covered Bridge, Schenectady, New York, USA *

Cernvir Covered Bridge, Pernstejn, South Moravia, Czech Republic

Chain Bridge, Nuremberg, Bavaria, Germany

Chateau Nové Mesto and Metuji, Krcin, Eastern Bohemia, Czech Republic

Chengyang Yongii (Wind and Rain) Bridge, Ma'an, Sanjian, Guangxi, China

Cobblers' Bridge, Ljubljana, Province of Ljubljana, Slovenia

Colossus Bridge, Philadelphia, Pennsylvania, USA*

Cornish-Windsor Covered Bridge, Cornish, New Hampshire, USA

Covered Bridge, Lovech, Lovech Province, Bulgaria

Covered Bridge, San Martino, Pavia, Italy

Duck Creek Aqueduct, Metamora, Indiana, USA

Hartland Covered Bridge, Saint-John, New Brunswick, Canada

Japanese Covered Bridge, Hoi An, Quang Nam, Vietnam

Koornbrug, Leiden, South Holland, Netherlands

Kubelbrücke, Appenzell Ausserrhoden, Herisau, Switzerland

Old Rhine Bridge, Vaduz, Oberland, Liechtenstein

Panzendorfer Bridge, Lienz, Tyrol, Austria

Pernstejn Castle Footbridges, Nedvedice, South Moravia, Czech Republic

Ponte degli Alpini, Bassano del Grappa, Vicenza, Italy

Powerscourt Covered Bridge, Hinchinbrooke, Quebec, Canada

Rhine Bridge, Bad Säckingen, Baden-Württemberg, Germany

Rulong (Like a Dragon) Bridge, Quingyuan, Sichuan, China

Schaffhausen Bridge, Schaffhausen, Switzerland

Smolen-Gulf Bridge, Ashtabula, Ohio, USA

Sulphite Railroad Bridge, Franklin, New Hampshire, USA

Timber Bridges (Spreuer and Chapel), Lucerne, Switzerland

Xidong Bridge, Sixi, Taishun, China

CRIB BRIDGES

A crib bridge is one in which a road deck is constructed above, or between a series of, an open criss-cross framework of stone or timbers. The entry below is for a rare granite cribstone bridge:

Bailey Island Bridge, Harpswell, Maine, USA

DEVIL'S BRIDGES

According to old legends, the builders of some early river bridges were forced to sacrifice people, supposedly in order to appease the river gods, before they would be able to complete their work. In other stories the Devil demanded in return the life of the first living creature to cross the new structure, but was outwitted by the cunning locals ensuring this was only a dog or cow. There are three so-called Devil's Bridges in Britain (see *AEBB*) and such bridges with entries in this book are:

Devil's Bridge, Ardino, Bulgaria

Devil's Bridge, Barcelona, Spain

Devil's Bridge, Martorell, Catalonia, Spain

Devil's Bridge, Park Kromlau, Germany

Epirus Bridges, Epirus, Greece

A less cold-blooded story from Korea, which tells of seven beautiful maidens descending from the heavens at night, is related to Seonim Bridge, Seogwipo, Jeju Volcanic Islands, South Korea.

DOUBLE-DECKED BRIDGES

Bridges where a single structure supports separate decks, one above the other, include:

Akashi Kaikyo Bridge, Kobe, Hyogo Prefecture, Japan

Alf-Bullay Bridge, Bullay, Rhineland-Palatinate, Germany

Carioca Aqueduct, Rio de Janeiro, Brazil

Chongqing Bridges, Chongqing, Sichuan, China

Craigavon Bridge, Londonderry, County Londonderry, Northern Ireland

Galata Bridge, Istanbul, Turkey

George Washington Bridge, New York, USA

International Bridge, Valenca, Viana de Castelo, Portugal

Kawazu-Nanadaru Loop Bridge, Shimoda, Shizuoka, Japan

Manhattan Bridge, New York, USA

Menai Straits Bridges, Menai Bridge, Isle of Anglesey, Wales

Metro Bridge, Kiev, Kievshchyna, Ukraine

Moving Bridges (Wells Street Bridge), Chicago, Illinois, USA

Nanjing Yangtze River Bridge, Nanjing, Jiangsu, China

Oberbaum Bridge, Berlin, Brandenburg, Germany

Old Yamuna Bridge, New Delhi, Punjab, India

Øresund Bridge, Malmö, Oresund Region, Denmark

Petronas Twin Towers Skybridge, Kuala Lumpur, Selangor, Malaysia

Pont Félix-Houphouët-Boigny, Abidjan, Ivory Coast

Putra Bridge, Putrajaya, Malaysia [triple-decked]

Richmond-San Rafael Bridge, Richmond, California, USA

Sai Van Bridge, Macau, Guangdong, China

Tsing Ma Bridge, Hong Kong, China

Twin River Bridges, Chongqing, China

Umshiang Double-Decker Root Bridge, Cherrapungi, Meghalaya State, India

Whirlpool Rapids Bridges, Niagara Falls, New York State, USA

Yangtze River Bridge, Nanking, Jiangsu, China

Yangtze River Bridge, Wuhan, Hubei, China

Zayanderud Bridges, Isfahan, Iran

Eight double-decked bridges in Britain have entries in *AEBB*.

EARLY CONCRETE BRIDGES

The Romans had used an early form of concrete, mixed using pozzolanic cement made from ground volcanic rock, but although Joseph Aspdin (1778–1855) invented Portland cement in 1824, it was nearly another half a century before modern forms of concrete began to be developed. Some early concrete bridges with entries in this book include:

Alvord Lake Bridge, San Francisco, California, USA [1899]

Châzelet Castle Moat Bridge, Centre-Val de Loire, France [1875]

Garden Bridge, Grenoble, France [1855]

The first mass lime concrete bridge in Britain was Cromwell Road Railway Bridge in London, built in 1870 (see *AEBB*).

EXTENDED CONSTRUCTION TIMES OF BRIDGES

Depending on factors such as ease of access to the construction site, the size of the structure and the complexity of its design, it typically takes between one and five years to build a bridge, although accidents, changes in design and political issues can lead to lengthy delays. Some of the longest recorded construction times are noted below:

6th October Bridge, Cairo, Egypt [27 years: 1969–1996]

Bogibeel Bridge, Dibrugarh, Assam, India [16 years: 2002–2018]

Carioca Aqueduct, Rio de Janeiro, Brazil [19 years: 1706–1723]

Forth Bridge, South Queensferry, City of Edinburgh, Scotland [17 years: 1873–1890]

Puente Nuevo, Ronda, Malaga, Spain [34 years: 1759–1793]

Valentré Bridge, Cahors, Lot, France [72 years: 1308–1380]

FLOATING AND PONTOON BRIDGES

Where a new crossing is to be constructed at a difficult site, pontoon bridges are often installed first to provide a temporary crossing while the permanent structure is built. However, pontoons are sometimes part of permanent bridges as in the following bridges with entries in this book:

Demerara Harbour Bridge, Georgetown, Demerara-Mahaica Region, Guyana

Dubai Creek Floating Bridge, Dubai, United Arab Emirates

Faidherbe Bridge, Saint-Louis, Saint-Louis Region, Senegal

Floating Bridge, Enshi City, Hubei Province, China

Galata Bridge, Istanbul, Marmara Region, Turkey

Governor Albert D. Rosellini Bridge, Seattle, Washington State, USA

Guangji / Xiangzi (Great Charity) Bridge, Chaozhou, Guangdong, China

Kristiansand to Trondheim Proposed Submerged Bridge, Kristiansand, Agder County, Norway

Lacey V. Murrow Memorial Bridge, Seattle, Washington, USA

Murinsel Bridge, Graz, Bundesland, Austria

Pontoon Railroad Bridge, Wabasha, Minnesota, USA *

Queen Emma Bridge, Willemstad, Curacao, Venezuela

Vytautas the Great Bridge, Kaunas, Kaunas County, Lithuania

FOLLY BRIDGES

Folly bridges are structures, often in a landscaped park or garden (and sometimes called a *fabrique*), that pretend to be what they are not, such as a hermit's grotto or a ruined Roman aqueduct, but that provide a focal point to a view and cannot be used as a crossing. There are several folly bridges in Britain (see *AEBB*) and those that are still standing and have entries in this book include:

Arkadia Park Aqueduct, Lowicz, Lodz, Poland [pseudo aqueduct]

Devil's Bridge, Gablenz, Saxony, Germany [bridge cannot be crossed as, with its reflection, it forms a perfect circle]

Peacock Island Palace Bridge, Potsdam, Brandenburg, Germany [wrought iron pretend link between towers]

FORTIFIED BRIDGES

The strategic importance of medieval crossings over rivers means they have often been fortified with defensive measures such as towers, from which protected archers could keep potential attackers at bay, and passageway over them has been restricted by drawbridges. Later, however, and reflecting those historic times, bridges were sometimes built with towers that were purely decorative – see **Bridges with Decorative Towers**). Bridges that still retain original fortified features such as castellated towers and that have entries in this book include:

Águas Livres Aqueduct, Lisbon, Estremadura, Portugal

Alcántara Bridge, Toledo, Castile-La Mancha, Spain

Anghel Saligny Bridge, Cernavoda, Romania

Attock Railway Bridge, Attock Khurd, Punjab Province, Pakistan

Bad Kreutznach Bridge, Mainz, Rhineland-Palatinate, Germany

Charles Bridge, Prague, Bohemia, Czech Republic

Chateau Moat Bridge, Fère-en-Tardenois, Aisne, France

Coulouvrenière Bridge, Geneva, Switzerland

Lomonosov Bridge, St Petersburg, Northwestern, Russia

London Bridge, London, England

Malan Bridge, Herat, Herat Province, Afghanistan

Nogat Bridge, Malbork, Zulawy Region, Poland

Oberbaum Bridge, Berlin, Brandenburg, Germany

Old Bridge, Albi, Tarn, France

Old Bridge, Pisek, South Bohemia, Czech Republic

Old Bridge, Sospel, Alpes-Maritimes, France

Old Stone Bridge, Regensburg, Bavaria, Germany

Pont des Trous, Tournai, Wallonia, Belgium

Pont Saint-Jacques, Parthenay, Deux-Sèvres, France

Pont Vieux, Orthez, Pyrénées-Atlantiques, France

Ponte della Maddalena, Borgo, Italy

Ponte delle Torri Aqueduct, Spoleto, Perugia, Italy

Ponte Salario, Rome, Italy

Ponte Scaligero, Verona, Italy

Regensburg Bridge, Regensburg, Bavaria, Germany

Roman Bridge, Frias, Burgos, Spain

Roman Bridge, Salamanca, Spain

Roman Bridge, Trier, Rhineland-Palatinate, Germany

San Martin Bridge, Toledo, Castile-La Mancha, Spain

Santa Teresa Bridge, Elche, Valencia, Spain

Shaharah Bridge, Shaharah, Yemen

Stari Most, Mostar, Herzegovina-Neretva Bosnia-Herzegovina

Stone Bridge, Zamora, Castile-León, Spain

Széchenyi Chain Bridge, Budapest, Hungary ???

Tauber Bridge, Rothenburg ob der Tauber, Germany

Treponti, Comacchio, Ferrara, Italy

Ucanha Bridge, Tarouca, Viseu, Portugal

Valentré Bridge, Cahors, Lot, France

Vienne Bridge, Confolens, Charente, France

Vistula Bridges, Dirschau, Eastern Pomerania, Poland

Two fortified bridges remain in Britain: Monnow and Warkworth (see *AEBB*).

GISCLARD SUSPENSION BRIDGES

A revolutionary suspension bridging system designed by Albert Gisclard (1844–1909), which was later licensed by Ferdinand Arnodin, consisted of a combination of diagonal stays near the main support towers and hangers from suspension cables for the central part of the span. Those still standing and that have entries in this book include:

La Cassagne Bridge, Planès, Pyrenees-Orientales, France

Lézardrieux Bridge, Lézardrieux, Côtes-d'Armor, France

Passerelle Marguerite, La Foa, South Province, New Caledonia

Pont de Cassagne, Planès, Eastern Pyrenees, France

Rochers Noirs Viaduct, Lapleau, Corrèze, France

Santa Fe Suspension Bridge, Santa Fe de la Vera Cruz, Santa Fe, Argentina

GLASS BRIDGES

A number of modern bridges, which have been built with glass decks and steps, include:

Aizhai Bridge, Jishou, West Hunan, China

Bridge of Peace, Tbilisi, Georgia

Campo Volantin Footbridge, Bilbao, Biscay, Spain

Constitution Bridge, Venice, Veneto, Italy

East Taiheng Glasswalk, Zhangjiajie, Hebei Province, China

Grand Canyon Skywalk, Peach Springs, Arizona, USA

Kinzua Viaduct, Westline, Pennsylvania, USA

Zhangjiajie Glass Bridge, Hunan, Hunan Province, China

GLULAM BRIDGES

Glulam consists of accurately planed thin strips of timber that are glued and laminated together under heat and pressure to make a single structural unit that is stronger, more stable and less liable to decay than a similarly-sized piece of natural timber. The technique can be used to make beams, posts, arches or decking. The following structures with entries in this book contain glulam members:

Dunajec Footbridge, Stromowce Nizne, Nowy Ttarg, Poland

Flisa Bridge, Flisa, Innlandet County, Norway

Leonardo Bridge, Ås, Viken County, Norway

Maronne Bridge, Saint-Geniez-ô-Merle, Corrèze, France

Vallikraavi Footbridge, Pärnu, Pärnu County, Estonia

Vézère Bridge, Les Eyzies-de-Tayac-Sireuil, Dordogne, France

Vihantasalmi Bridge, Mäntiharju, Åland, Finland

HIGHEST BRIDGES

Two of the major determining factors about a bridge are the height above ground of the deck (the tallest bridges) and the overall height of the structure itself above sea level (the highest bridges) (see also **longest bridges**). Entries in this book for some of what have been the highest bridges in the world, in date order of their opening, include the following:

1890: Malleco Railway Viaduct, Collipulli, Araucania, Chile: with its rails 102m above the bottom of the gorge, it was then the tallest railway bridge in the world]

1898: Kanoh Bridge, Kalka, Haryana, India: world's highest-located railway bridge above sea level: nearly 2,000m

1912: Cidi M'Cid Bridges, Constantine, Constantine Province, Algeria [1912; with its deck at 175m above water level, this became the highest bridge in the world]

1912: Sidi Rached Viaduct, Constantine, Constantine Province, Algeria [107m high]

1929: Royal Gorge Bridge, Canon City, Colorado, USA [955ft (291m) above the Arkansas River and was the highest bridge in the world]

1973: Mala Rijeka Viaduct, Podgorica, Montenegro [this bridge 200m above the Mala Rijeka River became the highest railway bridge in the world]

1979: Kochertal Viaduct, Schwäbisch Hall, Baden-Württemberg, Germany [highest bridge in the world at 185m (607ft)]

1987: Bridge of Immortals, Tangkouzhen, Anhui Province, China [world's highest-located bridge at 1,320m (4,330ft)]

2001: Beipan River Shuibai Bridge

2004: Millau Viaduct, Millau, Aveyron, France [the deck is a maximum of 245m (804ft) above the valley floor and the pylons, in which the cable stays are anchored, reach a further 87m (285ft) above the deck and are the tallest pylons in the world]

2005: Hegigio Gorge Pipeline Bridge, Komo Station, Komo-Magarima, Papua New Guinea [393m (1,289ft) above river level – world's highest pipeline structure]

2012: Titlis Suspension Bridge, Engelberg, Obwalden, Switzerland deck is 500m (1,640ft) above ground level and 3040m (9,974ft) above sea level]

2016: Duge Beipanjiang Bridge, Liupanshui, Guizhou Province, China [highest bridge in the world at 565m (1,854ft) above the river]

2020s: Chenab Rail Bridge, Sala, Jammu and Kashmir, India [highest railway bridge in the world at 322m (1,056ft) above water level]

HOTELS WITH BRIDGES

Several hotels are in buildings that include bridge structures or have bridges within their grounds, thus offering a particularly interesting experience for bridge lovers. Those with entries in this book include the following:

Ashford Castle Bridge, Clonbur, County Mayo, Ireland

Chateau Moat Bridge, Fère-en-Tardenois, Aisne, France

Paradise Island Royal Tower Hotel Bridge, Nassau, Grand Bahama, Bahamas

There are also sixteen hotel bridges in Britain (see *AEBB*).

INHABITED BRIDGES

Medieval bridges often had houses built on them to take advantage of the better air and simple way for disposing of waste, and later bridges have sometimes copied the concept. Inhabited bridges that have entries in this book include the following:

Bad Kreutznach Bridge, Mainz, Rhineland-Palatinate, Germany

London Bridge, London, England

Merchants' Bridge, Erfurt, Thuringia, Germany

Old Bridge, Albi, Tarn, France

LONGEST BRIDGES AND SPANS

Two of the major determining factors about a bridge are its overall length and the length of its longest span (but see also **highest bridges** and **widest bridges**). However, in order to keep things relatively simple, this entry is limited to a selection of the longest spans and bridges in the world for any type of structure, shown in date order of their opening, and a single list showing the longest span for each of the main type of bridge structure: arch, cable stay, cantilever, suspension and truss.

Some Historic Longest Overall Bridge Lengths

1700: Rhine Bridge, Bad Säckingen, Baden-Württemberg, Germany: longest timber covered bridge in Europe (668ft)

1762: Vanvitelli Aqueduct, Caserta, Italy: longest bridge structure in Europe 1,736ft (5696m)

1883: Lake Pontchartrain Rail Bridge, New Orleans, Louisiana, USA: world's longest bridge over water (21½ **miles**)

1968: Nanjing Yangtze River Bridge, Nanjing, Jiangsu, China: world's longest road:rail bridge 1,576m (5,170ft)

1978: Demerara Harbour Bridge, Georgetown, Guyana: world's longest floating bridge (6,074ft)

1999: Vihantasalmi Bridge, Mäntiharju, Åland, Finland: world's longest timber main road bridge (168m)

2006: Dunajec Footbridge, Stromowce Nizne, Nowy Ttarg, Poland: world's longest laminated timber deck (112m)

2011: Danyang-Kunshan Grand Bridge, Wuxi, Jiangsu, China: world's longest bridge (165km long)

2016: Governor Albert D. Rosellini Bridge, Seattle, Washington State, USA: world's longest floating bridge (7,710ft)

Some Historic Longest Span Lengths

1866: John A. Roebling Suspension Bridge, Covington, Kentucky, USA: longest span 322m (1,057ft)

1890: Firth of Forth Bridges, South Queensferry, City of Edinburgh, Scotland: longest span 521m (1,710ft)

1937: Golden Gate Bridge, San Francisco, California, USA: longest span 1280m (4200ft)

1977: New River Gorge Bridge, Fayetteville, West Virginia, USA: world's longest single-span bridge 1,700ft

1981: Humber Bridge, Barton-upon-Humber, North Lincolnshire, England: world's longest span 1,410m (4,626ft)

1988: Jundushan Aqueduct, Yanqing,Beijing, China: world's longest aqueduct span (126m)

1998: Akashi Kaikyo Bridge, Kobe, Hyogo Prefecture, Japan: world's longest span 1991m (6,532ft)

2003: Lupu Bridge, Shanghai, China: world's longest arch span 550m

2020s: Canakkale 1915 Bridge, Lapseki, Cannakkale, Turkey: world's longest suspension span 2,023m (6,637ft); length 4,608m (15,118ft)

Some Current Longest Span Lengths of Bridges by Type of Structure

Arch: Chaotianmen Bridge, Chongqing, Sichuan, China, 2009: 552m (1,811ft) long

Cable Stay: Russky Bridge, Vladivostok, Primorsky Krai, Russia, 2012: 1104m (3,622ft) long

Cantilever: Quebec Bridge, Quebec, Canada, 1919: 1,800ft (549m) long

Floating Bridge: Demerara Harbour Bridge, Georgetown, Demerara-Mahaica Region, Guyana, 1978: 6,074ft

Suspension: Akashi Kaikyo Bridge, Kobe, Hyogo Prefecture, Japan, 1998: 1,991m (6,532ft) long

Swing: El Ferdan Railway Bridge, Ismaelia, El Qantara, Egypt: twin balanced swinging cantilevers have 1,100ft span

Truss: Ikitsuki Bridge, Nagasaki, Japan, 1991: 400m (1,312ft) long

Vertical Lift: Arthur Kill Vertical Lift Bridge, Elizabeth, New Jersey, USA, 1959: 558ft long

LOUNGE, PAGODA, PAVILION AND TEAHOUSE BRIDGES

A special type of enclosed bridge in the Far East are the so-called lounge bridges, although they often also include temples. Bridges with buildings such as pavilions, where those crossing could rest and shelter, that have entries in this book include:

Anlan Suspension Bridge, Dujiangyan City, Sichuan, China

Anping (Peace) Bridge, Fujian, Hebei, China

Anshun (Peaceful and Favourable) Bridge, Chengdu, Sichuan, China

Baling Bridge, Weiyun, Gansun, China

Batuan Bridge, Sanjiang, Guangxi, China

Beijian Bridge, Taishun, Zhejiang, China

Black Dragon Pool Bridge, Lijiang, Yunnan, China

Bridge Pavilion, Zaragoza, Aragon, Spain

Buchan (Stepping Toad) Bridge, Qingyuan, Zhejiang, China

Chateau Veltrusy's Laudon Pavilion, Veltrusy, Bohemia, Czech Republic

Chengyang Yongji (Wind and Rain) Bridge, Ma'an, Sanjian, Guangxi, China

Dromana Bridge, Villierstown, Waterford, Ireland

Guangji / Xiangzi (Great Charity) Bridge, Chaozhou, Guangdong, China

Hangzhou Old Bridges, Hangzhou, Zhejiang Province, China

Haoshang Bridge, Leshan, Sichuan Province, China

Japanese Covered Bridge, Hoi An, Quang Nam, Vietnam

Jiemei (The Sisters') Bridges, Miamyang Xiaobei, Mianyang, Sichuan, China

Jingxing Qialoudian Bridge, Jingxing, Hebei, China

Lotus Lake Footbridges, Kaohsiung, Taiwan

Rama VIII Bridge, Bangkok, Thailand

Red Army Bridge, Taishun, Zhejiang, China

Rulong (Like a Dragon) Bridge, Quingyuan, Sichuan, China

Shahi Mughal Bridge, Jaunpur, Utter Pradesh, India

Slender West Lake Bridges, Yangzhou, Jiangsu, China

Stanley Bridge, Alexandria, Alexandria Governorate, Egypt

Tsarskoye Selo Park Bridges, Saint Petersburg, Northwestern, Russia

Tunca Bridge, Erdine, Turkey

U Bein Bridge, Amarapura, Myanmar

Wan'an (Eternal Peace) Bridge, Pinghan, Fujian, China

West Lake Pavilion Bridge, Hangzhou, China

Wuting (Five Pavilions) Bridge, Yangzhou, Jiangsu Province, China

Xidong Bridge, Sixi, Taishun, China

Yangtze River Bridge, Wuhan, China

Yongqing (Eternal Celebration) Bridge, Taishun, Zhejiang, China

Yu Yuan Bazaar / Garden Bridge, Shanghai, China

Yudai (Jade Belt) Bridge, Hangzhou, Zhejiang Province, China

Zayanderud Pol-e-Khaju and Siosopol Bridges, Isfahan, Iran

MEDIEVAL BRIDGES

For bridges built before 1000CE see **Ancient Bridges**. Bridges built between 1000 and 1500CE include:

Chapel Bridge, Lucerne, Switzerland [C14]

Citadel Bridge, Aleppo, Syria [C13]

Devil's Bridge, Martorell, Barcelona, Spain [C13]

Eurymedon Bridge, Aspendos, Antalya Province, Turkey [C13]

Geumcheongyo Bridge, Seoul, Gyeonggi Province, South Korea [1411]

Gothic Bridge, Vilomara, Barcelona, Spain [C12]

Guangji / Xiangzi (Great Charity) Bridge, Chaozhou, Guangdong, China [1170]

Khudafarin Bridges, Jabrayil, Hadrut, Azerbaijan [C11 or 12]

Loyang (Ten Thousand Peace) Bridge, Chuanchow, Fujian, China [1059]

Lucano Bridge, Villanova, Italy [1192]

Lugou / Lukou Bridge, Beijing, China [1192]

Magdalena Footbridge, Pamplona, Navarre, Spain [C12]

Malan Bridge, Herat, Herat Province, Afghanistan [C12]

Old Bridge, Albi, Tarn, France [C11]

Old Bridge, Béziers, Hérault, France [C12]

Old Bridge, Sospel, Alpes-Maritimes, France [C13]

Pont del Escalls, Escaldes-Engordany, Andorra [C13]

Pont des Trous, Tournai, Wallonia, Belgium [C13]

Ponte della Maddalena, Borgo a Mozzano, Lucca, Italy [C12]

Ponte delle Torri Aqueduct, Spoleto, Perugia, Italy [C13]

Queen's Bridge, Ourense, Navarre, Spain [C11]

Red Bridge, Meore Kesalo, Kvemo Kartli, Georgia [C12]

Regensburg Bridge, Regensburg, Bavaria, Germany [C12]

Roman Bridge, Frias, Burgos, Spain [C12]

San Martin Bridge, Toledo, Castile-La Mancha, Spain [1380]

Shahrestan Bridge, Isfahan, Iran [C11]

Spean Praptos Bridge, Chi Kiaeng, Siem Reap, Cambodia [C12]

Stone Bridge, Zamora, Castile-León, Spain [C13]

Vienne Bridge, Confolens, Charente, France

Yibna Bridge, Yavne, Central District, Israel [1274]

MEMORIAL BRIDGES

Important bridges have often been named after the sovereign or president of the country in which they stand and some bridge names honour their designer or promoter. However, a slightly unusual memorial to people who have died is a bridge, and bridges are sometimes also named to commemorate events in history. Memorial bridges in this book include:

4 de Abril Bridge, Benguela, Angola [end of civil war 2002]

25th April Bridge, Lisbon, Portugal [start of Carnation Revolution 1974]

6th October Bridge, Cairo, Egypt [start of Yom Kippur War in 1973]

Adolphe Bridge, Luxembourg City, Luxembourg [grand duke]

Alexandre III Bridge, Paris, France [czar]

Allenby (King Hussein) Bridge, Jordan [general (and king)]

Angus MacDonald Bridge, Halifax, Canada [state premier]

Atatürk Bridge and Atatürk Viaduct, Turkey [president]

Benjamin Franklin Bridge, Philadelphia, USA [president]

Bosphorus Bridge (15th July Martyrs Bridge), Istanbul, Turkey [250 people were killed in a failed coup in 2016]

Bridge of Appollodorus, Drobeta-Turnu, Romania [Roman military engineer]

Bridge of Boyacà, Tunja, Boyacà, Colombia [battle for independence 1819]

Bridge of Immortals, Tangkouzhen, Anhui Province, China [memorialises those who have fallen to their deaths]

Bridge of Remembrance, Christchurch, Canterbury, New Zealand [dedicated to the dead of WWI and later conflicts]

Canakkale 1915 Bridge, Lapseki, Cannakkale, Turkey [WWI campaign in Gallipoli]

Chamberlain Bridge, Bridgetown, Antigua and Barbuda [British Colonial Secretary]

Chateau Veltrusy's Laudon Pavilion, Veltrusy, Bohemia, Czech Republic [General Ernst Gideon von Laudon]

Delaware Memorial Bridge, Wilmington, Delaware, USA [local service people who died in WWII and afterwards]

Detroit-Superior (Veterans Memorial) Bridge, Cleveland, Ohio, USA [war veterans]

Eads Bridge, St Louis, USA [bridge designer James Buchanan Eads]

Firth of Forth Bridges, South Queensferry, City of Edinburgh, Scotland [Queen (later Saint) Margaret]

Francis Scott Key Bridge, Baltimore, Maryland, USA [writer, in 1812, of words to the *Star Spangled Banner*]

George P. Coleman Memorial Bridge, Yorktown, Virginia, USA [Head of Virginia Highway Commission]

George Washington Bridge, New York, USA [president]

Gorbaty Bridge, Minsk, Belarus [memorial to 800 soldiers killed in Afghanistan]

Haile Selassie Abay Bridge, GobaTsion, Ethiopia* [emperor]

Homer M. Hadley Memorial Bridge, Seattle, Washington, USA [Seattle bridge engineer]

Lacey V. Murrow Memorial Bridge, Seattle, Washington, USA [local director of highways]

Marco Polo Bridge, Beijing, China [Venetian traveller and merchant]

Mears Memorial Bridge, Nenana, Alaska, USA [chairman and chief engineer of the Alaska Engineering Commission]

Mohammed VI Bridge, Rabat, Morocco * [king]

Octavio Frias de Oliveira Bridge, Sao Paulo, Brazil [newspaper proprietor]

Oversteek Bridge, Nijmegan, Gelderland, Netherlands [American forced crossing of the river in 1944]

Red Army Bridge, Taishun, Zhejiang, China [Mao's forces successfully crossed the river here in 1937]

Red Women's Bridge, Huixian, Henan Province, China [built by 80 young peasant women]

Sonjuk Bridge, Kaesong, North Hwanghae, North Korea [assassination of statesman Jeong Mong-ju in 1392]

Unification Bridge, Rome, Italy [Italian Risorgimento]

Vasco da Gama Bridge, Lisbon, Estremadura, Portugal [explorer]

Vimy Memorial Bridge, Ottawa, Ontario, Canada [WWI battle]

Yavuz Sultan Selim Bridge, Istanbul, Turkey [Sultan Selim I]

Galata Bridge, Istanbul, Marmara Region, Turkey

Kingdom Centre Skywalk, Riyadh, Riyadh Region, Saudi Arabia

Oberbaum Bridge, Berlin, Brandenburg, Germany

Pinnacle Sky Bridges, Singapore

Royal Opera House Footbridge, Muscat, Oman

Sai Van Bridge, Macau, Guangdong, China

Santa Justa Lift, Lisbon, Estremadura, Portugal

Siosepol Bridge, Isfahan, Iran

State Castle Moat Bridge, Ceský Krumlov, South Bohemia, Czech Republic

Station Bridge, Beersheba, Northern Negev, Israel

Synod Hall Footbridge, Dublin, County Dublin, Ireland

Yongqing (Eternal Celebration) Bridge, Taishun, Zhejiang, China

Zhivopisny Bridge, Moscow, Russia

For bridges that include special areas which the public can access at the top of bridge towers or pylons, see **Observation Deck Bridges**.

MODERN ENCLOSED BRIDGES

Although most bridges are open-deck structures, there are a number in which the passageway across the deck is enclosed in some way. The early **covered bridges** were a type of enclosed bridge in which the barn-like superstructure served primarily to protect the structural parts from the weather and **Lounge, Pavilion, Pagoda and Teahouse Bridges** are all different types of historic enclosed bridge. On modern enclosed bridges, however, the enclosures serve to provide a large weather-proof deck area where people can assemble for dining or shopping. Such bridges include:

Bogdan Khmelnitsky Pedestrian Bridge, Moscow, Russia

Bridge Pavilion, Zaragoza, Aragon, Spain

Buchan (Stepping Toad) Bridge, Qingyuan, Zhejiang, China

Chateau Moat Bridge, Fère-en-Tardenois, Aisne, France

Chateau Veltrusy's Laudon Pavilion, Veltrusy, Bohemia, Czech Republic

MONO-CABLE SUSPENSION BRIDGES

Nearly all suspension bridges have two main cables, one above each side of the deck. Two bridges in this book are unusual for having only a single cable. These are:

College Footbridge, Kortrijk, West Flanders, Belgium

Peace Bridge, Derry, County Londonderry, Northern Ireland

MONORAIL BRIDGES

Bridges that were built to carry monorail transport systems and have entries in this book include:

Jacques Chaban-Delmas Lifting Bridge, Bordeaux, Nouvelle-Aquitaine, France

Monorail Suspension Bridge, Putrajaya, Malaysia

Putra Bridge, Putrajaya, Malaysia

Tokyo-Haneda Monorail Bridge, Kawasaki, Kanagawa, Japan

NATIONAL HERITAGE / MONUMENT BRIDGES

Many bridges are considered to be of international historic value and some have been officially registered by UNESCO (see **UNESCO World Heritage Bridges**). Bridges with entries in this book that have been listed as national treasures and cultural heritage monuments in their own countries are noted below:

Águas Livres Aqueduct, Lisbon, Estremadura, Portugal

Anghel Saligny Bridge, Cernavoda, Constanta, Romania

Blenheim Covered Bridge, Blenheim, New York, USA

Cabin John Aqueduct, Cabin John, Maryland, USA

Chamborigaud Viaduct, Chamborigaud, Gard, France

Devil's Bridge, Ardino, Smolyan, Bulgaria

Flat Bridge, St Catherine, Jamaica

Gignac Bridge, Gignac, Hérault, France

Gladesville Bridge, Sydney, New South Wales, Australia

Hartland Covered Bridge, Saint-John, New Brunswick, Canada

Hawksworth Suspension Bridge, San Ignacio, Western Belize, Belize

Hercilio Luz Bridge, Florianopolis, Santa Catarina, Brazil

Kintai Bridge, Iwakuni, Yamaguchi Prefecture, Japan

La Cassagne Bridge, Planès, Pyrenees-Orientales, France

Lima Bridge, Ponte de Lima, Norte, Portugal

Longteng Bridge, Sanyi Township, Miaoli County, Taiwan

Malleco Railway Viaduct, Collipulli, Araucania, Chile

Medjez-el-Bab Bridge, Béja, Tunisia

Musmeci Bridge, Potenza, Basilicata, Italy

Old Bridge, Béziers, Hérault, France

Old Bridge, Pisek, South Bohemia, Czech Republic

Panzendorfer Bridge, Lienz, Tyrol, Austria

Pompey's Bridge, Mtskheta, Mtskheta-Mtianeti, Georgia

Pont des Arts, Paris, Ile-de-France, France

Powerscourt Covered Bridge, Hinchinbrooke, Quebec, Canada

Puente Yayabo, Sancti Spiritus, Sancti Spiritus, Cuba

Richmond Bridge, Richmond, Tasmania, Australia

Rio Cobre Bridge, Spanish Town, Middlesex, Jamaica

Rochefort-Martrou Transporter Bridge, Rochefort-sur-Mer, Charente-Maritime, France

Rochers Noirs Viaduct, Lapleau, Corrèze, France

Roman Bridge, Vaison-la-Romaine, Vaucluse, France

San Martin Bridge, Toledo, Castile-La Mancha, Spain

Soca Railway Bridge, Solkan, Gorizia, Slovenia

Sonjuk Bridge, Kaesong, North Hwanghae, North Korea

Stadlec Chain Bridge, Stadlec, South Bohemia, Czech Republic

Starrucca Viaduct, Lanesboro, Pennsylvania, USA

Strimka Lan Bridge, Kamenets-Podolski, Ukraine

Swandbach Bridge, Hinterfultigen, Bern, Switzerland

Viaur Viaduct, Tanus, Tarn, France

Yayabo Bridge, Sancti Spiritus, Sancti Spiritus Province, Cuba

NATURAL FIBRE BRIDGES

Bridges have been built of natural fibres since the beginning of time and this book includes bridges made from palm fronds, feather grass, bamboo and climbing and twining plants:

Anlan Suspension Bridge, Dujiangyan City, Sichuan, China [first bridge here was of bamboo, 3rd century BC]

Bamboo Bridge, Surakarta, Central Java Province, Java [curved arches of bamboo]

Duy Xuyen Bamboo Bridge, Hoi An, Qang Nam, Vietnam [bamboo X-frame supports and deck]

Kazurabashi Vine Bridge, Shikoku, Ehime Prefecture, Japan [vines wrapped round and hanging from a steel cable core]

Keshwa Chaca Bridge, Huinchiri, Cuzco, Peru [woven from feathergrass]

Koh Pen Bamboo Bridge, Kampong Cham, Kampong Cham Province, Cambodia [bamboo stick structure and deck of bamboo strips]

Poubara Liana Bridge, Franceville, Haut-Ogooué, Gabon [woven from liana]

Vine Bridge, Koulé, Nzérékoré Prefecture, Guinea [56m-long bridge rebuilt every year]

Zuwulufu Bridge, Kandewe, Rumphi District, Malawi [palm frond suspension ropes and woven bamboo deck]

OBSERVATION DECK BRIDGES

Some bridges include special observation decks on the top of the bridge towers or pylons:

New Bridge, Bratislava, Slovakia [enclosed circular observation deck 'The Flying Saucer' on top of the pylon]

Panama Canal Fourth Bridge, Colon, Paraiso, Panama [an observation deck is planned for the top of one of the pylons]

Putra Bridge, Putrajaya, Malaysia [observation galleries]

Rama VIII Bridge, Bangkok, Thailand [observation deck on top of a pylon]

Tilikum Crossing, Portland, Oregon, USA [enlarged pedestrian belvederes around piers]

OPENING BRIDGES

Several bridges include a section that can be moved to allow traffic to pass along the route normally barred by all or part of the structure. (In Transporter Bridges, rather than a main part of the structure itself moving, the only moving part is a lightweight suspended carriage carrying foot passengers and a small number of vehicles; see the separate main entry **Transporter Bridges**.) The main types of opening bridge with example entries in this book include:

Bascule Bridges
A simple bascule bridge has an opening span, which is hinged near one end and is balanced by counterweights, that rotates like a seesaw with its free end rising up into the air. In a rolling lift bascule bridge, as the free end of the lifting deck section rises into the air, the whole section moves backwards over the abutments, in the manner of a rocking horse, thus leaving a wider opening for vessels.

Al Maktoum Bridge, Dubai, United Arab Emirates

Bizerte Bridge, Tunis, Tunisia

Butterfly Bridge, Copenhagen, Capital Region, Denmark

Chicago Bridges, Chicago, Illinois, USA

Dubai Creek Al Maktoum Bridge, Dubai, United Arab Emirates

Erasmus Bridge, Rotterdam, South Holland, Netherlands

Galata Bridge, Istanbul, Turkey

Glienicke Bridge, Potsdam, Brandenberg, Germany

International Railroad Bridge, Sault Ste Marie, Ontario, Canada

Jungfern Footbridge, Berlin, Brandenburg, Germany

Lake Pontchartrain Crossings, New Orleans, Louisiana, USA

Mountgarret Bridge, New Ross, South Leinster, Ireland

Pamban Bridge, Mandapam, Tamil Nadu, India

River Neva Bascule Bridges, St Petersburg, Northwestern, Russia

Slauerhoff Flying Drawbridge, Leeuwarden, Friesland, Netherlands

Wells Street Bridge, Chicago, Illinois, USA

Zeeland Bridge, Nieuwerkerk, South Holland, Netherlands

Defensive Drawbridges
The defensive drawbridge at the entrances to castles and cities was often an early type of bascule bridge. When the main part of the drawbridge was raised, it formed an additional barrier in front of the main gateway and the balancing back-span part of the deck would rotate into a pit which would itself then be a defensive measure. Such bridges in this book include:

Chateau Moat Bridge, Fère-en-Tardenois, Aisne, France

Chicago River Drawbridge, Chicago, Illinois, USA

London Bridge, London, England
Old Bridge, Albi, Tarn, France
Pernstejn Castle Footbridges, Nedvedice, South Moravia, Czech Republic
Sand Bridge, Wroclaw, Lower Silesia, Poland

Floating Bridges

Floating bridges can include a section that can be unlinked and towed aside to leave a clear passageway for vessels. Those in this book that had or have opening sections include:

Demerara Harbour Bridge, Georgetown, Guyana
Dubai Floating Bridge, United Arab Emirates
Galata Bridge, Istanbul, Turkey
Guangji Bridge, Chaozhou, China
Pontoon Railway Bridge, Wabasha, USA
Queen Emma Bridge, Willemstad, Curacao, Venezuela

Retracting Bridges

Retracting bridges move longitudinally back along the bridge axis as at:

Kiel-Hôrn Folding Bridge, Kiel, Schleswig-Holstein, Germany
Demerara Harbour Bridge, Georgetown, Demerara-Mahaica Region, Guyana

Submersible Bridges

The unusual submersible bridge is lowered to the bottom of the water channel so that vessels can pass over the sunken deck. The single entry in this book is for the Corinth Canal Submersible Bridges, Corinth, Peloponnese, Greece.

Swing Bridges

Swing bridges rotate horizontally like a turntable so that, when open for water traffic, the roadway is parallel to the water channel. Examples include:

Ampera Bridge, Palembang, Indonesia
Chicago Sanitary & Ship Canal Swing Bridge
Faidherbe Bridge, Saint-Louis, Saint-Louis Region, Senegal

Chamberlain Bridge, Bridgetown, Antigua and Barbuda
Five Circles Footbridge, Copenhagen, Denmark
George P. Coleman Memorial Bridge, Yorktown, Virginia, USA
Inner Harbour Bridge, Duisberg, Germany
River Neva Bascule Bridges, St Petersburg, Russia
Samuel Beckett Bridge, Dublin, County Dublin, Ireland
Seán O'Casey Bridge, Dublin, County Dublin, Ireland
Victoria & Alfred Harbour Bridge, Capetown, Western Cape, South Africa

Vertical Lift Bridges

In these structures the deck spans across the main water channel between towers which include motors that raise the deck section vertically leaving a clear passageway below for the water-borne traffic. The examples in this book include:

Ampera Bridge, Palembang, South Sumatra, Indonesia
Arthur Kill Vertical Lift Bridge, Elizabeth, New Jersey, USA
Buzzards' Bay Bridge, Bourne, Massachusetts, USA
Canal Street Bridge, Chicago, Illinois, USA
Double Lift Bridge, Portland, Oregon, USA
Duluth Lift Bridge, Duluth, Minnesota, USA
International Railroad Bridge, Sault Ste Marie, Ontario, Canada
Jacques Chaban-Delmas Lifting Bridge, Bordeaux, Nouvelle-Aquitaine, France
John T. Alsop Jr Bridge, Jacksonville, Florida, USA
Kattwyk Bridges, Hamburg, Lower Saxony, Germany
Puente Nicolas Avellaneda, Buenos Aires, Buenos Aires Province, Argentina
Vytautas the Great Bridge, Kaunas, Kaunas County, Lithuania

In an alternative form of vertical lift bridge, instead of each end of the movable section of deck being suspended from a tower, it stands on corner legs that can be jacked up, as at:

Pont Levant Notre Dame, Tornai, Hainaut, Belgium

PACKHORSE BRIDGES

Before transport by horse and cart became widespread, early trading of goods in remote parts even of well-developed countries was by packhorse or, in some countries, by camel. Bridges built for this traffic were generally simple arch structures, less than 6ft wide and with low parapet walls that would not obstruct the animals' loaded paniers. Such bridges that are described in this book include:

Aghakista Bridge, Castletownbere, County Cork, Ireland
Bridge of Elia, Kaminaria, Limassol District, Cyprus
Kelefos (Tzelefos) Bridge, Agios Nikolaos, Famagusta, Cyprus

There are still many extant packhorse bridges in England, Scotland and Wales (see *AEBB*).

PILGRIM ROUTES OVER BRIDGES

One of the reasons for building some medieval bridges was to enable the local town to benefit from the trade generated by pilgrim traffic. These bridges include the following:

International Bridge, Valenca, Viana do Castelo, Portugal [Santiago de Compostela]
Magdalena Footbridge, Pamplona, Navarre, Spain [Santiago de Compostela]
Merchants' Bridge, Erfurt, Thuringia, Germany [Via Regio Pilgrims' Way]
Pont Vieux, Orthez, Pyrénées-Atlantiques, France [Santiago de Compostela]
Ponte della Maddalena, Borgo a Mozzano, Lucca, Italy [Via Francigena Pilgrims' Way]
Queen's Bridge, Ourense, Navarre, Spain [Santiago de Compostela]
Tarn Bridge, Quézac, Lozère, France [Quézac's Black Madonna]
Valentré Bridge, Cahors, Lot, France [Santiago de Compostela]

PIPELINE BRIDGES

The slightly unusual structures built to carry an oil or gas pipeline over a major obstacle include:

Hegigio Gorge Pipeline Bridge, Komo Station, Papua New Guinea
LNG Pipeline Bridge, Bioko, Equatorial Guinea
Pipeline Bridge, Dargan Ata, Lebap, Uzbekistan

PONTOON BRIDGES

Some early river crossings consisted of linked boats until a fixed structure could be built and, later, a few bridges have included a pontoon section that could be pulled aside to leave a passageway for ships (see under Opening Bridges). Pontoon bridge entries include:

Afghan-Uzbek Friendship Bridge, Hairatan, Balkh, Afghanistan
Attock Railway Bridge, Attock Khurd, Punjab Province, Pakistan
Demerara Harbour Bridge, Georgetown, Demerara-Mahaica Region, Guyana
Dubai Creek Floating Bridge, Dubai, United Arab Emirates
Faidherbe Bridge, Saint-Louis, Saint-Louis Region, Senegal
Floating Bridge, Enshi City, Hubei Province, China
Galata Bridge, Istanbul, Turkey
Governor Albert D. Rosellini Bridge, Seattle, Washington State, USA
Guangji / Xiangzi (Great Charity) Bridge, Chaozhou, Guangdong, China
Lacey V. Murrow Memorial Bridge, Seattle, Washington, USA
Pontoon Railroad Bridge, Wabasha, Minnesota, USA *
Queen Emma Bridge, Willemstad, Curacao
Vytautas the Great Bridge, Kaunas, Lithuania

PRIZE-WINNING BRIDGES

The following is a short selection of some recent prize-winning bridges that have entries in this book, with the year of the award noted:

Abba G. Lichtenstein Medal

Civil engineer Abba G. Lichtenstein (1923–2015) was a leader in the preservation and restoration of historic bridges and the International Bridge Conference established a medal in his honour. This has been awarded to:

Frankford Avenue Bridge, Philadelphia, Pennsylvania, USA (2019)

Arthur G. Hayden Medal

Arthur G. Hayden was a civil engineer who specialised in the development of rigid frame bridges. The medal in his name recognises significant engineering innovation in non-traditional bridge structures. Award-winning bridges with entries in this book include:

Inner Harbour Bridge, Duisberg, North Rhine-Westphalia, Germany (2003)
Peace Bridge, Derry, Northern Island (2012)
Te Rewa Rewa Bridge, Taranaki, New Zealand (2011)
Zhangjiajie Glass Bridge, Hunan, Hunan Province, China (2018)

Eugene C. Figg Jr Medal for Signature Bridges

Eugene Figg (1936–2002) was an American structural engineer and the medal awarded in his memory is for signature bridges, those in this book including:

Dragon Bridge, Da Nang, Vietnam (2014)
Sanha Bridge, Shenyang, China (2009)
Triplets Bridge, La Paz, Bolivia (2012)

George S. Richardson Award

George S. Richardson (1896–1988) was an engineer who was involved in many major American highway projects. The award in his name has been given to the following bridge structures:

Confederation Bridge, Borden-Carleton, Prince Edward Island, Canada (1999)
Jiaozhou Bay Bridge, Hongdao, Shanxi, China (2013)
Rio-Antirrio Bridge, Patras, Greece (2005)
San Francisco-Oakland Bay Bridge, San Francisco, California, USA (2015)

Gustav Lindenthal Medal-Winning Bridges

The civil engineer Gustav Lindenthal (1850–1935) started his career working as an unqualified engineer on railways in Austria and Switzerland before emigrating to the USA in 1874. After initial employment as a mason, he joined the Keystone Bridge Company in Pittsburgh and then established his own consulting business in 1881. In 1902 he was appointed New York City's Commissioner of Bridges and was responsible for some of the greatest structures of his era. The Medal in his name is awarded for a single, recent outstanding achievement in bridge engineering, and the following winning bridges have entries in this book:

Akashi Kaikyo Bridge, Kobe, Hyogo Prefecture, Japan (1998)
Deh Cho Bridge, Fort Providence, Northwest Territories, Canada (2013)
Duge Beipanjiang Bridge, Liupanshui, China (2018)
George P. Coleman Memorial Bridge, Yorktown, Virginia, USA (1997)
Millau Viaduct, Millau, Aveyron, France (2005)
Vimy Memorial Bridge, Ottawa, Ontario, Canada (2015)
Xihoumen Bridge, Zhoushan, Zhejiang, China (2010)

John A. Roebling Medal

Roebling was the designer of the Brooklyn Bridge in New York but was mortally wounded when his foot was crushed as he was setting out the bridge's position. The medal is awarded for lifetime achievement in bridge engineering, including design, construction, research or teaching. Winners who have been responsible for bridges that have entries in this book include:

Menn, Christian (1997): Ganter Bridge, Brig, Valais, Switzerland
Troyano, Leonardo Fernandez (2008): Lerez River Bridge, Pontevedra, Galicia, Spain

Other International Prize-Winning Bridges

Other bridges that have won awards include:

Golden Dragon Bridge, Da Nang, Vietnam: Agora Prize for best architecture photo (2020)

High Bridge, Amsterdam, North Holland, Netherlands: International Footbridge Award (2002)

Juscelino Kubitschek Bridge, Brasilia, Brazil: International Bridge Conference award (2003)

Lego Bridge, Wuppertal, North Rhine-Westphalia, Germany: Deutscher Fassadenpreis Advancement Prize (2012)

Three Countries Bridge, Huningue, Haut-Rhin, France: German Bridge Construction Prize (2008)

Ushibuka Haiya Bridge, Kumamoto, Japan: Tanaker Prize (1998)

Webb Bridge, Melbourne, Australia: the Australian Joseph Reed Award for Urban Design (2005)

International Road Federation Prizes

Yavuz Sultan Selim Bridge [Global Road Achievement Award 2019]

World Architecture News Awards

Yavuz Sultan Selim Bridge, Istanbul, Marmara Region, Turkey [Best Bridge Award 2017]

RECORD-BREAKING BRIDGES

There are separate entries in this book listing bridges that are the **Highest**, **Longest** and **Widest** in the world. This entry lists a couple of other kinds of record-breaking bridges:

1967: Liteyny (Foundry) Bridge, St Petersburg, Russia [heaviest bascule 3,225 tons]

2005: Langkawi Sky Bridge, Langkawi Island, Malaysia [longest curved bridge in the world]

RE-LOCATED BRIDGES

Some bridges have been taken down and rebuilt on new sites:

Bizerte Bridge, Tunis, Tunis Governorate, Tunisia [first built 1898 over the canal in Tunis; re-erected in Brest in 1909]

Drum Bridge, San Francisco, California, USA [moved from Japan for re-erection in San Francisco in 1894]

Hradecky Bridge, Ljubljana, Province of Ljubljana, Slovenia [first erected 1897 and re-erected in 1931 and 2011]

Krasnoluzhsky Bridges, Moscow, Russia [first erected in 1907 as a railway bridge; re-erected as a footbridge in 2001]

Linn Branch Creek Bridge, Parkville, Missouri, USA [first erected in 1898 as a railway bridge; re-erected as a footbridge in 1981]

Lugard Footbridge, Kaduna, Kaduna State, Niger [first built at Zungeru in 1904; re-erected at Kaduna in 1920]

Seaka Bridge, Seaka, Quthing District, Lesotho [first erected 1882; moved to Seaka in 1950]

Stadlec Chain Bridge, Stadlec, South Bohemia, Czech Republic [first built in Podolsko in 1848; rebuilt in Stadlec in 1975]

RIBBON BRIDGES

All suspension bridges depend on chains or cables hanging originally in a catenary curve, from which the approximately level deck is hung, with the profile of the loaded chains or cables then becoming a compound curve between a catenary and a parabola. However, the term ribbon bridge (sometimes catenary bridge) is usually applied when the deck itself is laid on top of the suspension cables and therefore follows a true catenary curve when unloaded. In these structures, some or all of the supporting cables on which the concrete deck is placed are then post-tensioned in order to stiffen the whole structure. A constraint on the length of span that can be achieved is that the slope at the end of the deck as it reaches the bridge supports should not exceed 1 in 20. Ribbon bridges in this book include:

Airport Lighting Bridge, Bratislava, Slovakia

Emscher Landscape Park Bridges, Duisberg, North Rhine-Westphalia, Germany

Essinger Footbridge, Essing, Germany

Hureai Bridge, Nantan, Japan

Lake Vranov Footbridge, Znojmo, South Moravia, Slovakia

Lionel Viera Bridge, La Barra, Uruguay

Sacramento River Trail Footbridge, Redding, California, USA

Shiosai Bridge, Hamamatsu, Japan

Suransuns Bridge, Thusis, Graubünden, Switzerland

Swiss Bay Footbridge, Onsov, Czech Republic

There are also two ribbon bridges in Great Britain (see *AEBB*).

SEA BRIDGES

As well as a few bridges that have been built off sea cliffs, see **Cliff Walks**, there are also bridges that, instead of crossing rivers, streams and lakes, cross seas in order to link islands and connect them to the mainland. These include:

Akashi Kaikyo Bridge, Japan

Carrick-a-Rede Rope Bridge, Ballintoy, County Antrim, Northern Ireland

Cruit Island Sea Bridge, Ireland

Hangzhou Bay Bridge, Hangzhou, Zhejiang Province, China

Hong Kong-Zhuhai-Macau Bridge, Hong Kong, China

Jiaozhou Bay Bridge, Hongdao, Shanxi, China

LNG Pipeline Bridge, Bioko, Equatorial Guinea

Maslenica Bridges, Zadar, Zadar County, Croatia

Mizen Head Bridge, Crookhaven, Munster, Ireland

Öland Bridge, Kalmar, Sweden

Pamban Bridge, Mandapam, India

Paradise Island Bridges, Nassau, Bahamas

Rio-Antirrio Bridge, Patras, Greece

Second Penang Bridge, Bayan Lepas, Malaysia

Sinamalé Bridge, Malé, Maldives

Temburong Bridge, Bandar Seri Bagawan, Brunei

Tjörn Bridges, Tjörn, Sweden

Ushibuka Haiya Bridge, Kumamoto, Japan

Winding Patapat Viaduct, Pagadpud, Luzon, Philippines

SELF-ANCHORED SUSPENSION BRIDGES

In most suspension bridges, the ends of the supporting chains or wire cables are anchored into

solid rock or a concrete structure massive enough to withstand the tension forces. Occasionally, however, in tied or self-anchored bridges, the cables are terminated at the ends of the deck on the side spans, the deck itself then acting as a giant strut. Examples of such bridges in this book include:

Egongyan Bridge, Chongqing, Sichuan, China

Jiaozhou Bay Bridge, Hongdao, Shanxi, China

Peace Bridge, Derry, County Londonderry, Northern Ireland

San Francisco-Oakland Bay Bridge, San Francisco, California, USA

Seishun Footbridge, Tsumagoi, Gunma, Japan

Three Sisters Bridges, Pittsburgh, Pennsylvania, USA

STONE SLAB BRIDGES

Although in olden times the simplest bridges were formed from specially-felled trees or old logs, these would quickly rot or be swept away. Ancient bridges that could last for many years were formed from stone slabs (and, in the UK, these were called clam or clapper bridges – see *AEBB*.) There are few surviving examples of slab bridges, four of which have entries in this book:

Bunlahinch Clapper Footbridge, Louisburgh, County Mayo, Ireland

Loyang (Ten Thousand Peace) Bridge, Chuanchow, Fujian, China

Sonjuk Bridge, Kaesong, North Hwanghae, North Korea

Yudai (Jade Belt) Bridge, Hangzhou, Zhejiang Province, China

TALLEST BRIDGES

Bridges in which the deck is a considerable height above river level (or, sometimes, ground level) include the following that have entries in this book:

1884: Garabit Viaduct, St Flour, Auvergne, France [height of bridge deck 120m]

1909: Fades Viaduct, Les Fades, Puy-de-Dôme, France [height of bridge deck 133m]

1963: Europa Bridge, Innsbruck, Switzerland [height of bridge 192m]

2001: Beipan River Shuibai Bridge, Liupanshui, Guizhou Province, China [height of bridge deck 275m]

TRANSPORTER BRIDGES

The transporter principle was invented (but not patented) by Charles Smith (1843–82), a Scottish engineer, in 1873. Nevertheless, the French engineer Ferdinand Arnodin (1845–1924) was the first to implement the transporter bridge concept as a solution to the problem of enabling large numbers of pedestrians and relatively few vehicles to cross a narrow, busy shipping channel contained between low banks. However, only seventeen of these bridges were ever built, being completed in the period from 1893 to 1916, of which three remain in Britain (Middlesbrough, Newport and Warrington Transporter Bridges – see *AEBB*). This book has entries for seven of these structures (two of which marked * no longer exist) as shown below:

Bizerte Bridge, Tunis, Tunis Governorate, Tunisia*

Duluth Lift Bridge, Duluth, Minnesota, USA *

Osten Transporter Bridge, Osten, Lower Saxony, Germany

Rendsburg High Bridge, Rendsburg, Schleswig-Holstein, Germany

Rochefort-Martrou Transporter Bridge, Rochefort, France

Transborador del Riachuelo Nicolas Avellaneda, Buenos Aires, Argentina

Vizcaya Transporter Bridge, Portugalete, Spain

TRESTLE BRIDGES

Some bridges are simple trestle structures. Examples of such bridges that have entries in this book include:

Atamyrat-Kerkichi Bridges Atamyrat, Lebap Province, Turkmenistan

Banab River Bridge, Madang, Papua New Guinea

Dungeness River Bridge, Sequim, Washington, USA

Goat Canyon Trestle, San Diego, California, USA

Hardinge Bridge, Paksey, Pabna District, Bangladesh

Kintai Bridge, Iwakuni, Yamaguchi Prefecture, Japan

Kinzua Viaduct, Westline, Pennsylvania, USA

Lake Pontchartrain Crossings, New Orleans, Louisiana, USA

Lethbridge Viaduct, Lethbridge, Alberta, Canada

Pontoon Railroad Bridge, Wabasha, Minnesota, USA *

Spring Creek Trestle Bridge, Hanover, Montana, USA

Tokyo-Haneda Monorail Bridge, Kawasaki, Kanagawa, Japan

TRIUMPHAL ARCHWAYS AND GATEWAYS ON BRIDGES

Bridges described in this book that have (or had) triumphal archways across their approaches or decks include:

Alcántara Bridge, Alcántara, Cáceres, Spain

Alcántara Bridge, Toledo, Castile-La Mancha, Spain

Devil's Bridge, Martorell, Barcelona, Spain

Dromana Bridge, Villierstown, Waterford, Ireland

Huc Bridge, Hanoi, Red River Delta, Vietnam

Malabadi Bridge, Silvan, Diyarbakir, Turkey

Manhattan Bridge, New York, USA

Montauban Bridge, Montauban, Tarn-et-Garonne, France

Pont Flavien, Saint-Chamas, Bouches-du-Rhône, France

Pont Neuf, Toulouse, Haute-Garonne, France

Sinamalé Bridge, Malé, Northern, Maldives

Smithfield Street Bridge, Pittsburgh, Pennsylvania, USA

Stone Bridge, Zamora, Spain

UNESCO WORLD HERITAGE BRIDGES

The United Nations Educational, Scientific and Cultural Organization (UNESCO), founded in 1946, is an agency of the United Nations dedicated to promoting international collaboration in the protection of the world's shared heritage. More than 1,000 sites have been inscribed on UNESCO's World Heritage List because of their 'outstanding universal value', many of which are, or include, bridges. The forty-five World Heritage bridge structures with entries in this book are as follows:

Adolphe Bridge, Luxembourg City, Luxembourg

Águas Livres Aqueduct, Lisbon, Estremadura, Portugal

Alcántara Bridge, Toledo, Spain

Amoreira Aqueduct, Elvas, Alentejo, Portugal

Bridge of Immortals, Tangkouzhen, Anhui Province, China

Cahors, France

Canal du Midi Bridges, France

Celestial Bridge, Suzhou, China

Charles Bridge, Prague, Czech Republic

Citadel Bridge, Aleppo, Syria

Cividale Del Friuli Bridge, Udine, Italy

Geumcheongyo Bridge, Seoul, Gyeonggi Province, South Korea

Japanese Covered Bridge, Hoi An, Vietnam

Kalte Rinne Viaduct, Breitenstein, Lower Austria, Austria

Kanoh Bridge, Kalka, Haryana, India

Kromeríz Castle Gardens Iron Bridges, Czech Republic

Mehmed-Pasha Sokolovic Bridge, Visegrad, Bosnia and Herzegovina

Nine Arch Bridge, Hortobágy, Hungary

Old Bridge, Albi, France

Old Bridge, Mostar, Bosnia-Herzegovina

Old Stone Bridge, Regensburg, Germany

Padre Templeque Aqueduct, Hidalgo, Mexico

Pompey's Bridge, Mtskheta, Georgia

Pont de Diable, Aniane, France

Pont du Gard, Nîmes, France

Ponte Pietra, Verona, Italy

Querétaro Aqueduct, Santiago de Querétaro, Mexico

Red Gate Aqueduct, Augsburg, Bavaria, Germany

Regensburg Bridge, Regensburg, Germany

Rhaetian Railway Bridges, Filisur, Graubünden, Switzerland

Rimac River Bridge, Rimac, Lima, Peru

Roman Bridge, Mérida, Spain

Roman Bridge, Trier, Rhineland-Palatinate, Germany

San Martin Bridge, Toledo, Castile-La Mancha, Spain

Segovia Aqueduct, Segovia, Spain

Seonim Bridge, Seogwipo, Jeju Volcanic Islands, South Korea

Shadorvan Bridge, Shushtar, Khuzestan Province, Iran

Shinkyo (Sacred) Bridge, Nikko, Tochigi Prefecture, Japan

Shuangshi Bridge, Lijiang, China

Sonjuk Bridge, Kaesong, North Korea

Speicherstadt Island Bridges, Hamburg, Lower Saxony, Germany

State Castle Moat Bridge, Cesky Krumlov, Czech Republic

Suspension Bridge (T E Clark), Budapest, Hungary

Tarragona Aqueduct, Tarragona, Catalonia, Spain

Tensift Bridge, Marrakesh, Morocco

Tsarskoye Selo Park Bridges, Saint Petersburg, Northwestern, Russia

Valens Aqueduct, Istanbul, Turkey

Valentré Bridge, Cahors, France

Vanvitelli Aqueduct, Caserta, Italy

Vizcaya Transporter Bridge, Portugalete, Spain

Vizegrad Bridge, Bosnia-Herzegovina

West Lake Bridges, Hangzhou, Zhejiang, China

Western Bridge, Colombia [application submitted]

Wörlitz Garden Bridges, Dessau-Wörlitz, Saxony-Anhalt, Germany

Yiheyuan Summer Palace Bridges, Beijing, China

There are twenty-eight UNESCO bridge sites in Britain (see *AEBB*).

UNUSUAL BRIDGE FEATURES

Many bridges have unusual or distinctive features, including:

Decorative animals

Apart from bridges with statues of famous people, some bridges are decorated with creatures. These are mainly carvings or castings of heraldic-style lions or eagles, although mythical creatures are also depicted, all as in the following examples:

Anichkov Bridge, St Petersburg, Northwestern, Russia [bronze horses, mermaids and sea horses adorn this structure]

Bank Footbridge, St Petersburg, Northwestern, Russia [cast iron winged griffins hold the suspension chains]

Egyptian Bridge, St Petersburg, Northwestern, Russia [sphinxes sit on the abutments]

Fangsheng (Liberate Living Things) Bridge, Zhujiajiao, Shanghai, China [four carved stone lions on parapet]

Geumcheongyo Bridge, Seoul, Gyeonggi Province, South Korea [carvings of imaginary animals]

Jisr Jindas (Baybars) Bridge, Zeitan, Tulkarm Governorate, Palestine [a pair of lions flank a carved inscription]

Lions' Footbridge, St Petersburg, Northwestern, Russia [cast iron lions hold the suspension chains]

Lowen Bridge, Berlin, Brandenburg, Germany [the suspension chains emerge from the mouths of bronze lions]

Marco Polo (Lugou) Bridge, Beijing, China [originally 627 carved lions stood on the parapets]

Pont Flavien, Saint-Chamas, Bouches-du-Rhône, France [a pair of carved lions stand on the triumphal arch]

Salzach Bridge, Oberndorf, Salzburg, Austria [decorative eagles stand on pylons each side of the river]

Sanahin Bridge, Alawerdi, Lori, Armenia [a carved lion terminates the parapet]

Spean Praptos Bridge, Chi Kiaeng, Siem Reap, Cambodia [mythical creatures stand on the balustrade ends]

Stauffacher Bridge, Zurich, Switzerland [corner pylons are capped by seated lions]

Vittoria Emanuele II Bridge, Rome, Italy [bronze winged victories]

Yiheyuan Summer Palace Shiqigong (Seventeen Arches) Bridge, Beijing, China [544 lions stand on the parapets]

Single-pylon stay or suspension bridges

Although the vast majority of stay and suspension bridges have two main piers, a few have just one, as in the examples below:

Ada Bridge, Belgrade, Serbia

Captain William Moore Bridge, Skagway, Alaska, USA

Langebrug, Haarlem, Netherlands

Lerez River Bridge, Pontevedra, Spain

Manning Crevice Bridge, Riggins, Idaho, USA

Rama VIII Bridge, Bangkok, Thailand

Odd-shaped decks

Bridges with unconventional decks that are not simply rectangular include:

Abdoun Bridge, Amman, Jordan [S-shaped deck]

Aveiro Circular Footbridge, Aveiro, Centro, Portugal [circular ring deck]

Bamako Third Bridge, Bamako, Mali [S-shaped deck]

Brusio Spiral Viaduct, Filisur, Graubünden, Switzerland [spiral deck]

College Footbridge, Kortrijk, West Flanders, Belgium [S-shaped deck]

Emscher Landscape Park Bridges, Duisberg, North Rhine-Westphalia, Germany [S-shaped deck]

Eurymedon Bridge, Aspendos, Antalya Province, Turkey [cranked deck]

Hiyoshi Dam Footbridge, Nantan, Kyoto, Japan [circular ring deck]

Kawazu-Nanadaru Loop Bridge, Shimoda, Japan [helical deck]

Laguna Garzón Bridge, José Ignacio, Rocha, Uruguay [circular ring deck]

Maputo-Katembe Bridge, Maputo, Mozambique [S-shaped approach viaduct]

Nanpu Bridge, Shanghai, China [spiral deck]

Neckar River Footbridge, Stuttgart, Baden-Württemberg, Germany [S-shaped deck]

Peace Bridge, Derry, County Londonderry, Northern Ireland [S-shaped deck]

San Juanico Bridge, Tacloban, Eastern Visayas, Philippines [S-shaped deck]

Second Penang Bridge, Bayan Lepas, Penang, Malaysia [double S-curved deck]

Senbonmatsu Bridge, Osaka, Japan [circular deck]

Sultan Omar Ali Saifuddien Pedestrian Bridge, Bandar Seri Begawan, Brunei [S-shaped deck]

Yu Yuan Bazaar / Garden Bridge, Shanghai, China [ten-legged zigzag deck]

Zarasas Bridge, Ditkūnai, Lithuania [nearly circular ring deck]

Zhivopisny Bridge, Moscow, Russia [S-shaped deck]

Decks with Non-Traffic-Bearing Uses
The normal function of a bridge deck is simply to carry vehicles or people across the obstacle spanned by the structure. On early covered bridges (q.v.) the opportunity was sometimes taken to include shops. Occasionally, however, other bridge structures have also been used for purposes outside simple transportation, such as:

Galata Bridge, Istanbul, Turkey [the bridge has a second deck that carries cafes and restaurants]

New Bridge, Bratislava, Slovakia [the support pylon is topped by an observation deck]

Rama VIII Bridge, Bangkok, Thailand [this bridge has a glass observation deck]

Rialto Bridge, Venice, Veneto, Italy [this footbridge is famous for its parade of shops]

Zhivopisny Bridge, Moscow, Russia [a suspended drum-shaped enclosure was planned to house a restaurant]

UNUSUAL TRUSS DESIGNS

Although there are many types of truss bridge (such as the eponymous Burr, Howe, Long, Pratt, Town, Warren and others with their different arrangements of vertical, horizontal, diagonal and arched members), the unusual Bollman and Fink truss are sufficiently rare to be worth listing separately. The Fink truss is effectively an inter-connected series of basic king post trusses and the Bollman truss could be described as a series of overlapping and inverted king post trusses. A further different design is that by Hans Ulrich Grubenmann, of which only one still stands, that consists of a series of rectangular timber frames supported by multi-coursed arched timber ribs. Bridges of these three types that have entries in this book include:

Fink Deck Truss Bridge, Lynchburg, Virginia, USA [Fink Truss]

Kubelbrücke, Appenzell Ausserrhoden, Herisau, Switzerland [Grubenmann]

Olstgracht Bridge, Almere, Netherlands [Fink Truss]

Savage Mills Bridge, Savage, Maryland, USA [Bollman Truss]

WIDEST BRIDGES

Some of the widest bridges in the world, in date order of their opening, include the following:

1818: Blue Bridge, St Petersburg, Northwestern, Russia [widened 1844; 97m wide]

1866: Cornish-Windsor Covered Bridge, Cornish, New Hampshire, USA [50ft wide]

1880: O'Connell Bridge, Dublin, County Dublin, Ireland [150ft wide]

1883: Brooklyn Bridge, New York, New York, USA [85ft wide (26m)]

1900: Alexandre III Bridge, Paris, Ile-de-France, France [40m wide (131ft)]

1909: Manhattan Bridge, New York, New York, USA [120ft (37m) wide]

1932: Sydney Harbour Bridge, Sydney, New South Wales, Australia [160ft (49m) wide]

1959: Auckland Harbour Bridge, Auckland, New Zealand [13m (43ft) wide; widened 1969 to 26m (85ft)]

1964: Port Mann Bridge, Vancouver, British Colombia, Canada [65m (213ft) wide]

1990: San Francisco-Oakland Bay Bridge, San Francisco, California, USA [258ft (79m) wide]

1997: Wadi Laban Bridge, Riyadh, Riyadh Province, Saudi Arabia [85m (279ft) wide]

2008: Ramstor Bridge, Nur Sultan, Central Kazakhstan, Kazakhstan [50m wide]

2009: Chaotianmen Bridge, Chongqing, Sichuan, China [37m wide]

2010: Bandra-Worli Sea Link, Mumbai, India: [40m wide]

2011: Ada Bridge, Belgrade, Serbia [45m wide]

2015: Port Mann Bridge, Vancouver, British Colombia, Canada [45m wide]

(Glanworth Bridge, Ballyquane, County Cork, Ireland, with a deck that is only 9ft wide, is considered to be the narrowest and oldest public bridge still in everyday use in Europe.)

WORLD MONUMENTS FUND BRIDGES

The World Monuments Fund was founded in 1996 to call attention to important cultural heritage sites that are at risk from the ravages of conflict, natural disasters and absence of local support and produces a biennial Watch List. Since then more than 800 such sites have been identified, those containing bridges include the following that have entries in this book:

Chain Bridge, Bagni di Lucca, Italy
Chains Bridge, Fomoli, Italy
Dalal Bridge, Zakho, Iraq
Iron Bridge, Spanish Town, Jamaica
Karl-Theodor Bridge, Heidelberg, Germany
Lucano Bridge, Villanova, Italy
Mehmed-Pasha Sokolovic Bridge, Visegrad, Republika Srnka, Bosnia-Herzegovina
Merritt Parkway Bridges, Greenwich, Connecticut, USA
Lucano Bridge, Villanova, Padua, Italy
Segovia Aqueduct, Segovia, Spain
Stari Most, Mostar, Bosnia-Herzegovina

WORLD'S BEST BRIDGES

Many people have made lists of the most famous bridges in the world. The following fourteen bridges appear regularly on most of these lists:

Akashi-Kaikyo Bridge, Kobe, Japan
Alcántara Bridge, Alcántara, Spain
Brooklyn Bridge, New York, USA
Chapel Bridge, Lucerne, Switzerland
Charles Bridge, Prague, Czech Republic
Chengyang Bridge, Guangxi, China
Golden Gate Bridge, San Francisco, USA
Great Belt Bridge, Korsør, Denmark
Millau Viaduct, Millau, France
Rialto Bridge, Venice, Italy
Siosepol Bridge (Bridge of 33 Arches), Isfahan, Iran
Stari Most, Mostar, Bosnia and Herzegovina
Sydney Harbour Bridge, Australia
Tower Bridge, London, England (see *AEBB*)

ZIGZAG BRIDGES

Both China and Japan have a heritage of zigzag bridges, supposedly built originally in order to protect people using them from evil spirits, which, it was believed, could travel only in straight lines. Some of these bridges that have entries in this book include:

Lotus Lake Footbridges, Kaohsiung, Taiwan
Nine-Turn Bridge, Yangzhou, Jiangsu Province, China
Yu Yuan Bazaar / Garden Bridge, Shanghai, China

Bibliography

Adams, Kramer A. *Covered Bridges of the West: A History and Illustrated Guide*. Howell-North 1963

Agnelli, Claude. *37 Bridges of Paris*. Magellan & Cie 2006

Aguiló, Miguel. *The Character of Spanish Bridges*. Actividades De Construcción Y Servicios (Acs) 2007

Allen, Richard Saunders. *Covered Bridges of the Middle Atlantic States*. Stephen Greene Press 1957

Allen, Richard Saunders. *Covered Bridges of the Northeast*. Stephen Greene Press (2nd) 1974

Allen, Richard Saunders. *Covered Bridges of the South*. Bonanza Books 1970

Allen, Richard Saunders. *Old North Country Bridges in Upstate New York*. North Country Books 1983

American Wooden Bridges. American Society of Civil Engineers 1976

Andrić, Ivo. *The Bridge on the Drina*. University of Chicago Press 1977

Archaeology of Bridges. Verlag Friedrich Pustet 2011

Architectural Design of Concrete Bridges. Chicago, Portland Cement Association (nd 1930s)

Arcila, Martha Torres. *Puentes*. Atrium 2002

ASCE. *List of Historic Engineering Landmarks: Bridges*. American Society of Civil Engineers (nd)

Balk, Jaap. *Shell-Journaal van Nederlandse Bruggen*. Shell Nederland 1980

Bank, Gretchen G. and Kees Moerbeek. *Master Builder: Bridges*. Thunder Bay Press 2008

Barna, Ed. *Covered Bridges of Vermont*. The Countryman Press 1996

Barnham, Kay and Laszlo Veres. *Brilliant Bridges*. Collins 2012

Barow, Horst. *Roads and Bridges of the Roman Empire*. Edition Axel Menges 2013

Barry, Michael. *Across Deep Waters: Bridges of Ireland*. Frankfort Press 1985

Bascove, I. *Stone & Steel: Paintings & Writings Celebrating the Bridges of New York City*. David R Godine 1998

Baus, Ursula and Mike Schlaich. *Footbridges: Structure Design History*. Birkhauser Verlag 2008

Beck, Haig and Jackie Cooper. *Kurilpa Bridge*. The Images Publishing Group 2012

Beckett, Derrick. *Great Buildings of the World: Bridges*. Paul Hamlyn 1969

Bennett, David. *The Architecture of Bridge Design*. Thomas Telford 1997

Bennett, David. *The Creation of Bridges*. Aurum Press 1999

Beren, Peter and Morton Beebe. *The Golden Gate: San Francisco's Celebrated Bridge*. Earth Aware Editions 2011

Berfield, Rick L. *Covered Bridges of New York State*. Syracuse University Press 2003

Berg, Christa van den and Gerhard Nijenhuis. *Bridging the Dutch Landscape; Design Guide for Bridges*. BIS Publishers 2006

Beyer, Peter and Jurgen Stritzke. *Die Goltzschtalbrucke*. Bundesingenieurkammer 2009

Bill, Max. *Robert Maillart: Bridges and Constructions*. Pall Mall Press (3rd) 1969

Billings, Henry. *Bridges*. New York, Viking 1956

Billington, David P. *Robert Maillart's Bridges: The Art of Engineering*. Princeton University Press 1979.

Billington, David P. *The Tower and the Bridge: The New Art of Structural Engineering*. Princeton University Press 1985

Billington, Robert and Sarah Billington. *The Bridge*. Peribo 1999

Binney, Marcus. *Bridges Spanning the World*. Pimpernel Press 2017

Bishop, Peter. *Bridge*. Reaktion Books 2008

Black, Annette and Michael B. Barry. *Bridges of Dublin: The Remarkable Story of Dublin's Liffey Bridges*. Dublin City Council 2015

Black, Archibald. *The Story of Bridges*. McGraw Hill 1936

Blakely, Joe R. *Lifting Oregon Out of the Mud: Building the Oregon Coast Highway*. CraneDance Publications 2010.

Blakstad, Lucy. *Bridge: The Architecture of Connection.* Birkhauser 2002

Blankenstein, Elisabeth van et al. *De Nederlandse Brug; 40 MarkanteVoorbeelden.* Uitgeverjii Thoth 2012

Blockley, David. *Bridges: The Science and Art of the World's Most Inspiring Structures.* Oxford University Press 2010

Bloise, Remo Capra and Pat Fahey. *Bridge over Niger: The True Story of the J.F. Kennedy Bridge.* Writer's Showcase 2001

Boff, Charles. *Boys' Book of Bridges.* George Routledge 1938

Bonatz, Paul and Fritz Leonhardt. *Brücken.* Karl Robert Langewiesche 1953

Bosman, Francoise, Martine Mille, and Gersende Piernas. *L'art du Vide: Ponts d'Ici et d'Ailleurs; Trois siècles de genie francais, XVIII-XX.* Somogy Editions d'Art, Paris & Archives National du Monde du Travail, Roubaix 2010

Bosporus Bridge, The. Freeman Fox & Partners 1973

Bottenberg, Ray. *Bridges of the Oregon Coast.* Arcadia Publishing 2006

Bottenberg, Ray. *Bridges of Portland.* Arcadia Publishing 2007

Bradfield J. J. C. *Sydney Harbour Bridge.* H. Phillips nd (1932?)

Brake, Mike and Francisco Javier Gil Oreja. *Bridges.* Parkstone International 2010

Brangwyn, Frank and Christian Barman. *The Bridge.* The Bodley Head 1926

Brangwyn, Frank and Walter Shaw Sparrow. *A Book of Bridges.* The Bodley Head (2nd) 1920

Break, Kevin. *Bridges of Downtown Los Angeles.* Arcadia Publishing 2015

Bridge Aesthetics Around the World. (Transportation Research Board 1991)

Bridges: A Few Examples of the Work of a Pioneer Firm in the Manufacture of Steel and Steelwork. Dorman Long and Company 1930

Bridge: The Heritage of Connecting Places and Cultures. Ironbridge International Institute for Cultural Heritage 2018

Bridging Normandy to Berlin. Printing and Stationery Service, British Army of the Rhine 1945

Broos, James. *Bridges of Venice Walking Tours.* James Broos 2008

Brown, David J. *Bridges: Three Thousand Years of Defying Nature.* Mitchell Beazley (2nd) 2005

Browne, Lionel. *Bridges; Masterpieces of Architecture.* Todtri Productions 1996

Bullard, Stephan G.,Bridget J. Gromek, Martha Fout, Ruth Fout, and the Point Pleasant River Museum. *The Silber Bridge Disaster of 1967.* Arcadia Publishing 2012

Burk, John S. *Massachusetts Covered Bridges.* Arcadia Publishing 2010

Burke, Kathryn W. *Hudson River Bridges.* Arcadia Publishing 2007

Butterfield, Ben. *Bridges.* Nelson Doubleday 1962

Byrd, Ty. *Lantau Link: Tsing Ma Bridge.* New Civil Engineer 1997

Canon Van De Nederlandse Brug. Nederlandse Bruggenstichting 2016

Caravan, Jill. *American Covered Bridges: A Pictorial History.* Courage Books 1990

Casson, Sir Hugh (ed). *Bridges.* Chatto and Windus 1963

Caswell, William S. Jnr. *Connecticut and Rhode Island Covered Bridges.* Arcadia Publishing 2011

Cazneaux, H. *The Bridge Book.* Art in Australia 1930

Cerver, Francisco Asensio (ed). *Bridges; New Architecture 1.* Ediciones Atrium 1993

Chadraba, Rudolf. *Charles Bridge.* Odeon 1974

Cleary, Richard L. *Bridges.* W.W. Norton & Co 2007

Cockrell, Bill. *Oregon's Covered Bridges.* Arcadia Publishing 2008

Congdon, Herbert Wheaton. *The Covered Bridge.* (5th) Vermont Books 1979

Conwill, Joseph D. *Covered Bridges.* Shire Publications 2014

Conwill, Joseph D. *Maine's Covered Bridges.* Arcadia Publishing 2003

Conwill, Joseph D. *New England Covered Bridges Through Time.* America Through Time 2014

Conzett, Jurg and Martin Linsi. *Landscape and Structures.* Verlag Scheidegger & Spiess 2010

Cook, Richard J. *The Beauty of Railroad Bridges in North America – Then and Now.* Golden West Books 1987

Corbett, Scott. *Bridges.* Four Winds Press 1978

Cornille, Didier. *Who Built That? Bridges: An Introduction to Ten Great Bridges and their Designers.* Princeton Architectural Press 2016

Cortright, Robert S. *Bridging: Discovering the Beauty of Bridges.* Bridge Ink 1998

Cortright, Robert S. *Bridging the World.* Bridge Ink 2003

Covered Bridges: A Close-Up Look. Fox Chapel Publishing. 2011

Cox, Ronald and Philip Donald. *Ireland's Civil Engineering Heritage.* The Collins Press 2013

Cox, R.C. and M.H. Gould. *Civil Engineering Heritage: Ireland*. Thomas Telford 1998

Cox, Ronald and Michael Gould. *Ireland's Bridges*. Wolfhound Press 2003

Creighton, Jeff. *Bridges of Spokane*. Arcadia Publishing 2013

Cruikshank, Dan. *Bridges: Heroic Designs that Changed the World*. Collins 2010

Daidalos 57: Brücken/Bridges. Bertelsmann Fachzeitschriften GmbH 1955

D'Acres, Lilia and Donald Luxton. *Lion's Gate*. Talonbooks 1999

Dale, Frank T. *Bridges over the Delaware River: A History of Crossings*. Rutgers University Press 2003

Dance, H.E. *Sydney Harbour Bridge*. Thomas Nelson and Sons Ltd 1946

Dechau, Wilfried. *Nossa Punt; Bridges Landscape*. Scheidegger & Spiess 2018

Dechau, Wilfried. *Traversinersteg*. Ernst Wasmuth Verlag 2006

Dechau, Wilfried. *Trutg dil Flem: Seven Bridges by Jurg Conzett*. Scheidegger & Spiess 2013

De Jong, Henk and Nico Muyen. *Bruggen in Nederland*. Atrium 2001

DeLony, Eric. *Context for World Heritage Bridges*. TICCIH 1996

DeLony, Eric. *Landmark American Bridges*. American Society of Civil Engineers 1993

Demeude, Hugues and Patrick Escudero. *Ponts de Paris*. Editions Flammarion 2003

Dennison, Edward and Ian Stewart. *How to Read Bridges: A Crash Course Spanning the Centuries*. Herbert Press 2012

Desveaux, Delphine. *Le Pont Jacques-Chaban-Delmas; An urban vertical-lift bridge*. Archibooks + Sautereau Editeur 2013

Dethier, Jean and Ruth Eaton. *48 Rassegna (Inhabited Bridges)*. CIPLA 1991

Dragoun, Zdeněk and Jirina Šebková. *Charles Bridge*. Oswald 1991

Du Camp, Maxime. *Paris, Seine et Ponts*. Sepia III (nd)

Dubly, Henry-Louis. *Ponts de Paris à Travers les Siècles*. Editions des Deux Mondes 1957

Dubreuil, Julia. *The Millau Viaduct Portfolio*. Editions CEVM 2005

Dunn, Andrew. *Structures: Bridges*. Wayland 1993

Dupré, Judith. *Bridges: A History of the World's Most Famous and Important Spans*. Konemann 1997

Dupré, Judith. *Bridges: A History of the World's Most Spectacular Spans*. Black Dog & Leventhal Publishers 2017

Elborough, Travis. *London Bridge in America: The Tall Story of a Transatlantic Crossing*. Jonathan Cape 2013

Elrick, Wil and Kelly Kazek. *Covered Bridges of Alabama*. The History Press 2018

English, Michael. *The Ha'penny Bridge Dublin*. Dublin City Council 2016

Erlacher, Gisela. *Skies of Concrete*. Park Books 2015

Evans, Benjamin D. and R. June. *New England's Covered Bridges: A Complete Guide*. University Press of New England 2004

Evans, Benjamin D. and R. June. *Pennsylvania's Covered Bridges: A Complete Guide*. University of Pittsburgh Press 1993

Fabre, Guilhem et al. *The Pont Du Gard: Water and the Roman Town*. Caisse Nationale des Monuments Historiques et des Sites 1992

Feather, Carl E. *The Covered Bridges of Ashtabula County, Ohio*. The History Press 2014

Featherstone, Donald. *Bridges of Battle: Famous Battlefield Actions at Bridges and River Crossings*. Arms and Armour 1998

Feichtinger Architectes. *Passserelle Simone-de-Beauvoir Paris*. Archives d'Architecture Moderne 2006

Flaga, Andrzej. *Mosaty dla pieszych*. WKL Warsaw 2011

Fornes, Mike. *Mackinac Bridge*. Arcadia Publishing 2007

Fortier-Kriegel, Anne. *Ouvrage d'Art Remarquables et leur Sites*. Editions Villes et Territoires 1996

Foster, Norman and Thomas Leslie. *Millau Viaduct*. Prestel 2012

Frampton, Kenneth, Anthony C. Webster and Anthony Tischhauser. *Calatrava Bridges*. Artemis 1993

Fromont, Francois. *Solferino Bridge, Paris*. Birkhauser 2001

Gaillard, Marc. *The Quays and Bridges of Paris: An Historical Guide*. Martelle Editions (nd)

Galbreath, David, Carolyn Temple, Lucile Estell and Joy Graham. *Historic Bridges of Milam County*. Arcadia Publishing 2017

Gannon, Todd. *UN Studio/Erasmus Bridge*. Princeton Architectural Press 2004

Gardner, Denis P. *Wood, Concrete, Stone and Steel: Minnesota's Historic Bridges*. University of Minnesota Press 2008

Garlipp, Richard J., Jr. *New Jersey's Covered Bridges*. Arcadia Publishing 2014

Gies, Joseph. *Bridges and Men*. Cassell 1963

Gottemoeller, Frederick. *Bridgescape: The Art of Designing Bridges*. John Wiley & Sons 1998

Graf, Bernhard. *Bridges that Changed the World*. Prestel 2002

Grattesat, Guy. *Ponts de France*. Presses de l'école nationale des ponts chausees (2nd) 1984

Great East River Bridge 1883–1983, The. The Brooklyn Museum (2nd) 1983

Greater Astoria Historical Society, The and The Roosevelt Island Historical Society. *The Queensboro Bridge*. Arcadia Publishing 2008

Greenhill, Ralph. *Spanning Niagara: The International Bridges 1848–1962*. Niagara University 1984

Gurrieri, Francesco, Lucia Bracci and Giancarlo Pedreschi. *I Ponti Sull' Arno dal Falterone al Mare*. Edizione Polistampa Firenze 1998

Gyukics, Peter *Danube-Bridges: from the Black Forest to the Black Sea*. Yuki Studio 2010

Hadlow, Robert W. *Elegant Arches, Soaring Spans: C. B. McCullough, Oregon's Master Bridge Builder*. Oregon's State University Press 2010

Hall, Donald M. *A Collection of Bridges*. Donald M Hall 1995

Hannavy, John. *Transporter Bridges; An Illustrated History*. Pen & Sword Transport 2020

Harper, Douglas. *River, Railway and Ravine; Foot Suspension Bridges for Empire*. The History Press 2015

Harrington, Lyn and Richard Harrington. *Covered Bridges of Central and Eastern Canada*. McGraw-Hill Ryerson 1976

Hartley, H. A. *Famous Bridges and Tunnels of the World*. Muller (2nd) 1962

Haw, Richard. *Art of the Brooklyn Bridge: A Visual History*. Routledge 2008

Haw, Richard. *The Brooklyn Bridge: A Cultural History*. Rutgers University Press 2005

Hayden, Martin. *The Book of Bridges: The History, Technology and Romance of Bridges and their Builders*. Marshall Cavendish 1976

Heritage Bridges of County Cork. Cork County Council 2013

Holstine, Craig and Richard Hobbs. *Spanning Washington; Historic Highway Bridges of the Evergreen State*. Washington State University Press 2005

Holth, Nathan. *Chicago's Bridges*. Shire Publications 2012

Hopkins, H. J. *A Span of Bridges: An Illustrated History*. David & Charles 1970

Horton, Tom and Baron Wolman. *Superspan: The Golden Gate Bridge*. (2nd) Squarebooks 1997

Hourcadette, Claudine, Sophie Marguerite and Serge Montens. *Paris de Pont en Pont: le long de la Seine et du canal Saint-Martin*. Christine Bonneton 2010

Howard, Michael and Maureen Howard. *The Benjamin Franklin Bridge*. Arcadia Publishing 2009

Humar, Gorazd (ed). *Civil Engineering Heritage in Europe*. Grafika Soca, Slovenia 2009

Humar, Goradz (ed). *Footbridges: Small is Beautiful*. European Council of Civil Engineers 2014

Hurley, Michael. *The World's Most Amazing Bridges*. Raintree 2013

Idelberger, Klaus. *The World of Footbridges: From the Utilitarian to the Spectacular*. Ernst & Sohn 2011

Jackson, Donald C. *Great American Bridges and Dams*. The Preservation Press 1988

Jacobs, David and Anthony E. Neville. *Bridges, Canals and Tunnels: The Engineering Conquest of America*. American Heritage Publishing 1968

James, J.G. *The Cast-Iron Bridges of Thomas Wilson*. The Newcomen Society 1979

Jodidio, Philip. *Santiago Calatrava: Complete Works 1979 – Today*. Taschen 2015

Johnson, Stephen and Roberto T. Leon. *Encyclopaedia of Bridges and Tunnels*. Checkmark Books 2002

Joiner, J. H. *One More River to Cross*. Leo Cooper 2001

Josef, Dusan. *Nase Mosty Historicke a Soucasne [historic bridges]*. Nadas Prague 1984

Kane, Bob and Trish Kane. *New York State's Covered Bridges*. Arcadia Publshing 2014

Kawada, Tadaki. *History of the Modern Suspension Bridge: Solving the Dilemma Between Economy and Stiffness*. ASCE Press 2010

Keith, L. Bruce and Lewis Matheson. *Bridgescapes*. Dunnotar Productions 2017

Kidney, Walter C. *Pittsburgh's Bridges: Architecture and Engineering*. Pittsburgh History & Landmarks Foundation 1999

Knapp, Ronald G. *Chinese Bridges*. Oxford University Press 1993

Knapp, Ronald G. and A. Chester Ong. *Chinese Bridges: Living Architecture from China's Past*. Tuttle Publishing 2008

Knoblock, Glenn A. *New Hampshire Covered Bridges*. Arcadia Publishing 2002

Kranakis, Eda. *Constructing a Bridge: An Exploration of Engineering Culture, Design and Research in Nineteenth Century France and America*. The MIT Press 1997

Lacaze, Marc. *European Bridges*. Castor & Pollux 2007

Lambert, Guy. *Les Ponts de Paris*. Action Artistique de la Ville de Paris 1999

Lane, Oscar F. (ed). *World Guide to Covered Bridges*. The National Society for the Preservation of Covered Bridges Inc (3rd) 1972

Larbodière, Jean-Marc. *Ponts de Paris; Découverte & Histoire*. Éditions Massin 2013

Laughlin, Robert W.M. and Melissa C. Jurgensen. *Kentucky's Covered Bridges*. Arcadia Publishing 2007

Leonhardt, Fritz. *Brücken / Bridges: Aesthetics and Design*. Architectural Press 1982

Lesbros, Dominique. *Paris Promenades au bord de l'eau; 12 itinéraires de charme le long de la Seine et des canaux*. Parigramme 2015

Lewis, Tom. *By Derwent Divided: The Story of Lake Illawarra, the Tasman Bridge and the 1975 Disaster*. Tall Stories 1999

Liu, Jie and Weiping Shen. *Lounge Bridges in Taishun*. Shanghai People's Fine Arts Publishing House 2005

Locke, Tim and Anne Locke. *Bridges of the World: An Illustrated History*. AA Publishing 2008

Macauley, David. *Building Big*. Houghton Mifflin 2000

MacDonald, Copthorne. *Bridging the Strait: The Story of the Confederation Bridge Project*. Dundurn Press 1997

MacDonald, Donald and Ira Nadel. *Golden Gate Bridge: History and Design of an Icon*. Chronicle Books 2008

MacDonald, Donald and Ira Nadel. *Tilikum Crossing, Bridge of the People; Portland's Bridges and a New Icon*. Overcup Press 2018

Mackaness, Caroline (ed). *Bridging Sydney*. Historic Houses Trust of New South Wales 2006

Maggi, Angelo and Nicola Navone. *John Soane and the Wooden Bridges of Switzerland: Architecture and the Culture of Technology from Palladio to the Grubenmanns*. Sir John Soane's Museum 2003

Mair, George. *Bridge Down*. Stein and Day 1982

Mak, Geert. *The Bridge: A Journey between Orient and Occident*. Vintage 2009

Mao, Yi-Shen. *Bridges in China, Old and New: From the Ancient Chaochow Bridge to the Modern Nanking Bridge over the Yangtse*. Foreign Languages Press, Peking 1978

Maristany, Manuel. *Els Ponts de Pedra de Catalunya*. Generalitat de Catalunya 1998

Marking, Tonja Koob and Jennifer Snaper. *Huey P. Long Bridge*. Arcadia Publishing 2013

Marrey, Bernard. *Les Ponts Modernes 20ᵉ siècle*. Picard Éditeur 1995

Masi, Antonio and Joan Marans Dims. *New York's Golden Age of Bridges*. Fordham University Press 2012

Maxwell, Yolanda. *Famous Bridges of the World; Measuring Length, Weight and Volume*. Rosen Publishing Group 2010

McBriarty, Patrick T. *Chicago River Bridges*. University of Illinois Press 2013

McCormack, Michael J. *Timeless Crossings: Vermont's Covered Bridges*. Schiffer Publishing 2011

McFetrich, David. *An Encyclopaedia of British Bridges*. Pen and Sword Transport 2019

McKee, Alexander. *The Race for the Rhine Bridges: 1940, 1944, 1945*. Souvenir Press 1971

McKee, Brian J. *Historic American Covered Bridges*. American Society of Civil Engineers 1997

McKee, Harley J. *Original Bridges on the National Road in Eastern Ohio*. The Ohio Historical Society 1972

Mesqui, Jean. *Les Vieux Ponts*. Arthaud 1998

Mettem, Christopher J. *Timber Bridges*. Spon Press 2011

Middleton, William D. *Landmarks on the Iron Road: Two Centuries of North American Railroad Engineering*. Indiana University Press 2011

Middleton, William D. *The Bridge at Québec*. Indiana University Press 2001

Mikesell, Stephen D. *A Tale of Two Bridges: The San Francisco-Oakland Bay Bridges of 1936 and 2013*. University of Nevada Press 2017

Miller, David. *Bridges*. Chartwell Books 2006

Miller, Terry E. and Ronald G. Knapp. *America's Covered Bridges: Practical Crossings: Nostalgic Icons*. Tuttle Publishing 2013

Mock, Elizabeth. *The Architecture of Bridges*. The Museum of Modern Art 1949

Mohr, Marsha Williamson. *Indiana Covered Bridges*. Quarry Books 2012

Moll, Fred. *Pennsylvania's Covered Bridges*. Arcadia Publishing 2012

Moll, Fred J. *Pennsylvania's Historic Bridges*. Arcadia Publishing 2007

Montens, Serge. *Les Plus Beaux Ponts de France*. Bonneton 2001

Montens, Serge. *Les 500 Plus Beaux Ponts de France*. Christine Bonneton 2009

Moore, Elma Lee. *Ohio's Covered Bridges*. Arcadia Publishing 2010

Morley, S. Griswold. *The Covered Bridges of California*. University of California 1938

Morse, Victor. *Windham County's Famous Covered Bridges*. The Book Cellar Brattleboro, Vermont 1960

Mort, Mike. *A Bridge Worth Saving; A Community Guide to Historic Bridge Preservation*. Michigan State University Press 2008

Murray, Peter and Mary Anne Stevens. *Living Bridges: The Inhabited Bridge, Past, Present and Future*. Royal Academy of Arts 1996

Myer, Donald Beekman. *Bridges and the City of Washington*. The Commission of Fine Arts 1974

Myerscough, Matthew. *Suspension Bridges: Past and Present*. The Structural Engineer July 2013

Ness, Wolfgang, Chritine Onnen and Dirk J. Peters. *Die Schwebefähre Osten-Menmoor*. Bundesingnieurkammer 2009

O'Connor, Colin. *Roman Bridges*. Cambridge University Press 1993

O'Connor, Colin. *Spanning Two Centuries: Historic Bridges of Australia*. University of Queensland Press 1985

O'Keefe, Peter and Tom Simington; revised by Rob Goodbody. *Irish Stone Bridges: History and Heritage*. Irish Academic Press (2nd) 2016

Ostrow, Steven A. *Bridges*. Metrobooks 1997

Outerbridge, Graeme. *Bridges*. Harry N Abrams Inc 1989

Oxlade, Chris. *Superstructures: Bridges*. Belitha Press 1996

Parkyn, Neil. *Superstructures: The World's Greatest Modern Structures*. Merrell 2004

Paulik, Peter. *Bridges in Slovakia*. Jaga Group 2014

Pearce, Martin and Richard Jobson. *Bridge Builders*. Wiley-Academy 2002

Peirson, J. Gordon. *The Work of the Bridge Builders*. Pen-in-Hand 1948

Penberthy, Ian. *Man-Made Wonders: Bridges – 75 Most Spectacular Bridges*. Grange Books 2008

Perino, Angia and Georgio Faraggiana. *Bridges*. White Star Publishers 2004

Peters, Tom F. *The Development of Long-Span Bridge Building*. ETH Zurich (3rd) 1981

Petroski, Henry. *Engineers of Dreams: Great Bridge Builders and the Spanning of America*. Vintage Books 1996

Phillips, Valmai. *Bridges and Ferries of Australia*. Bay Books 1983

Pliukhin, E. and A. Punin. *And Bridges Spanned the Water's Width … .* Aurora Art Publishers 1975

Plowden, David. *Bridges: The Spans of North America*. Viking Press 1974

Polano, Sergio. *Santiago Calatrava Complete Works*. Ginko 1996

Prade, Marcel. *Les Grands Ponts du Monde Hors d'Europe*. Brissaud 1992

Prade, Marcel. *Les Ponts Monuments Historiques*. Brissaud 1988

Prade, Marcel. *Ponts et Viaducs Remarquables d'Europe*. Brissaud 1990

Priwer, Shana and Cynthia Phillips. *Bridges and Spans*. Sharpe Focus 2009

Rastorfer, Darl. *Six Bridges: The Legacy of Othmar H. Ammann*. Yale University Press 2000

Reed, Henry Hope, Robert M. McGee, Esther Mipaas. *Bridges of Central Park*. Greensward Foundation 1990

Reed, Robert. *Indiana's Covered Bridges*. Arcadia Publishing 2004

Reier, Sharon. *The Bridges of New York*. Quadrant Press 1977

Ricciuti, Edward. *America's Top 10 Bridges*. Blackbirch Press 1998

Richman, Stephen M. *The Bridges of New Jersey: Portraits of Garden State Crossings*. Rutgers University Press 2005

Rickard, Graham. *Bridges*. Wayland 1986

Riehle, Tomas. *Rheinebrücken / Rhine Bridges*. Edition Axel Menges 2015

Robert, Hugues. *Viaduc de Millau: Le Mini Mag*. Office du Tourisme de Millau (nd) c. 2006

Robertson, E.B. and D.K. Robertston. *Covered Bridges in the Saco River Valley in Maine and New Hampshire*. Robertston Books 1984

Robins, F.W. *The Story of the Bridge*. Cornish 1948

Robinson, John V. *Al Zampa and the Bay Area Bridges*. Arcadia Publishing 2005

Robinson, John V. *Carquinez Bridge 1927–2007*. Fonthill Media 2017

Roig, Joan. *New Bridges*. Editorial Gustavo Gili 1996

Ross, David. *Bridges*. Amber Books 2018

Rundall, J.W. *Bridges*. R & R Clark 1890

Sadler, Heiner. *Brücken*. Harenburg Kommunikation (2nd) 1987

Savet, John-Marie. *Les Ponts d'Hier et d'Aujourd'hui*. Mae Editeurs 2006

Scharer, Caspar and Christian Menn. *Brücken Bridges*. Scheidegger & Spiess 2006

Scheer, Joachim. *Failed Bridges: Case Studies, Causes and Consequences*. Ernst & Sohn 2010

Schlaich, Mike and Ardnt Goldack, (eds). *The World's Footbridges for Berlin*. Jovis Verlag 2017

Scott, Quinta and Howard S. Miller. *The Eads Bridge*. University of Missouri Press 1979

Scott, Richard. *In the Wake of Tacoma: Suspension Bridges and the Quest for Aerodynamic Stability*. ASCE Press 2001

Sealey, Anthony. *Bridges and Aqueducts*. Hugh Evelyn 1976

Seine sous ses Ponts de Paris à Honfleur, La. Anthèse 1995

Semur, Francois-Christian and Emmanuel Pain. *L'Ile de la Cité et ses Ponts.* Editions Ouest-France 2010

Shank, William H. *Historic Bridges of Pennsylvania.* American Canal & Transportation Center (3rd) 1980

Shapiro, Mary J. *A Picture History of the Brooklyn Bridge.* Dover Publications 1983

Shirley-Smith, H. *The World's Great Bridges.* Phoenix House (2nd) 1964

Siebel, George A. *Bridges over the Niagara Gorge; Rainbow Bridge 50 Years 1941–1991.* Niagara Falls Bridge Commission 1991

Sitler, Nevin D. and Richard N. Sitler. *The Sunshine Skyway Bridge Spanning Tampa Bay.* The History Press 2013

Sloane, Eric. *American Barns and Covered Bridges.* Wilfred Funk 1954

Soloman, Brian. *North American Railroad Bridges.* Voyageur Press 2008

Spangenburg, Ray and Diane K. Moser. *Connecting a Continent: The Story of America's Bridges.* Replica Books 1999

Stamp, Robert M. *Bridging the Border: The Structures of Canadian-American Relations.* Dundurn Press 1992

Starr, Kevin. *Golden Gate: The Life and Times of America's Greatest Bridge.* Bloomsbury Press 2010

Steel Bridges Span Europe. European Convention for Constructional Steelwork 1996

Steinman, David B. *Famous Bridges of the World.* Dover 1961

Steinman, David B. and Sara Ruth Watson. *Bridges and their Builders.* (2nd) Dover 1957

Story of the Sydney Harbour Bridge, The. Roads & Traffic Authority (nd – c. 1989?)

Straub, Hans. *A History of Civil Engineering: An Outline from Ancient to Modern Times.* Leonard Hill 1952

Sutherland, Cara A. *Bridges of New York City.* Barnes & Noble 2003

Sweetman, John. *The Artist and the Bridge 1700–1920.* Ashgate Publishing 1999

Symes, Michael. *The English Landscape Garden in Europe.* Historic England 2016

Tassotti, Giorgio. *Bassano Il Ponte degli Alpini.* Tassotti Editore 2014

Thiemann, Eckhard and Dieter Desczyk. *Berlinner Brucken; Gestaltung und Schmuk.* Lukas Verlag 2012

Thomas, W.N. *The Development of Bridges.* Geoffrey Parker & Gregg 1920

Thornton, Geoffrey. *Bridging the Gap: Early Bridges in New Zealand 1830–1939.* Reed Books 2001

To Bridge – The Danish Way. The Danish State Railways and the Road Directorate 1986

Tomasi, Peter J. and Sara Duvall. *The Bridge: How the Roeblings Connected Brooklyn to New York.* Abrams ComicArts 2018

Trachtenberg, Alan. *Brooklyn Bridge: Facts and Symbol.* University of Chicago Press (2nd) 1979

Trimble, Paul C. and John C. Alioto, Jr. *The Bay Bridge.* Arcadia Publishing 2004

Troitsky, M.S. *Planning and Design of Bridges.* John Wiley & Sons 1994

Troyano, Leonardo Fernandez. *Bridge Engineering: A Global Perspective.* Thomas Telford 2003

Tsipis, Yanni. *Boston's Bridgers.* Arcadia Publishing 2004

Tzonis, Alexander and Rebeca Caso Donadei. *Calatrava Bridges.* Thames & Hudson 2005

Van Deputte, Jocelyne. *Ponts de Paris.* Editions Sauret 1994

Van Uffelen, Chris. *Link It! Masterpieces of Bridge Design.* Braun Publishing 2005

Van Uffelen, Chris. *Masterpieces: Bridge Architecture + Design.* Braun Publishing 2010

Viaduc de Millau. Eiffage 2005

Warren, John. *Chesapeake Bay Bridge-Tunnel.* Arcadia Publishing 2015

Watson, Wilbur J. *Bridge Architecture: Containing 200 Illustrations of the Most Notable Bridges of the World, Ancient and Modern with Descriptive, Historical and Legendary Text.* New York, Helburn 1927

Watson, Wilbur J. and Sarah Ruth Watson. *Bridges in History and Legend.* J H Jansen 1937

Wells, Matthew. *30 Bridges.* Laurence King Publishing 2002

White, Joseph and M.W. Von Bernewitz. *The Bridges of Pittsburgh.* Cramer Printing & Publishing Co 1928

White, Warren H. *Covered Bridges in the Southeastern United States.* McFarland 2003

Whitney, Charles S. *Bridges; A Study in their Art, Science and Evolution.* Rudge 1929

Wiese, Eigel. *Die Brucken von Hamburg.* Die Hanse 2008

Wilson, Forrest. *Bridges Go from Here to There.* National Trust for Historic Preservation, Washington DC 1993

Wittfoht, Hans. *Building Bridges: History, Technology, Construction.* Beton-Verllag 1984

Wong, John B. *Battle Bridges: Combat River Crossings World War II*. Trafford Publishing 2004

Wortman, Sharon Wood. *The Portland Bridge Book*. Oregon Historical Society Press 2001

Xiang, Haifan (ed). *Bridges in China*. Tongi University Press and A&U Publication (HK) 1993

Zaidenberg, Arthur. *How to Draw Historic and Modern Bridges*. Abelard-Schuman 1962

Ziesel, Wolfdietrich. *Dream Bridges*. Springer-Verlag 2004

Zuercher, Gary. *The Glow of Paris; The Bridges of Paris by Night*. Marcop Editions 2015

Zuk, William. *Notable Unbuilt Bridges*. Virginia Transportation Research Council 1988

Geographic Index

Country	State / Province / Region Department / County	City, Town or Village	Bridge Name (* indicates bridge no longer exists)
Afghanistan	Balkh	Hairatan	Afghan-Uzbek Friendship Bridge
Afghanistan	Herat Province	Herat	Malan Bridge
Albania	Berat County	Berat	Gorica Bridge
Albania	Shkodër County	Shkodër	Mes Bridge
Algeria	Constantine Province	Constantine	Cidi M'Cid Bridges
Algeria	Constantine Province	Constantine	El-Kantara Bridge
Algeria	Constantine Province	Constantine	Sidi Rached Viaduct
Andorra		Andorra la Vella	Paris Bridge
Andorra		Escaldes-Engordany	Pont del Escalls
Andorra		Fontaneda	Pont de la Margineda
Angola	Benguela Province	Benguela	4 de Abril Bridge
Angola	Cunene Province	Xangongo	Cunene River Bridge
Antigua and Barbuda		Bridgetown	Chamberlain Bridge
Argentina	Anta Fe Province	Rosario	Rosario-Victoria Bridge
Argentina	Buenos Aires Province	Buenos Aires	Movable Bridges
Argentina	Buenos Aires Province	Buenos Aires	Puente de la Mujer (Women's Bridge)
Argentina	Corrientes	Corrientes	General Belgrano Bridge
Argentina	Santa Fe	Santa Fe de la Vera Cruz	Santa Fe Suspension Bridge
Armenia	Aragatsotn	Ashtarak	Ashtarak Bridge
Armenia	Ararat	Yerevan	Red Bridge
Armenia	Ararat	Yerevan	Victory (Haghtanak) Bridge
Armenia	Lori	Alawerdi	Sanahin Bridge
Armenia	Syunik	Tatey	Khndzoresk Swinging Bridge
Armenia	Yerevan	Oshakan	Kasakh Bridge
Australia	New South Wales	Sydney	Gladesville Bridge
Australia	New South Wales	Sydney	Sydney Harbour Bridge
Australia	Queensland	Brisbane	Goodwill Bridge
Australia	Queensland	Brisbane	Kurilpa Bridge
Australia	Tasmania	Richmond	Richmond Bridge
Australia	Tasmania	Ross	Ross Bridge
Australia	Tasmania	Sidmouth	Batman Bridge
Australia	Victoria	Melbourne	Webb Bridge
Austria	Bundesland	Graz	Murinsel Bridge
Austria	Bundesland	Linz	Danube Bridge
Austria	Lower Austria	Breitenstein	Kalte Rinne Viaduct
Austria	Salzburg	Obendorf	Salzach Bridge
Austria	Tyrol	Innsbruck	Europa Bridge

Country	State / Province / Region Department / County	City, Town or Village	Bridge Name (* indicates bridge no longer exists)
Austria	Tyrol	Lienz	Panzendorfer Bridge
Austria	Tyrol	Stams	Kanzler-Dollfuss Footbridge
Austria	Vienna	Vienna	Augartenbrücke
Azerbaijan	Hadrut	Jabrayil	Khudafarin Bridges
Bahamas	Grand Bahama	Nassau	Paradise Island Bridges
Bahrain		Bahrain	Sitra Causeway Bridges
Bangladesh	Chittagong	Chittagong	Shah Amanat Bridge
Bangladesh	Dhaka	Sonargaon	Panam Bridge
Bangladesh	Dhaka Division	Mawa	Padma Bridge
Bangladesh	Pabna District	Paksey	Hardinge Bridge
Bangladesh	Rajshahi Division	Sirajganj	Bangabandhu Bridge
Belarus	Minsk	Minsk	Gorbaty Bridge
Belgium	Hainaut	La Louvière	Sart Aqueduct
Belgium	Hainaut	Mons	Montignies-St-Christophe Roman Bridge
Belgium	Hainaut	Tornai	Pont Levant Notre Dame
Belgium	Liège	Plombières	Moresnet Viaduct
Belgium	Wallonia	Tournai	Pont des Trous
Belgium	Walloon	Liège	Lanaye Bridge
Belgium	West Flanders	Kortrijk	College Footbridge
Belize		Belize City	Belize City Swing Bridge
Belize	Western Belize	San Ignacio`	Hawkesworth Suspension Bridge
Bhutan	Paro Province	Paro	Tachogang Lhakang Bridge
Bolivia	Chuquisaca	Villa Azurduy	La Marca Suspension Bridge
Bolivia	Obispo Santistevan	Montero	Banegas Bridge
Bolivia	Pedro Domingo Murillo	La Paz	Triplets Bridges
Bosnia-Herzegovina	Herzegivina-Neretva	Mostar	Stari Most
Bosnia-Herzegovina	Republika Srpska	Visegrad	Mehmed-Pasha Sokolovic Bridge
Bosnia-Herzegovina	Sarajevo	Sarajevo	Latin Bridge
Botswana	Central District	Mathathane	Platian Bridge
Botswana	North-West District	Mohembo	Okavango River Bridge
Brazil	Amazonas	Manaus	Rio Negro Bridge
Brazil	Central-West Region	Brasilia	Juscelino Kubitschek Bridge
Brazil	Rio de Janeiro	Gavea	Favela de Rocinha Footbridge
Brazil	Rio de Janeiro	Rio de Janeiro	Carioca Aqueduct
Brazil	Rio de Janeiro	Rio de Janeiro	Rio-Niteroi Bridge
Brazil	Santa Catarina	Florianopolis	Hercilio Luz Bridge
Brazil	Southeast Region	Sao Paulo	Octavio Frias de Oliveira Bridge
Brunei	Brunei-Muara	Bandar Seri Bagawan	Temburong Bridge
Brunei	Brunei-Muara	Bandar Seri Begawan	Sultan Omar Ali Saifuddien Pedestrian Bridge
Brunei	Brunei-Muara	Bandar Seri Begawan	Sungai Kebun Bridge
Bulgaria	Haskova Province	Svilengrad	Svilengrad Bridge
Bulgaria	Lovech Province	Lovech	Covered Bridge
Bulgaria	Smolyan	Ardino	Devil's Bridge

Country	State / Province / Region Department / County	City, Town or Village	Bridge Name (* indicates bridge no longer exists)
Bulgaria	Varna	Varna	Asparuhov Bridge
Bulgaria	Yambol Province	Elhovo	Elhova Footbridge
Burkina Faso	Balé	Boromo	Mouhoun Bridge
Cambodia	Kampong Cham	Steung Trang	Mekong River Crossing
Cambodia	Kampong Cham Province	Kampong Cham	Koh Pen Bamboo Bridge
Cambodia	Siem Reap	Chi Kiaeng	Spean Praptos Bridge
Cameroon	Littoral Province	Besseke	Bonaberi Bridge
Cameroon	Littoral Province	Edea	Japoma Bridge
Cameroon	West Region	Sakbayémé	Sakbayémé Bridge
Canada	Alberta	Banff National Park	Land Bridge
Canada	Alberta	Calgary	Peace Bridge
Canada	Alberta	Edmonton	High Level Bridge
Canada	Alberta	Lethbridge	Lethbridge Viaduct
Canada	British Colombia	Vancouver	Sky Bridge
Canada	British Columbia	Castlegar	Columbia River Bridge
Canada	British Columbia	Vancouver	Capilano Suspension Bridges
Canada	British Columbia	Vancouver	Port Mann Bridge
Canada	New Brunswick	Saint John	Hartland Covered Bridge
Canada	Northern Manitoba	Hudson Bay	Port Nelson Bridge
Canada	Northwest Territories	Fort Providence	Deh Cho Bridge
Canada	Ontario	Niagara Falls	Rainbow Bridge
Canada	Ontario	Ottawa	Vimy Memorial Bridge
Canada	Ontario	Peterborough	Hunter Street Bridge
Canada	Ontario	Sault Ste Marie	International Railroad Bridge
Canada	Prince Edward Island	Borden-Carleton	Confederation Bridge
Canada	Quebec	Hinchinbrooke	Powerscourt Covered Bridge
Canada	Quebec	Quebec	Quebec Bridge
Canada	Quebec	Saguenay	Arvida Bridge
Chile	Auraucania	Collipulli	Malleco Railway Viaduct
Chile	Chiloé Province	Chiloé	Chacao Channel Bridge
Chile	Valdivia	Torobayo	Rio Cruces Bridge
China	Anhui	Tankhouzhen	Bridge of Immortals
China	Anhui Province	Huangshan City	Hongcun Moon Bridge
China	Beijing	Beijing	Marco Polo (Lugou) Bridge
China	Beijing	Beujing	Yiheyuan Summer Palace Bridges
China	Beijing	Yanqing	Jundushan Aqueduct
China	Fujian	Chuanchow	Loyang (Ten Thousand Peace) Bridge
China	Fujian	Pinghan	Wan'an (Eternal Peace) Bridge
China	Fujian	Xiadang	Luanfeng (Mythical Bird Peak) Bridge
China	Guangdong	Chaozhou	Guangi / Xiangzi (Great Charity) Bridge
China	Guangdong	Macau	Sai Van Bridge
China	Guangxi	Sanjiang	Batuan Bridge
China	Guangzhou	Guandong	Dongguan City Aqueduct Bridge
China	Guanxi	Ma'an, Sanjian	Chengyang Yongji (Wind and Rain) Bridge

Country	State / Province / Region Department / County	City, Town or Village	Bridge Name (* indicates bridge no longer exists)
China	Guizhou Province	Liupanshui	Beipan River Shuibai Bridge
China	Guizhou Province	Liupanshui	Duge Beipanjiang Bridge
China	Hebei	Fujian	Anping (Peace) Bridge
China	Hebei	Jingxing	Jingxing Qialoudian Bridge
China	Hebei	Tianjin	Yongle Bridge
China	Hebei	Zhaozhao	Zhaozhou Bridge
China	Hebei Province	Zhangjiajie	East Taiheng Glasswalk
China	Henan	Huixian	Red Women's Bridge
China	Hong Kong	Hong Kong	Hong Kong-Zhuhai-Macau Bridge
China	Hong Kong	Hong Kong	Tsing Ma Bridge
China	Hubei	Enshi City	Floating Bridge
China	Hubei	Wuhan	Yangtse River Bridge
China	Hunan	Changsa	Lucky Knot Bridge
China	Hunan	Zhangjiajie	Zhangjiajie Glass Bridge
China	Inner Mongolia	Jingpen Pass	Symingyi Viaduct
China	Jiangsu	Nanjing	Nanjing Yangtze River Bridge
China	Jiangsu	Nanking	Yangtse River Bridge
China	Jiangsu	Suzhou	Baodai (Precious Belt) Bridge
China	Jiangsu	Wuxi	Danyang-Kunshan Grand Bridge
China	Jiangsu	Yangzhou	Slender West Lake Bridges
China	Liaoning	Shenyang	Sanhao Bridge
China	Shanghai	Shanghai	Donghai Bridge
China	Shanghai	Shanghai	Lupu Bridge
China	Shanghai	Shanghai	Nanpu Bridge
China	Shanghai	Shanghai	Xu Pu Bridge
China	Shanghai	Shanghai	Yangpu Bridge
China	Shanghai	Shanghai	Yu Yuan Bazaar / Garden Bridge
China	Shanghai	Zhujiajiao	Fansheng (Liberate Living Things) Bridge
China	Shanxi	Hongdao	Jiaozhou Bay Bridge
China	Sichuan	Chengdu	Anshun (Peaceful and Favourable) Bridge
China	Sichuan	Chongqing	Chaotianmen Bridge
China	Sichuan	Chongqing	Chongqing Bridges
China	Sichuan	Chongqing	Wanxian Bridge
China	Sichuan	Dujiangyan City	Anlan Suspension Bridge
China	Sichuan	Garze	Luding Bridge
China	Sichuan	Leshan	Haoshang Bridge
China	Sichuan	Mianyang, Xiaobei	Jiemi (The Sisters') Bridges
China	Sichuan	Quingyuan	Rulong (Like a Dragon) Bridge
China	Taishun	Sixi	Xidong Bridge
China	Taiwan	Baguashan to Zuoying	Changhua-Kaohsiung Viaduct
China	West Hunan	Jishou	Aizhai Bridge
China	Yunan	Yoingpin	Jihong (Rainbow in the Clear Sky) Bridge
China	Yunnan	Lijiang	Black Dragon Pool Bridge
China	Zhejiang	Hangzhou	West Lake Bridges
China	Zhejiang	Qingyuan	Buchan (Stepping Toad) Bridge

Country	State / Province / Region Department / County	City, Town or Village	Bridge Name (* indicates bridge no longer exists)
China	Zhejiang	Taishun	Beijian Bridge
China	Zhejiang	Taishun	Red Army Bridge
China	Zhejiang	Taishun	Yonqing (Eternal Celebration) Bridge
China	Zhejiang	Taishun	Yuwen (Nourish Culture) Bridge
China	Zhejiang	Yiwu	Guyue Bridge
China	Zhejiang	Zhoushan	Xihoumen Bridge
China	Zhejiang	Hangzhou	Gongchen Bridge
China	Zhejiang	Hangzhou	Hangzhou Bay Bridge
China	Zhejiang	Hangzhou	Hangzhou Old Bridges
Colombia	Boyacá	Tunja	Bridge of Boyacà
Colombia	Risaralda Department	Pereira	César Gaviria Trujillo Viaduct
Colombia	Western Antioquia	Olaya	Bridge of the West
Costa Rica	Alajuela	La Fortuna	Arenal Hanging Bridges
Costa Rica	Puntarenas	Quepos	Quepos Bridge
Croatia	Dubrovnik-Neretva	Dubrovnik	Ploče Gate Footbridges
Croatia	Dubrovnik-Neretva	Kentafig	Franjo Tudjman Bridge
Croatia	Osijek-Baranja	Osijek	Drava Footbridge
Croatia	Primorge-Gorski Kotar	Rijeka	Krk Bridge
Croatia	Zadar	Kosinj	Lika River Bridge
Croatia	Zadar County	Zadar	Harbour Footbridge
Croatia	Zadar County	Zadar	Maslenica Bridges
Cuba	Mayabeque Province	Santa Cruz del Norte	Bacunayagua Bridge
Cuba	Sancti Spiritus	Sancti Spiritus	Puente Yayabo
Cuba	Sancti Spiritus Province	Sancti Spiritus	Yayabo Bridge
Cyprus	Famagusta	Agios Nikolaos	Kelefos (Tzelefos) Bridge
Cyprus	Limassol District	Akapnou	Akapnou Bridge
Cyprus	Limassol District	Kaminaria	Bridge of Elia
Cyprus	Limassol District	Trimiklini	Trimiklini Double Bridge
Czech Republic	Bohemia	Prague	Charles Bridge
Czech Republic	Bohemia	Veltrusy	Chateau Veltrusy's Laudon Pavilion
Czech Republic	Central Bohemia	Mirejovice	Mirejovice Bridge
Czech Republic	Central Bohemian Region	Borovnice	Borovsko Bridge
Czech Republic	Eastern Bohemia	Krcin	Chateau Nové Mesto and Metuji
Czech Republic	South Bohemia	Ceský Krumlov	State Castle Moat Bridge
Czech Republic	South Bohemia	Pisek	Old Bridge
Czech Republic	South Bohemia	Stadlec	Stadlec Chain Bridge
Czech Republic	South Moravia	Ivancice	Ivancice Railway Viaducts
Czech Republic	South Moravia	Nedvedice	Pernstein Castle Footbridges
Czech Republic	South Moravia	Onsov	Swiss Bay Footbridge
Czech Republic	South Moravia	Pernstein	Cernvir Covered Bridge
Democratic Republic of Congo	Central Province	Matadi	Matadi Bridge
Denmark	Capital Region	Copenhagen	Butterfly Bridge

Country	State / Province / Region Department / County	City, Town or Village	Bridge Name (* indicates bridge no longer exists)
Denmark	Capital Region	Copenhagen	Five Circles Footbridge
Denmark	Öresund Region	Malmö	Øresund Bridge
Denmark	Zealand	Stege	Queen Alexandrine Bridge
Denmark	Zealand Region	Korsør	Great Belt East Bridge
Dominican Republic	Distrito Nacional Province	Santo Domingo	Ozama River Bridge
Dominican Republic	San Pedro de Marcoris	San Pedro de Marcoris	Mauricio Báez Bridge
Ecuador	Manabí	Bahia de Caráquez	Los Caros Bridge
Ecuador	Pichincha	Quito	Chiche Bridges
Egypt	Alexandria Governorate	Alexandria	Stanley Bridge
Egypt	Cairo Governorate	Cairo	6th October Bridge
Egypt	Ismaelia	El Qantara	Al Salam Peace Bridge
Egypt	Ismaelia	El Qantara	El Ferdan Railway Bridge
El Salvador	La Unión Department	El Amatillo	Rio Goascoran Bridge
El Salvador	San Miguel Department	San Miguel	Cuscatlán Bridge
England	Greater London	London	London Bridge
Equatorial Guinea	Bioko Sur Province	Bioko	LNG Pipeline Bridge
Equatorial Guinea	Littoral Province	Mbini	Benito River Bridge
Eritrea	Gash-Barka	Molki	Obel Railway Bridge
Estonia	Lääne County	Matsalu	Kasari Old Bridge
Estonia	Pärnu County	Pärnu	Vallikraavi Footbridge
Ethiopia	Amhara	Motta	Sebara Dildiv (Broken Bridge)
Ethiopia	Oromia	Goha Tsion	Blue Nile Bridges
Ethiopia	Senafe	Debre Libanos	Portuguese Bridge
Ethiopia	Somali	Bombas-Jijiga	Bombas Bridge
Fiji	Nadroga-Navosa	Sigatoka	Nakabuta Bridge
Fiji	Nausori	Nausori	Rewa Bridge
Fiji	Rewa Province	Suva	Vatuwaqa Bridge
Finland	Åland	Mäntiharju	Vihantasalmi Bridge
Finland	Lapland	Rovaniemi	Lumberjack's Candle Bridge
Finland	Ostrobothnia	Korsholm	Replot Bridge
France	Ain	Leaz	Longeray Viaduct
France	Aisne	Fère-en-Tardenois	Chateau Moat Bridge
France	Allier	Coutansouze	Bellon Viaduct
France	Allier	Gannat	Ronzat Viaduct
France	Allier	Vichy	Boutiron Bridge
France	Alpes Maritimes	Castillon	Caramel Viaduct
France	Alpes-Maritimes	Sospel	Old Bridge
France	Alpes-Maritimes	Sospel	Bevera Railway Bridge
France	Ardennes	Chooz	Meuse Footbridge
France	Ariege	St Girons	Saint-Girons Footbridge
France	Auvergne	St Flour	Garabit Viaduct
France	Aveyron	Belcastel	Aveyron Bridge

Country	State / Province / Region Department / County	City, Town or Village	Bridge Name (* indicates bridge no longer exists)
France	Loir-et-Cher	Blois	Jacques Gabriel Bridge
France	Lot	Cahors	Valentré Bridge
France	Lot Department	Souillac	Souillac Bridge
France	Lot-et-Garonne	Clairac	Clairac Bridge
France	Lot-et-Garonne	Tonneins	Tonneins Bridge
France	Lozère	Quézac	Tarn Bridge
France	Maine-et-Loire	Montjean-sur-Loire	Montjean-sur-Loire Bridge
France	Nouvelle-Aquitaine	Bordeaux	Jacques Chaban-Delmas Lifting Bridge
France	Paris	Juvisy-sur-Orge	Pont des Belles Fontaines
France	Puy-de-Dôme	Les Fades	Fades Viaduct
France	Pyrénées-Atlantiques	Orthez	Pont Viuex
France	Pyrenees-Orientales	Planès	La Cassagne Bridge
France	Pyrénées-Orientales	Céret	Devil's Bridge
France	Pyrénées-Orientales	Fontpédrouse	Séjourné Viaduct
France	Rhône	Lyon	College Footbridge
France	Rhône	Lyon	Pont Galliéni
France	Sâone-et-Loire	Saint-Marcel	Pont des Echavannes
France	Seine-et-Marne	Luzancy	Luzancy Bridge
France	Seine-Maritime	Le Havre	Normandie Bridge
France	Seine-Maritime	Le Havre	Tancarville Bridge
France	St Flour	Clermont Ferrand	Clermont Ferrand Viaduct
France	Tarn	Albi	Old Bridge
France	Tarn	Daranche	Cize-Bolozon Viaduct
France	Tarn	Tanus	Viaur Viaduct
France	Tarn-et-Garonne	Montauban	Montaubon Bridge
France	Val-d'Oise	Argenteuil	Argenteuil Bridges
France	Vaucluse	Avignon	Avignon Bridge
France	Vaucluse	Bonnieux	Pont Julien
France	Vaucluse	Vaison-la-Romaine	Roman Bridge
France	Vaucluse	Vaucluse	Carpentras Aqueduct
France	Yonne	Sens	Grand Maître Aqueduct
Gabon	Haut-Ogooué	Franceville	Poubara Liana Bridge
Gambia	North Bank Division	Farafenni	Senegambia Bridge
Georgia	Kvemo Kartli	Meore Kesalo	Red Bridge
Georgia	Mtskheta-Mtianeti	Mtskheta	Pompey's Bridge
Georgia	Tbilisi	Tbilisi	Bridge of Peace
Germany	Baden-Württemberg	Bad Säckingen	Bad Säckingen Covered Bridge
Germany	Baden-Württemberg	Bad Säckingen	Rhine Bridge
Germany	Baden-Württemberg	Bietigheimer	Enz Viaduct
Germany	Baden-Württemberg	Ditzingen	Trumpf Footbridge
Germany	Baden-Württemberg	Heidelberg	Old Bridge (Karl Theodor Bridge)
Germany	Baden-Württemberg	Schwäbisch Hall	Kochertal Viaduct
Germany	Baden-Württemberg	Stuttgart	Neckar River Footbridge
Germany	Baden-Württemberg	Stuttgart	Nesenbachtal Bridge
Germany	Bavaria	Augsburg	Red Gate Aqueduct
Germany	Bavaria	Essing	Essinger Footbridge
Germany	Bavaria	Füssen	Queen Mary's Bridge

Country	State / Province / Region Department / County	City, Town or Village	Bridge Name (* indicates bridge no longer exists)
Germany	Bavaria	Leipheim	Leipheim Bridge
Germany	Bavaria	Nantenbach	Main Viaduct
Germany	Bavaria	Nuremberg	Chain Bridge
Germany	Bavaria	Regensburg	Old Stone Bridge
Germany	Bavaria	Regensburg	Regensburg Bridge
Germany	Bavaria	Regensburg	Schwabelweis Bridge
Germany	Bavaria	Rothenburg ob der Tauber	Tauber Bridge
Germany	Bavaria	Würzburg	Old Main Bridge
Germany	Brandenburg	Berlin	Britzer Garden Footbridges
Germany	Brandenburg	Berlin	Jungfern Footbridge
Germany	Brandenburg	Berlin	Lowen Bridge
Germany	Brandenburg	Berlin	Oberbaum Bridge
Germany	Brandenburg	Potsdam	Glienicke Bridge
Germany	Brandenburg	Potsdam	Peacock Island Palace Bridge
Germany	Hesse	Friedberg	Fibreglass Bridge
Germany	Hesse	Schlitz	Rombachtal Bridge
Germany	Lower Saxony	Hamburg	Kattwyk Bridges
Germany	Lower Saxony	Hamburg	Speicherstadt Island Bridges
Germany	Lower Saxony	Osten	Osten Transporter Bridge
Germany	North Rhine-Westphalia	Cologne	Hohenzollern Bridge
Germany	North Rhine-Westphalia	Duisburg	Emscher Landscape Park Bridges
Germany	North Rhine-Westphalia	Duisburg	Inner Harbour Bridge
Germany	North Rhine-Westphalia	Gelsenkirchen	Double-Arched Bridge
Germany	North Rhine-Westphalia	Lünen	Schwansbell Bridge
Germany	North Rhine-Westphalia	Wuppertal	Lego Bridge
Germany	Rhineland-Palatinate	Bullay	Alf-Bullay Bridge
Germany	Rhineland-Palatinate	Mainz	Bad Kreutznach Bridge
Germany	Rhineland-Palatinate	Trier	Roman Bridge
Germany	Saxonhy-Anhalt	Dessau-Wörlitz	Wörlitz Garden Bridges
Germany	Saxony	Bad Muskau	Muskauer Park Bridges
Germany	Saxony	Dresden	Augustus Bridge
Germany	Saxony	Gablenz	Devil's Bridge
Germany	Saxony	Kurort Rathen	Saxon Switzerland National Park Bastei Bridge
Germany	Saxony	Lohmen	Bastion Bridge
Germany	Saxony	Mylau	Göltzschtal Viaduct
Germany	Saxony	Plauen	Peace Bridge
Germany	Saxony-Anhalt	Dessau-Wörlitz	Dessau-Wörlitz Garden Kingdom Bridges
Germany	Saxony-Anhalt	Magdeburg	Magdeburg Water Bridge
Germany	Schleswig-Holstein	Kiel	Kiel-Hörn Folding Bridge
Germany	Schleswig-Holstein	Rendsburg	Rendsburg High Bridge

Country	State / Province / Region Department / County	City, Town or Village	Bridge Name (* indicates bridge no longer exists)
Germany	Thuringia	Erfurt	Merchants' Bridge
Ghana	Ashanti	Wawase	Kakum Canopy Walk
Ghana	Asuogyaman	Atimpoko	Adomi Bridge
Ghana	Volta	South Tongu	Lower Volta Bridge
Greece	Epirus	Arta, Plaka and Tymfi	Epirus Bridges
Greece	Epirus	Konitsa	Konitsa Bridge
Greece	Patras	Patras	Rio-Antirrio Bridge
Greece	Peloponnese	Arkadiko	Kazarma Bridge
Greece	Peloponnese	Corinth	Corinth Canal Submersible Bridges
Guatemala	Guatemala Department	Guatemala City	National Post Office Bridge
Guatemala	Santa Rosa	Cuilapa	Slaves Bridge
Guinea	Kankan	Kankan	Milo Bridge
Guinea	Nzérékoré Prefecture	Koulé	Vine Bridge
Guinea-Bissau	Cacheu Region	São Vicente	São Vicente Bridge
Guyana	Demerara-Mahaica Region	Georgetown	Demerara Harbour Bridge
Guyana	Potaro-Siparuni Region	Mahdia	Denham Suspension Bridge
Guyana	Upper Takutu-Essequibo	Lethem	Takutu River Bridge
Haiti	Centre	Cange	Ti Peligre Bridge
Honduras	Choluteca Province	Choluteca	Choluteca Bridge
Hungary	Central Hungary	Budapest	Elisabeth Bridge
Hungary	Central Hungary	Budapest	Liberty Bridge
Hungary	Central Hungary	Budapest	Margaret Bridge
Hungary	Central Hungary	Budapest	Megyeri Bridge
Hungary	Central Hungary	Budapest	Széchenyi Chain Bridge
Hungary	Fejer	Dunaijvaros	Pentele Bridge
Hungary	Hajdú-Bihar	Hortobágy	Nine-Hole Bridge
Hungary	Jász-Nagykun-Szolnok	Szolnok	Mayfly Footbridge
Iceland	Eastern Region	Egilsstaðir	Jökulsá á Dal Canyon Bridge
Iceland	Eastern Region	Kirkjubæjarklaistur	Skeiðarárbrú Bridge *
Iceland	North Iceland	Fosshóll	Skjálfandafljót Bridge
Iceland	Reykjanes	Midlina	Bridge Between Two Continents
Iceland	Western Region	Hvanneyn	Hvitá Bridge
India	Assam	Dibrugarh	Bogibeel Bridge
India	Haryana State	Kalka	Kanoh Bridge
India	Jammu and Kashmir	Sala	Chenab Rail Bridge
India	Maharashtra State	Mumbai	Bandra-Worli Sea Link
India	Meghalaya State	Cherrapungi	Umshiang Double-Decker Root Bridge
India	Punjab	New Delhi	Old Yamuna Bridge
India	Tamil Nadu State	Mandapam	Pamban Bridge
India	Utter Pradesh	Jaunpur	Shahi Mughal Bridge
India	West Bengal State	Calcutta	Howrah Bridge
Indonesia	Bali	Nusa Lembongan	Yellow Bridge

Country	State / Province / Region Department / County	City, Town or Village	Bridge Name (* indicates bridge no longer exists)
Indonesia	Bali	Ubud	Dragon Bridge
Indonesia	East Java	Surabaya	Suramadu National Bridge
Indonesia	East Kalimantan	Tenggarong	Kutai Kartanegara Bridge
Indonesia	Java	Surakarta	Bamboo Bridge
Indonesia	South Sumatra	Palembang	Ampera Bridge
Iran	Fars	Marydasht	Band-e-Amir Bridge
Iran	Isfahan	Isfahan	Zayanderud Bridges
Iran	Khuzestan Province	Shushtar	Shadorvan Bridge
Iraq	Kurdistan	Zakho	Delal Bridge
Iraq	Kurdistan	Zakho	Pira Delal
Iraq	West Azerbaijan	Tello	Girsu Bridge
Ireland	County Clare	Lisdoonvarna	Spectacle Bridge
Ireland	County Clare	Spanish Point	Bealaclugga Bridge
Ireland	County Cork	Ballyquane	Glanworth Bridge
Ireland	County Cork	Castletownbere	Aghakista Bridge
Ireland	County Cork	Cork	St Patrick's Bridge
Ireland	County Dublin	Dublin	Father Mathew Bridge
Ireland	County Dublin	Dublin	Ha'penny Bridge
Ireland	County Dublin	Dublin	James Joyce Bridge
Ireland	County Dublin	Dublin	Lucan Bridge
Ireland	County Dublin	Dublin	O'Connell Bridge
Ireland	County Dublin	Dublin	Samuel Beckett Bridge
Ireland	County Dublin	Dublin	Seán O'Casey Bridge
Ireland	County Dublin	Dublin	Synod Hall Footbridge
Ireland	County Kerry	Dingle	Lispole Viaduct
Ireland	County Limerick	Limerick	Pedestrian Living Bridge
Ireland	County Longford	Abbeyshrule	Whitworth Aqueduct
Ireland	County Louth	Drogheda	Boyne Viaduct
Ireland	County Louth	Drogheda	Mary McAleese Boyne Valley Bridge
Ireland	County Mayo	Clonbur	Ashford Castle Bridge
Ireland	County Mayo	Louisburgh	Bunlahinch Clapper Footbridge
Ireland	County Meath	Trim	Trim Bridge
Ireland	County Waterford	Dungarvan	Devonshire Bridge
Ireland	County Wexford	Scarawalsh	Scarawalsh Bridge
Ireland	County Wicklow	Britonstown	Pollaphuca Bridge
Ireland	Munster	Crookhaven	Mizen Head Bridge
Ireland	Munster	Limerick City	Barrington Bridge
Ireland	North Leinster	Drogheda	Obelisk Bridge
Ireland	South Leinster	New Ross	Mountgarret Bridge
Ireland	Waterford	Villierstown	Dromana Bridge
Israel	Central District	Yavne	Yibna Bridge
Israel	Haifa	Caesarea	High Level Aqueducts
Israel	Jerusalem	Jerusalem	Chords Bridge
Israel	Northern Negev	Beersheba	Station Bridge
Italy	Basilicata	Potenza	Musmeci Bridge
Italy	Belluno	Caralte	Cadore Viaduct
Italy	Calabria	Scilia	Favazzina Bridge

Country	State / Province / Region Department / County	City, Town or Village	Bridge Name (* indicates bridge no longer exists)
Italy	Caserta	Caserta	Vanvitelli Aqueduct
Italy	Emilia	Bologna	A13 Motorway Footbridge
Italy	Ferrara	Comacchio	Treponti
Italy	Lombardy	Genoa	Polcevera Viaduct
Italy	Lucca	Borgo a Mozzano	Ponte della Maddalena
Italy	Lucca	Fornoli	Chains Bridge
Italy	Padua	Villanova	Lucano Bridge
Italy	Pavia	San Martino	Covered Bridge
Italy	Perugia	Spoleto	Ponte delle Torri Aqueduct
Italy	Rimini	Rimini	Tiberius Bridge
Italy	Rome	Rome	Ancient Roman Bridges
Italy	Rome	Rome	Ponte del Risorgimento
Italy	Rome	Rome	Ponte Salario
Italy	Rome	Rome	Unification Bridge
Italy	Rome	Rome	Vittoria Emanuele II Bridge
Italy	Sicily	Adrano	Saracens Bridge
Italy	Sicily	Termine Imerese	Ponte San Leonardo
Italy	South Tyrol	Bozen	Bolzano Museum Footbridge
Italy	Terni	Narni	Bridge of Augustus
Italy	Tuscany	Florence	Santa Trinita Bridge
Italy	Varese	Varese	Bevera Viaduct
Italy	Veneto	Venice	Bridge of Sighs
Italy	Veneto	Venice	Constitution Bridge
Italy	Veneto	Venice	Rialto Bridge
Italy	Veneto	Venice	Three-Arched Bridge
Italy	Verona	Verona	Ponte Scaligero
Italy	Vicenza	Bassano del Grappa	Ponte degli Alpini
Ivory Coast	Abidjan	Abidjan	Pont Félix-Houphouët-Boigny
Ivory Coast	Gbôklé	Sassandra	Weygand Bridge
Jamaica	Middlesex	Spanish Town	Rio Cobre Bridge
Jamaica	Middlesex	St Catherine	Flat Bridge
Japan	Chiba	Katori	Tonegawa Bridge
Japan	Ehime Prefecture	Imabiri	Kurishima-Kaikyö Bridge
Japan	Ehime Prefecture	Shikoku	Kazurabashi Vine Bridge
Japan	Gunma	Tsumagoi	Seishun Footbridge
Japan	Hiroshima	Hiroshima	Aioi Bridge
Japan	Hokkaido	Hokkaido	Boukei Bridge
Japan	Honshu	Shizuoka	Shiosai Bridge
Japan	Hyogo Prefecture	Kobe	Akashi Kaikyo Bridge
Japan	Kagawa	Takamatsu	Ritsurin Garden Bridges
Japan	Kanagawa	Kawasaki	Tokyo-Haneda Monorail Bridge
Japan	Kumamoto	Kumamoto	Ushibuka Haiya Bridge
Japan	Kyoto	Nantan	Hiyoshi Dam Footbridge
Japan	Nagasaki	Nagasaki	Ikitsuki Bridge
Japan	Osaka	Osaka	Senbonmatsu Ohashi Megane Bridge
Japan	Saitama	Chichibu	Tomoegawa Bridge
Japan	Shiga	Shigoraki	Miho Museum Bridge

Country	State / Province / Region Department / County	City, Town or Village	Bridge Name (* indicates bridge no longer exists)
Japan	Shimane	Matsue	Eshima Grand Bridge
Japan	Shizuoka	Shimoda	Kawazu-Nanadaru Loop Bridge
Japan	Tochigi Prefecture	Nikko	Shinkyo (Sacred) Bridge
Japan	Tokyo	Tokyo	Koishikawa Korakuen Garden Bridges
Japan	Yamaguchi Prefecture	Iwakuni	Kintai Bridge
Jordan	Amman Governorate	Amman	Abdoun Bridge
Jordan	Jericho Governorate	Jericho	Allenby (King Hussein) Bridge
Kazakhstan	Astana	Astana	Atyrau Bridge
Kazakhstan	Central Kazakhstan	Nur Sultan	Ramstor Bridge
Kazakhstan	East Kazakhastan	Semey	Semey Bridge
Kenya	Busia County	Sigiri	Sigiri Bridge
Kenya	Mombasa	Likoni	Mombasa Gate Bridge
Kenya	Nairobi	Athi River	Athi River Super Bridge
Kuwait		Kuwait City	Sheikh Jaber Causewaay
Laos	Bokeo Province	Ban Houayxay	Fourth Thai-Lao Friendship Bridge
Laos	Luang Prabang Province	Wat Luang	Pakse (Lao-Nippon) Friendship Bridge
Laos	Sekong Province	Sekong	Sekong Bridge
Latvia	Courland	Kurzeme	Kuldiga Bridge
Latvia	Riga	Riga	Southern Bridge
Latvia	Zemgale	Jelgava	Mitava Footbridge
Lebanon	Aintoura	Mazraat El Ras	Nahr al-Kalb Bridge
Lebanon	Mount Lebanon	Sawfar	Mudeirej Bridge
Lebanon	Tyre	Maameltein	Maameltein Roman Bridge
Lesotho	Qacha's Nek District	Mohlapiso	Mohlapiso Bridge
Lesotho	Quthing District	Seaka	Seaka Bridge
Liberia	Montserrado	Monrovia	Gabriel Tucker Bridge
Libya	Benghazi Province	Benghazi	Giuliana Bridge
Libya	Cyrenaica	Al Bayda	Wadi El Kuf Bridge
Liechtenstein	Oberland	Vaduz	Old Rhine Bridge
Lithuania	Kaunas County	Kaunas	Vytautas the Great Bridge
Lithuania	Utena County	Ditkūnai	Zarasas Bridge
Lithuania	Vilnius County	Vilnius	Mindaugas Bridge
Luxembourg		Luxembourg City	Adolphe Bridge
Madagascar	Betsiboka	Maevatanana	Betsiboka Bridge
Madagascar	Sava	Fanambana	Fanambana Bridge
Malawi	Chikwawa	Chikwawa	Chapananga Bridge
Malawi	Rumphi District	Kandewe	Zuwulufu Bridge
Malaysia	Kedar	Langkawi Island	Langkawi Sky Bridge
Malaysia	Kedar	Pulau Langkawi	Sky Bridge
Malaysia	Pahang	Kuala Tahan	Taman Negara Canopy Walkway
Malaysia	Penang	Bayan Lepas	Second Penang Bridge
Malaysia	Putrajaya	Putrajaya	Monorail Suspension Bridge
Malaysia	Putrajaya	Putrajaya	Putra Bridge
Malaysia	Putrajaya	Putrajaya	Seri Wawasan Bridge
Malaysia	Selangor	Kuala Lumpur	Petronas Twin Towers Sky Bridge

Country	State / Province / Region Department / County	City, Town or Village	Bridge Name (* indicates bridge no longer exists)
Maldives	Northern	Malé	Sinamalé Bidge
Mali	Koulikoro	Bamako	Bamako Third Bridge
Mali	Koulikoro	Bamako	King Fahd Bridge
Mali	Koulikoro	Bamako	Martyrs Bridge
Malta	Northern	Mosta	Mosta Bridge
Malta	South Eastern	Valetta	Wignacourt Aqueduct
Mauritius	Rodrigues Island	La Ferme	Suspension Bridge
Mauritius	Savanne District	Souillac	Senneville Bridge
Mexico	Durango	Mapimi	Ojuela Bridge
Mexico	Durango	Témoris	Chinipas Bridge
Mexico	Eastern	Hidalgo	Padre Templeque Aqueduct
Mexico	Nuevo León	San Pedro Garza Garcia	National Unity Bridge
Mexico	Oaxaca	Huajuapan de Leon	Colonial Bridge of Tequixtepec
Mexico	Querétaro	Santiago de Querétaro	Querétaro Aqueduct
Moldova	Bessarabia	Lipcani	Lipcani-Rădăuți Bridge
Moldova	Ungheni District	Ungheni	Eiffel Bridge
Mongolia	Khövsgöl Province	Jaargalant	Jaargalant Wooden Bridge
Montenegro	Bar Municipality	Stari Bar	Bar Aqueduct
Montenegro	Central Region	Danilovgrad	Adzija's Bridge
Montenegro	Central Region	Podgorica	Adži-Paša's Bridge
Montenegro	Central Region	Podgorica	Mala Rijeka Viaduct
Montenegro	Central Region	Podgorica	Millennium Bridge
Montenegro	Crmnica Region	Rijeka Crnojevica	Danilo's Bridge
Montenegro	Niksic Municipality	Mostanica	Roman Bridge
Montenegro	Northern Region	Zabljak	Djurdjevica Tara Bridge
Morocco	Khenifra	Ismail	Kantara Moulay Bridge
Morocco	Larache	Ksar el Kebir	Faisal Bridge
Morocco	Marrakesh-Safi	Marrakesh	Dar-Tahar-ben-Abbou Bridge
Morocco	Rabat-Salé-Kénitra	Rabat	Mohammed VI Bridge
Morocco	Rabat-Salé-Kénitra	Rabat	Rabat Bridges
Mozambique	Maputo Province	Matola	Maputo-Katembe Bridge
Mozambique	Tete Province	Tete City	Kassuende Bridge
Mozambique	Zambézia Province	Chimuara	Armando Emilio Guebuza Bridge
Myanmar	Mandalay	Amarapura	U Bein Bridge
Myanmar	Mandalay	Mandalay	Ayeyarwady Bridge
Myanmar	Mandalay	Sagaing	Yadanabon Bridge
Namibia	Hardap Region	Groot Aub	Groot Aub Bridge
Nepal	Kanchanpur	Bhimdatta	Dodhara Chandani Bridge
Nepal	Mustang District	Lete	Hanging Bridge of Ghasa
Nepal	Parbat	Kushma	Kushma-Gyadi Bridge
Netherlands	Flevoland	Almere	Olstgracht Bridge
Netherlands	Friesland	Leeuwarden	Slauerhoff Flying Drawbridge
Netherlands	Gelderland	Nijmegan	Oversteek Bridge
Netherlands	Limburg	Maastricht	St Servatius Bridge
Netherlands	North Holland	Amsterdam	Amstel Park Footbridge
Netherlands	North Holland	Amsterdam	Enneüs Heerma Bridge
Netherlands	North Holland	Amsterdam	High Bridge

Country	State / Province / Region Department / County	City, Town or Village	Bridge Name (* indicates bridge no longer exists)
Netherlands	North Holland	Haarlem	Langebrug
Netherlands	South Holland	Leiden	Koornbrug
Netherlands	South Holland	Nieuwerkerk	Zeeland Bridge
Netherlands	South Holland	Rotterdam	Erasmus Bridge
Netherlands	South Holland	Vlaardingen	Vlaardingse Vaart Bridge
Netherlands	Zeeland	Westenschouwen	Eastern Scheldt Bridge
New Caledonia	South Province	La Foa	Passerelle Marguerite
New Zealand	Auckland	Auckland	Auckland Harbour Bridge
New Zealand	Canterbury	Christchurch	Bridge of Remembrance
New Zealand	Central Otago	Alexandra	Alexandra Suspension Bridge
New Zealand	Northland	Panguru	Kohukohu Bridge
New Zealand	Otago	Queenstown	Kawarau Bridge
New Zealand	Taranaki	Taranaki	Te Rewa Rewa Bridge
Nicaragua	Nueva Segovia	Pantasma	Arenales Suspension Bridge
Niger	Kaduna State	Kaduna	Lugard Footbridge
Nigeria	South West	Lagos	Lekki-Ikoyi Bridge
Nigeria	South West	Lagos	Third Mainland Bridge
North Korea	North Hwanghae	Kaesong	Bridge of No Return
North Korea	North Hwanghae	Kaesong	Sonjuk Bridge
North Korea	North Pyongan	Sinuiji	Sino-Korean Friendship Bridge
North Macedonia	Greater Skopje	Skopje	Skopje Aqueduct
North Macedonia	Greater Skopje	Skopje	Skopje City Centre Footbridges
North Macedonia	Greater Skopje	Skopje	Stone Bridge
Northern Ireland	County Antrim	Antrim	Greenisland Railway Viaducts
Northern Ireland	County Antrim	Ballintoy	Carrick-a-Rede Rope Bridge
Northern Ireland	County Armagh	Bessbrook	Craigmore Viaduct
Northern Ireland	County Londonderry	Derry	Foyle Bridge
Northern Ireland	County Londonderry	Derry	Peace Bridge
Northern Ireland	County Londonderry	Londonderry	Craigavon Bridge
Northern Ireland	Ulster	Castledawson	Moyola Park Bridge
Norway	Agder County	Kristiansand	Kristiansand to Trondheim Submerged Bridge
Norway	Greater Oslo	Sandvika	Lokke Bridge
Norway	Innlandet County	Flisa	Flisa Bridge
Norway	Møre og Romsdal	Kristiansund	Storseisundet Bridge
Norway	Nordland	Sandnessjøen	Helgeland Bridge
Norway	Viken	Fredrikstad	Fredrikstad Bridge
Norway	Viken County	Ås	Leonardo Bridge
Oman	Ash Sharqiyah South	Sur	Ayjay (Khor Al Batah) Bridge
Oman	Muscat	Muscat	Royal Opera House Footbridge
Pakistan	Gwalior	Marena	Noorabad Bridge
Pakistan	Khyber Pakhtunkhwa	Peshawar	Chuha Gujar Mughal Bridge
Pakistan	Punjab	Rawalpindi	Dhangali Bridge
Pakistan	Punjab Province	Attock Khurd	Attock Railway Bridge
Pakistan	Sindh	Karachi	Malir River Bridge
Pakistan	Sindh Province	Sukkur	Ayub Bridge
Pakistan	Sindh Province	Sukkur	Lansdowne Bridge
Pakistan	Upper Hunza	Passu	Hussaini Hanging Bridge

Country	State / Province / Region Department / County	City, Town or Village	Bridge Name (* indicates bridge no longer exists)
Palau		Koror	Japan-Palau Friendship Bridge
Palestine	Tulkarm Governorate	Zeltan	Jisr Jindas (Baybars) Bridge
Panama	Panama City	Balboa	Bridge of the Americas
Panama	Paraiso	Colon	Panama Canal Bridges
Papua New Guinea	Komo-Magarima	Komo Station	Hegigio Gorge Pipeline Bridge
Papua New Guinea	Madang	Madang	Banab River Bridge
Paraguay	Alto Paraná Department	Ciudad del Este	Friendship Bridge
Peru	Canas Province	Huinchiri	Keshwa Chaca Bridge
Peru	Cuzco	Machu Pichu	Inca Bridge
Peru	Lima	Rimac	Rimac River Bridge
Philippines	Central Visayas	Cebu City	Marcelo Fernan Bridge
Philippines	Eastern Visayas	Tacloban	San Juanico Bridge
Philippines	Luzon	Pagadpud	Winding Patapat Viaduct
Poland	Eastern Pomerania	Dirschau	Vistula Bridges
Poland	Kuyavian-Pomerania	Grudziadz	Bronislaw Malinowski Bridge
Poland	Lesser Poland	Cracow	Father Bernatek Footbridge
Poland	Lodz	Lowicz	Arkadia Park Aqueduct
Poland	Lower Silesia	Wroclaw	City Centre Bridges
Poland	Lower Silesia	Wroclaw	Redzinski Bridge
Poland	Nowy Ttarg	Stromowce Nizne	Dunajec Footbridge
Poland	Wielkopolska	Opatówek	Park Footbridge
Poland	Zulawy Region	Malbork	Nogat Bridge
Portugal	Alentejo	Elvas	Amoreira Aqueduct
Portugal	Coimbra	Coimbra	Pedro & Inês Footbridge
Portugal	Coimbra	Figueira da Foz	Edgar Cardosa Bridge
Portugal	Estramadura	Lisbon	Aguas Livres Aqueduct
Portugal	Estremadura	Lisbon	25th April Bridge
Portugal	Estremadura	Lisbon	Santa Justa Lift
Portugal	Estremadura	Lisbon	Vasco da Gama Bridge
Portugal	Norte	Ponte de Lima	Lima Bridge
Portugal	Norte	Porto	Dom Luis I Bridge
Portugal	Norte	Porto	Maria Pia Bridge
Portugal	Viana do Castelo	Valenca	International Bridge
Portugal	Viseu	Tarouca	Ucanha Bridge
Qatar		Doha	Doha Sharq Crossing
Republic of the Congo	Brazzaville	Brazzaville	Brazzaville-Kinshasa Bridge
Romania	Constanta	Cernavoda	Anghel Saligny Bridge
Romania	Constanta	Cernavoda	Cernavoda Bridge
Romania	Mehedinti	Droberta-Turnu	Bridge of Appollodorus *
Romania	Transylvania	Sibiu	Bridge of Lies
Russia	Leningrad Oblast	Saint Petersburg	Blue Bridge
Russia	Leningrad Oblast	Saint Petersburg	City Bridges
Russia	Leningrad Oblast	Saint Petersburg	River Neva Bascule Bridges
Russia	Leningrad Oblast	Saint Petersburg	Tsarkoye Selo Park Bridges

Country	State / Province / Region Department / County	City, Town or Village	Bridge Name (* indicates bridge no longer exists)
Russia	Moscow Oblast	Moscow	Bogdan Khmelnitsky Pedestrian Bridge
Russia	Moscow Oblast	Moscow	Krasnoluzhsky Bridges
Russia	Moscow Oblast	Moscow	Tsaritsyno Park Bridges
Russia	Moscow Oblast	Moscow	Zhivopisny Bridge
Russia	Primorsky Krai	Vladivostok	Russky Bridge
Russia	Siberia	Kuandinsky	Vitim Bridge
Rwanda	Eastern Province	Rusumo	Rusumo Bridge
Samoa		Apia	Loto Samasoni Bridge
San Marino	Borgo Maggiore	Valdragone	Fontevecchia Bridge
Saudi Arabia	Eastern Province	Al Khobar	King Fahd Causeway Bridge
Saudi Arabia	Riyadh Province	Riyadh	Wadi Laban Bridge
Saudi Arabia	Riyadh Region	Riyadh	Kingdom Centre Skywalk
Scotland	City of Edinburgh	South Queensferry	Firth of Forth Bridges
Senegal	Saint-Louis Region	Saint-Louis	Faidherbe Bridge
Serbia	Belgrade	Belgrade	Ada Bridge
Serbia	Belgrade	Belgrade	Gazela Bridge
Serbia	Vojvodina	Novi Sad	Zezelj Bridge
Sierra Leone	Pujehun District	Moa	Moa Bridge
Singapore		Singapore	Henderson Waves Bridge
Singapore		Singapore	Pinnacle Sky Bridges
Singapore		Singapore	Singapore River Bridges
Slovakia	Bratislava Region	Bratislava	Airport Lighting Bridge
Slovakia	Bratislava Region	Bratislava	New Bridge
Slovakia	South Moravia	Znojmo	Lake Vranov Footbridge
Slovenia	Gorizia	Solkan	Soca Railway Bridge
Slovenia	Koper	Stepani	Crni Kal Viaduct
Slovenia	Province of Ljubljana	Ljubljana	Borovnica Railway Viaduct *
Slovenia	Province of Ljubljana	Ljubljana	Cobblers' Bridge
Slovenia	Province of Ljubljana	Ljubljana	Dragon Bridge
Slovenia	Province of Ljubljana	Ljubljana	Harp Motorway Bridge
Slovenia	Province of Ljubljana	Ljubljana	Hradecky Bridge
Slovenia	Styria	Ptuj	Puch Bridge
Solomon Islands	Makira-Ulawa Province	Makira	Makira Bridges
Somalia	Benadir	Mogadishu	Old Railway Viaduct
South Africa	Eastern Cape	Storms River	Paul Sauer Bridge
South Africa	Gauteng Province	Johannesburg	Nelson Mandela Bridge
South Africa	Kwa-Zulu-Natal	Port Edward	C.H. Mitchell Bridge
South Africa	Western Cape	Albertinia	Gouritz River Bridges
South Africa	Western Cape	Capetown	Foreshore Freeway Bridge
South Africa	Western Cape	Capetown	Victoria & Alfred Harbour Bridge
South Africa	Western Cape	Knysna	Bloukrans Bridge
South Africa	Western Cape	Knysna	Tsitsikamma Suspension Bridge
South Korea	Gyeonggi Province	Daejeon	Daedunsan Cloud Bridge
South Korea	Gyeonggi Province	Seoul	Banpo-Jaamsu Bridge
South Korea	Gyeonggi Province	Seoul	Geumcheongyo Bridge

Country	State / Province / Region Department / County	City, Town or Village	Bridge Name (* indicates bridge no longer exists)
South Korea	Gyeonggi Province	Seoul	Seonyu Footbridge
South Korea	Jeju Volcanic Islands	Seogwipo	Seonimgyo Bridge
South Sudan	Central Equatoria	Juba	Juba Nile (Freedom) Bridge
Spain	Andalusia	Almuñécar	Rio Seco Aqueduct
Spain	Andalusia	Cordoba	Roman Bridge
Spain	Andalusia	Nerja	Aguila Aqueduct
Spain	Andalusia	Seville	Alamillo Bridge
Spain	Andalusia	Seville	Barqueta Bridge
Spain	Aragon	Lascellas-Ponzano	Lascellas Bridge
Spain	Aragon	Zaragoza	Bridge Pavilion
Spain	Asturios	Mieres del Camino	Caudal Bridge
Spain	Badajoz	Mérida	Lusitania Bridge
Spain	Badajoz	Mérida	Roman Bridge
Spain	Barcelona	Cardona	Cardona Bridge
Spain	Barcelona	Martorell	Devil's Bridge
Spain	Barcelona	Vilomara	Gothic Bridge
Spain	Biscay	Bilbao	Campo Volantin Footbridge
Spain	Biscay	Bilbao	Galindo Bridge
Spain	Biscay	Bilbao	Zubizuri Bridge
Spain	Biscay	Portugalete	Vizcaya Transporter Bridge
Spain	Burgos	Frias	Roman Bridge
Spain	Cáceres	Alcántara	Alcántara Bridge
Spain	Castile-La Mancha	Toledo	Alcántara Bridge
Spain	Castile-La Mancha	Toledo	San Martin Bridge
Spain	Castile-León	Salamanca	Roman Bridge
Spain	Castile-León	Segovia	Segovia Aqueduct
Spain	Castile-León	Zamora	Stone Bridge
Spain	Catalonia	Barcelona	Bac de Roda-Felipe II Bridge
Spain	Catalonia	Girona	Eiffel Bridge
Spain	Catalonia	Sant Celoni	Llinars Bridge
Spain	Catalonia	Tarragona	Tarragona Aqueduct
Spain	Galicia	Pontevedra	Burgo Bridge
Spain	Galicia	Pontevedra	Lerez River Bridge
Spain	Gipuzkoa	San Sebastian	Marina Cristina Bridge
Spain	Gipuzkoa	Segura	Roman Bridge
Spain	Girona	Besalu	Besalu Bridge
Spain	Huelva	Ayamonte	Guadiana International Bridge
Spain	La Rioja	Logrono	Sagasta Bridge
Spain	Madrid	Madrid	Arganzuela Footbridge
Spain	Malaga	Ronda	Puente Nuevo
Spain	Navarre	Alloz	Alloz Aqueduct
Spain	Navarre	Ourense	Queen's Bridge
Spain	Navarre	Pamplona	Magdalena Footbridge
Spain	Ourense	Ourense	Millennium Bridge
Spain	Valencia	Alameda	Alameda Bridge
Spain	Valencia	Elche	Santa Teresa Bridge
Spain	Zaragoza	Osera	Ebro Bridge
Sri Lanka	Uva	Badulla	Bogoda Wooden Bridge

Country	State / Province / Region Department / County	City, Town or Village	Bridge Name (* indicates bridge no longer exists)
Sri Lanka	Uva	Demodara	Nine Arch Bridge
Sudan	Khartoum State	Khartoum	MacNimir Bridge
Sudan	Khartoum State	Khartoum	White Nile (Omdurman) Bridge
Suriname	Coronie District	Jenny	Coppename Bridge
Suriname	Suriname District	Paramiribo	Jules Wijdenbosch Bridge
Sweden	Bohuslän	Tjörn	Tjörn Bridges
Sweden	Småland	Kalmar	Öland Bridge
Sweden	Västerbotten	Skellefteå	Lejonströmsbron Bridge
Sweden	Västernorrland County	Kramfors	Sandö Bridge
Sweden	Västra Götaland County	Håverud	Håverud Aqueduct
Sweden	Västra Götaland County	Svinesund	Svinesund Bridge
Switzerland	Appenzell Ausserrhoden	Herisau	Kubelbrücke
Switzerland	Bern	Gadmen	Trift Bridge
Switzerland	Bern	Hinterfultigen	Swandbach Bridge
Switzerland	Fribourg Canton	Fribourg	Grandfey Viaduct
Switzerland	Geneva	Geneva	Coulouvrenière Bridge
Switzerland	Geneva	Geneva	Vessy Bridge
Switzerland	Graubünden	Filisur	Rhaetian Railway Bridges
Switzerland	Graubünden	St Moritz	Traversina Footbridge
Switzerland	Graubünden	Thusis	Suransuns Bridge
Switzerland	Graubünden	Zuoz	Zuoz Bridge
Switzerland	Lucerne	Lucerne	Timber Bridges (Spreuer and Chapel)
Switzerland	Obwalden	Enelberg	Titlis Suspension Bridge
Switzerland	Schaffhausen Canton	Schaffhausen	Schaffhausen Bridge
Switzerland	Ticino	Lavertezzo	Ponte dei Salti
Switzerland	Uri	Andermatt	Schöllenen Gorge Devil's Bridge
Switzerland	Valais	Brig	Ganter Bridge
Switzerland	Valais	Sierre	Sierre Footbridge
Switzerland	Valais	Zermatt	Charles Kuonen Suspension Bridge
Switzerland	Vaud	Veytaux	Chillon Viaduct
Switzerland	Zurich	Zurich	Stauffacher Bridge
Syria	Al Hasakah Governorate	Ain Diwar	Ain Diwar Bridge
Syria	Aleppo Governorate	Aleppo	Citadel Bridge
Syria	Deir ez-Zor Province	Deir ez-Zor	Deir ez-Zor Suspension Bridge
Taiwan	Kaohsiung County	Kaohsiung	Lotus Lake Footbridges
Taiwan	Miaoli County	Sanyi Township	Longteng Bridge
Taiwan	Xinya District	Taipei	Danjiang Bridge
Tajikistan	Gorno-Bakakshan Province	Vang	Vang Bridge
Tanzania	Dar es Salaam	Kurasini	Kigamboni Bridge
Tanzania	Ikwiriri	Mkana	Mkapa Bridge

Country	State / Province / Region Department / County	City, Town or Village	Bridge Name (* indicates bridge no longer exists)
Tanzania	Mtwara Region	Mtambaswala	Unity Bridge
Thailand	Kanchanaburi Province	Kanchanaburi	Ma Kham Bridge
Thailand	Krung Thep Maha Nakhon	Bangkok	Bhumibol Bridges
Thailand	Krung Thep Maha Nakhon	Bangkok	Rama VIII Bridge
Timor-Leste	Timor Timur	Oecusse	Noefefan Bridge
Togo	Lacs	Aneho	Aneho Bridge
Tonga	Tongatapu	Nuku'alofa	Fanga'uta Lagoon Bridge
Trinidad & Tobago	Saint George County	Piarco	Caroni Delta Bridge
Tunisia	Béja Governorate	Béja	Medjez-el-Bab Bridge
Tunisia	Bizerte Governorate	Kantarat Binzart	Protville Bridge
Tunisia	Tunis Governorate	Tunis	Bizerte Bridge
Turkey	Adana Province	Adana	Adana Roman Bridge
Turkey	Adana Province	Karaisali	Varda Viaduct
Turkey	Adivaman Province	Arsameia	Severan Bridge
Turkey	Alacahan District	Yeslikale	Halil Rifat Pasha Bridge
Turkey	Antalya	Aspendos	Eurymedon Bridge
Turkey	Antalya	Aspendos	Roman Aqueducts
Turkey	Antalya	Manaygat	Naras Bridge
Turkey	Antalya Province	Limyra	Kirkgoz Kemeri Bridge
Turkey	Canakkale	Lapseki	Canakkale 1915 Bridge
Turkey	Diyarbakir	Silvan	Malabadi Bridge
Turkey	East Thrace	Erdine	Tunca Bridge
Turkey	Harabesehir District	Ahlat	Emir Bayindir Bridge
Turkey	Izmir Province	Gazientep	Atatürk Viaduct
Turkey	Izmir Province	Smyrna	Kızılçullu Aqueducts
Turkey	Marmara Region	Istanbul	Atatürk Bridge
Turkey	Marmara Region	Istanbul	Bosphorus Bridge
Turkey	Marmara Region	Istanbul	Galata Bridge
Turkey	Marmara Region	Istanbul	Mağlova Kemer Aqueduct
Turkey	Marmara Region	Istanbul	Valens Aqueduct
Turkey	Marmara Region	Istanbul	Yavuz Sultan Selim Bridge
Turkey	Smyrna	Izmir	Caravan Bridge
Turkmenistan	Lebap Province	Atamyrat	Atamyrat-Kerkichi Bridges
Uganda	Buikwe District	Njeru	Source of the Nile Bridge
Ukraine	Crimea	Yalta	Ai-Petrie Bridge
Ukraine	Dnipropetrovsk	Dnipro	Kaidatsky Bridge
Ukraine	Khmelnytskyi	Kamianets-Podilskyi	Castle Bridge
Ukraine	Khmelnytskyi	Kamianets-Podilskyi	Strimka Lan Bridge
Ukraine	Kievshchyna	Kiev	Darnytskyi Bridge
Ukraine	Kievshchyna	Kiev	Metro Bridge
Ukraine	Kievshchyna	Kiev	Parkovy Bridge
United Arab Emirates		Abu Dhabi	Sheikh Zayed Bridge
United Arab Emirates		Dubai	Al Garhoud Bridge

Country	State / Province / Region Department / County	City, Town or Village	Bridge Name (* indicates bridge no longer exists)
United Arab Emirates		Dubai	Dubai Creek Bridges
Uruguay	Maldonado	La Barra	Lionel Viera Bridge
Uruguay	Rocha	José Ignacio	Laguna Garzón Bridge
USA	Alaska	Anchorage	Hurricane Gulch Railroad Bridge
USA	Alaska	Nenana	Mears Memorial Bridge
USA	Alaska	Skagway	Captain William Moore Bridge
USA	Arizona	Peach Springs	Grand Canyon Skywalk
USA	Arkansas	Cotter	White River Concrete Arch Bridge
USA	California	Bridgeport	Bridgeport Bridge
USA	California	Keddie	Wye Bridge
USA	California	Redding	Sacramento River Trail Footbridge
USA	California	Redding	Sundial Bridge
USA	California	Richmond	Richmond-San Rafael Bridge
USA	California	San Diego	Goat Canyon Trestle
USA	California	San Francisco	Alvord Lake Bridge
USA	California	San Francisco	Drum Bridge
USA	California	San Francisco	Golden Gate Bridge
USA	California	San Francisco	San Francisco-Oakland Bay Bridge
USA	Colorado	Canon City	Royal Gorge Bridge
USA	Colorado	Keenesburg	Mile into the Wild Walkway
USA	Connecticut	Greenwich	Merritt Parkway Bridges
USA	Delaware	Wilmington	Delaware Memorial Bridge
USA	Florida	Jacksonville	John T. Alsop Jr. Bridge
USA	Georgia	Savannah	South Viaduct
USA	Idaho	Riggins	Manning Crevice Bridge
USA	Illinois	Chicago	Chicago Bridges
USA	Illinois	Chicago	Moving Bridges
USA	Indiana	Metamora	Duck Creek Aqueduct
USA	Iowa	Boone	Boone Viaduct
USA	Kentucky	Covington	John A. Roebling Suspension Bridge
USA	Louisiana	New Orleans	Lake Pontchartrain Crossings
USA	Louisiana	Shreveport	Cross Bayou Bridge
USA	Maine	Harpswell	Bailey Island Bridge
USA	Maryland	Baltimore	Carrollton Viaduct
USA	Maryland	Baltimore	Francis Scott Key Bridge
USA	Maryland	Cabin John	Cabin John Aqueduct
USA	Maryland	Savage	Savage Mills Bridge
USA	Massachusetts	Bourne	Buzzards' Bay Bridge
USA	Massachusetts	Ipswich	Choate Bridge
USA	Massachusetts	Newburyport	Chain Bridge
USA	Michigan	Mackinaw City	Mackinac Straits Bridge
USA	Minnesota	Duluth	Duluth Lift Bridge

Country	State / Province / Region Department / County	City, Town or Village	Bridge Name (* indicates bridge no longer exists)
USA	Minnesota	Minneapolis	Irene Hixon Whitney Bridge
USA	Minnesota	Wabasha	Pontoon Railroad Bridge *
USA	Missouri	Parkville	Linn Branch Creek Bridge
USA	Missouri	St Louis	Eads Bridge
USA	Montana	Hanover	Spring Creek Trestle Bridge
USA	New Hampshire	Cornish	Cornish-Windsor Covered Bridge
USA	New Hampshire	Franklin	Sulphite Railroad Bridge
USA	New Jersey	Elizabeth	Arthur Kill Vertical Lift Bridge
USA	New York	Binghamton	South Washington Street Bridge
USA	New York	Blenheim	Blenheim Covered Bridge
USA	New York	New York	Brooklyn Bridge
USA	New York	New York	Hell Gate Bridge
USA	New York	New York	Manhattan Bridge
USA	New York	New York	Tappan Zee Bridge
USA	New York	New York	Verrazzano Narrows Bridge
USA	New York	Poughkeepsie	Poughkeepsie Railroad Bridge
USA	New York	Schenectady	Burr Covered Bridge *
USA	New York State	Niagara Falls	Whirlpool Rapids Bridges
USA	Ohio	Ashtabula	Ashtabula Bridge
USA	Ohio	Ashtabula	Smolen-Gulf Bridge
USA	Ohio	Cleveland	Detroit-Superior Bridge
USA	Ohio	Sciotoville	Sciotoville Bridge
USA	Oregon	Eugene	Cape Creek Bridge
USA	Oregon	Portland	Double Lift Bridge
USA	Oregon	Portland	St John's Bridge
USA	Oregon	Portland	Tilikum Crossing
USA	Oregon	Waldport	Alsea Bay Bridge
USA	Pennsylvania	Harrisburg	Rockville Bridge
USA	Pennsylvania	Lackawaxen	Delaware Aqueduct
USA	Pennsylvania	Lanesboro	Starrucca Viaduct
USA	Pennsylvania	Muncy	Reading-Halls Station Bridge
USA	Pennsylvania	Nicholson	Tunkhannock Viaduct
USA	Pennsylvania	Philadelphia	Colossus Bridge *
USA	Pennsylvania	Philadelphia	Frankford Avenue Bridge
USA	Pennsylvania	Pittsburgh	Smithfield Street Bridge
USA	Pennsylvania	Pittsburgh	Three Sisters Bridges
USA	Pennsylvania	Pittsburgh	West End Bridge
USA	Pennsylvania	Stalker	Kellams Bridge
USA	Pennsylvania	Tuckerton	Peacock's Lock Viaduct
USA	Pennsylvania	Westline	Kinzua Viaduct
USA	Virginia	Lynchburg	Fink Deck Truss Bridge
USA	Virginia	Yorktown	George P. Coleman Memorial Bridge
USA	Washington	Seattle	Lacey V. Murrow Memorial Bridge
USA	Washington	Sequim	Dungeness River Bridge
USA	Washington	Yakima	Rock Island Railroad Bridge

Country	State / Province / Region Department / County	City, Town or Village	Bridge Name (* indicates bridge no longer exists)
USA	Washington State	Seattle	Governor Albert D. Rosellini Bridge
USA	Washington State	Seattle	Tacoma Narrows Bridges
USA	West Virginia	Fayetteville	New River Gorge Bridge
Uzbekistan	Lebap	Dargan Ata	Pipeline Bridge
Uzbekistan	Samarquand	Samarkand	Bridge of Shadman Malik
Vanuata	Shefa Province	Port Vila	Bridges of Eden
Venezuela	Bolivar State	Ciudad Bolivar	Angostura (Narrows) Bridge
Venezuela	Curacao	Willemstad	Queen Emma Bridge
Venezuela	Curacao	Willemstad	Queen Juliana Bridge
Venezuela	Greater Caracas	Caracas	Caracas-La Guaira Highway Viaducts
Venezuela	Zulia	Maracaibo	General Rafael Urdaneta Bridge
Vietnam	Hai Chau	Da Nang	Dragon Bridge
Vietnam	Hai Chau	Da Nang	Golden Bridge
Vietnam	Quang Nam	Hoi An	Duv Xuven Bamboo Bridge
Vietnam	Quang Nam	Hoi An	Japanese Covered Bridge
Vietnam	Red River Delta	Hanoi	Huc Bridge
Vietnam	Red River Delta	Hanoi	Long Biên Bridge
Wales	Isle of Anglesey	Menai Bridge	Menai Straits Bridges
Yemen	Amran Governorate	Shaharah	Shaharah Bridge
Zambia	Lusaka Province	Kafue	Kafue Railway Bridge
Zambia	Southern Province	Kazungula	Kazungula Bridge
Zambia	Western Province	Sesheke	Katimo Mulilo Bridge
Zimbabwe	Manicaland	Chipinge	Birchenough Bridge
Zimbabwe	Matabeleland North	Victoria Falls	Victoria Falls Bridge
Zimbabwe	Matabeleland South	Beitbridge	Beitbridge Bridges

Analysis of Number of Entries By Country

Afghanistan	2	England (part of UK)	1
Albania	2	Equatorial Guinea	2
Algeria	3	Eritrea	1
Andorra	3	Estonia	2
Angola	2	Ethiopia	5
Antigua & Barbuda	1	Fiji	3
Argentina	5	Finland	3
Armenia	6	France	89
Australia	8	Gabon	1
Austria	8	Gambia	1
Azerbaijan	1	Georgia	3
Bahamas	1	Germany	52
Bahrain	1	Ghana	3
Bangladesh	5	Greece	5
Belarus	1	Guatemala	2
Belgium	7	Guinea	2
Belize	2	Guinea-Bissau	1
Bhutan	1	Guyana	3
Bolivia	3	Haiti	1
Bosnia-Herzegovina	3	Honduras	1
Botswana	2	Hungary	8
Brazil	7	Iceland	5
Brunei	3	India	9
Bulgaria	5	Indonesia	6
Burkino Faso	1	Iran	3
Cambodia	3	Iraq	3
Cameroon	3	Ireland	29
Canada	19	Israel	4
Chile	3	Italy	30
China	68	Ivory Coast	2
Colombia	3	Jamaica	2
Costa Rica	2	Japan	21
Croatia	7	Jordan	2
Cuba	3	Kazakhstan	3
Cyprus	4	Kenya	3
Czech Republic	12	Kuwait	1
Democratic Rep Congo	1	Laos	3
Denmark	5	Latvia	3
Dominican Republic	2	Lebanon	3
Ecuador	2	Lesotho	2
Egypt	4	Liberia	1
El Salvador	2	Libya	2

Lichtenstein	1	Singapore	3
Lithuania	3	Slovakia	3
Luxembourg	1	Slovenia	8
Madagascar	2	Solomon Islands	1
Malawi	2	Somalia	1
Malaysia	8	South Africa	8
Maldives	1	South Korea	5
Mali	3	South Sudan	1
Malta	2	Spain	44
Mauritius	2	Sri Lanka	2
Mexico	6	Sudan	2
Moldova	2	Suriname	2
Mongolia	1	Sweden	6
Montenegro	8	Switzerland	20
Morocco	5	Syria	3
Mozambique	3	Taiwan	3
Myanmar	3	Tajikistan	1
Namibia	1	Tanzania	3
Nepal	3	Thailand	3
Netherlands	13	Timor-Leste	1
New Caledonia	1	Togo	1
New Zealand	6	Tonga	1
Nicaragua	1	Trinidad & Tobago	1
Niger	1	Tunisia	3
Nigeria	2	Turkey	21
North Korea	3	Turkmenistan	1
North Macedonia	3	Uganda	1
Northern Ireland (part of UK)	7	Ukraine	7
Norway	8	United Arab Emirates	3
Oman	2	Uruguay	2
Pakistan	8	USA	87
Palau	1	Uzbekistan	2
Palestine	1	Vanuata	1
Panama	2	Venezuela	5
Papua New Guinea	2	Vietnam	6
Paraguay	1	Wales (part of UK)	1
Peru	3	Yemen	1
Philippines	3	Zambia	3
Poland	9	Zimbabwe	3
Portugal	12		
Qatar	1		
Republic of the Congo	1		
Romania	4		
Russia	10		
Rwanda	1		
Samoa	1		
San Marino	1		
Saudi Arabia	3		
Scotland (part of UK)	1		
Senegal	1		
Serbia	3		
Sierra Leone	1		

General Index

NOTE The heading below for **Individual Bridges** lists bridges that do not have separate main entries under their own names but are noted in descriptions under other bridge names (such as Ponte Milvio in the entry named Ancient Roman Bridges).

International Exhibitions and Olympic Games Sites

National Parks

Non-Railway Organisations

Streams, Lakes, Rivers and Seas

Acknowledgements

First, I must thank my family and friends who have let me use holiday photographs they have taken of bridges that have entries in this book. In addition, I am indebted to various enthusiasts with whom I have had contact over the years, who have also let me use their photos. Their names, together with those of the relevant bridge(s) for which they have provided images, are listed below. However, these photos covered only a very small proportion of the total number of bridge entries and it has been necessary to find illustrations from the internet. My sister-in-law Donna McFetrich has put in an enormous amount of time in searching for and finding many of these other photos and I am most grateful for all her help. The formal acknowledgements recording the names of each website and of the photographer are listed in the table below.

IMAGES FROM FAMILY AND FRIENDS

Freddie Corrin: Campo Volantin Footbridge, Bilbao, Spain
Lucy Corrin: Rio-Antirrio Bridge, Patras, Greece
Neil McFetrich: Segovia Aqueduct, Segovia, Spain
Ray Paulson: Dom Luis I Bridge, Lisbon; Santa Justa Lift, Lisbon

IMAGES COURTESY OF BRIDGE ENTHUSIASTS

Emily Beck: Central of Georgia South Viaduct, Savannah, Georgia, USA
Engineer Girmu Worku: Blue Nile Bridges and Bombas Bridge, Ethiopia
Stephen Jones: Chain Bridge, Newburyport, Massachusetts, USA
The Happy Pontist: Bercy Bridge, Paris, Ile-de-France, France; Hvita Bridge, Hvanneyn, Western Regions, Iceland; Jokulsa a Dal Canyon Bridge, Egilsstaoir, Eastern Region, Iceland; Skjalfandafljot Bridge, Fossholl, North Iceland, Iceland

IMAGES COURTESY OF MABEY BRIDGE

Loto Samasoni Bridge, Samoa
Nakabuta Bridge, Fiji

ACKNOWLEDGEMENTS FOR PICTURES COPIED FROM WIKIMEDIA COMMONS

25th April Bridge, Lisbon, Estremadura, Portugal (ceiling)

Ada Bridge, Belgrade, Serbia (Михајло Анђелковић)
Adana Roman Bridge, Adana, Adana Province, (Mustafa Tor)
Afghan-Uzbek Friendship Bridge, Hairatan, Balkh, Afghanistan (Bradley Lail)
Águas Livres Aqueduct, Lisbon, Estramadura, Portugal (Juntas)
Águila Aqueduct, Nerja, Andalusia, Spain (Adan Fernandez Berrocal)

ACKNOWLEDGEMENTS

Ain Diwar Bridge, Ain Diwar, Al Hasakah Governortae, Syria (Zoeperkoe)

Akashi Kaikyo Bridge, Kobe, Hyogo Prefecture, Japan (Tysto)

Alamillo Bridge, Seville, Spain (Pablo Pérez)

Alcántara Bridge, Alcántara, Cáceres, Spain (Alonso de Mendoza)

Alexandre III Bridge, Paris, Ile-de-France, France (Dimitri Destugues)

Alf-Bullay Bridge, Bullay, Rhineland-Palatinate, Germany (Holger Weinandt)

Amoreira Aqueduct, Elvas, Alentejo, Portugal (Concierge.2C)

Ampera Bridge, Palembang, South Sumatra, Indonesia (Achmad Rabin Taim)

Aného Bridge, Aného, Lacs, Togo (Jeff Attaway)

Anghel Saligny Bridge, Cernavoda, Constanta, Romania (vizualizare.stradala)

Anping (Peace) Bridge, Fujian, Fujian, China (Vmenkov)

Anshun (Peaceful and Favourable) Bridge, Chengdu, Sichuan, China (Charlie Fong)

Arenal Hanging Bridges, La Fortuna, Alajuela, Costa (Peter Sheik)

Arganzuela Footbridge, Madrid, Spain (Tamorlan)

Arkadia Park Aqueduct, Lowiez, Poland (Tomasz Kuran aka Meteor2017)

Ashtarak Bridge, Ashtarak, Aragatsotn, Armenia (Travis K. Witt)

Atamyrat-Kerkichi Bridge, Atamyrat, Lebap Province, Turkmenistan (Elena8109)

Attock Railway Bridge, Attock Khurd, Punjab Province, Pakistan (Annemarie Schwarzenbach)

Auckland Harbour Bridge, Auckland, Auckland, New Zealand (GPS 56)

Aveiro Circular Footbridge, Aveiro, Centro, Portugal (Domoreira~commonswiki)

Aveyron Bridge, Belcastel, Aveyron, France (Kajimoto)

Avignon Bridge, Avignon, Vaucluse, France (Henk Monster)

Ayub Bridge, Sukkur, Sindh Province, Pakistan (Ibaadnaqvi)

Bad Kreutznach Bridge, Mainz, Rhineland-Palatinate, Germany (Rainer Lippert)

Bad Säckingen Covered Bridge, Bad Säckingen, Baden-Württemberg, Germany (BlueBreezeWiki)

Baling Bridge, Weiyun, Gansu, China (Unknown Photographer)

Bank Footbridge, St Petersburg, Leningrad Oblast, Russia (Vlad&Mirom)

Banpo-Jaamsu Bridge, Seoul, Gyeonggi Province, South Korea (이명석)

Baodai (Precious Belt) Bridge, Suzhou, Jiangsu, China (Rolf Müller)

Barqueta Bridge, Seville, Andalusia, Spain (Auricaria)

Batman Bridge, Sidmouth, Tasmania (Rowan.M.McDonald)

Beijian Bridge, Taishun, Zhejiang, China (Unknown Photographer)

Beipan River Shuibai Bridge, Liupanshui, Guizhou Province, China (Glabb)

Belize City Swing Bridge, Belize City, Belize (Haakon S. Krohn)

Besalu Bridge, Besalu, Girona, Spain (Rafael Lluis Comamala)

Bévera Bridge, Sospel, Alpes-Maritimes, France (Markus Schweiss)

Bhumibol Bridge, Bangkok, Thailand (Nik Cyclist)

Birchenough Bridge, Chipinge, Manicaland, Zimbabwe (Kjdhambuza)

Bizerte Bridge, Tunis, Tunisia (khaled abdelmoumen)

Black Dragon Pool Bridge, Lijiang, Yunnan, China (Chris from Canada)

Blenheim Covered Bridge, Blenheim, New York, USA (Cg-realms)

Bloukrans Bridge, Knynsa, South Africa (NJR ZA)

Bogibeel Bridge, Dibrugarh, Assam, India (Vikramjit Kakati)

Bogoda Wooden Bridge, Badulla, Uva, Sri Lanka (Ebaran) Bosphorus Bridge, Istanbul, Marmara Region, Turkey (Jorge1767)

Boyne Viaduct, Drogheda, County Louth, Ireland (Trounce)

Briare Aqueduct, Briare, Loiret, France (SovalValtos)

Bridge of Boyacà, Tunja, Boyacà, Colombia (Néstor Daza)

Bridge of Immortals, Tangkouzhen, Anhui, China (Stephane.janel)

Bridge of Peace, Tbilisi, Georgia (Dudva)

Bridge of Shadman Malik, Samarkand, Samaquand, Uzbekistan (Bogaevskii, Nikolai V)

Bridge of Sighs, Venice, Veneto, Italy (Didier Descouens)
Bridge of the Americas, Balboa, Panama (Gualberto107)
Bridge Pavilion, Zaragoza, Aragon, Spain (Grez)
Brooklyn Bridge, New York, New York, USA (Suiseiseki)
Bunlahinch Clapper Footbridge, Louisburgh, County Mayo, Ireland (Egbert Polski)
Butterfly Bridge, Copenhagen, Capital Region, Denmark (Dietmar Rabich)

Cadore Viaduct, Caralte, Belluno, Italy (Antonio De Lorenzo)
Cape Creek Bridge, Eugene, Oregon, USA (Bonnie Moreland)
Capilano Suspension Bridges, Vancouver, British Columbia, Canada (Loadmaster (David R. Tribble))
Caracas-La Guaira Highway Viaducts, Caracas, Greater Caracas, Venezuela (Unknown Photographer)
Caramel Viaduct, Castillon, Alpes Maritimes, France (Jean Gilletta)
Carioca Aqueduct, Rio de Janeiro, Rio de Janeiro, Brazil (Wolfhardt)
Carpentras Aqueduct, Vaucluse, Vaucluse, France (Véronique Pagnier)
César Gaviria Trujillo Viaduct, Pereira, Risaralda Department, Colombia (ledpup)
Chain Bridge, Nuremberg, Bavaria, Germany (Dr Bernd Gross)
Chamberlain Bridge, Bridgetown, Antigua and Barbuda (regani)
Chaotianmen Bridge, Chongqing, Sichuan, China (Graeme Bray)
Chapel Bridge, Lucerne, Switzerland (Simon Koopmann)
Charles Bridge, Prague, Bohemia, Czech Republic (Sergey Ashmarin)
Château de Chenonceau, Chenonceaux, Indre-et-Loire, France (Taxiarchos228)
Chateau Nové Mesto and Metuji, Krcin, Eastern Bohemia, Czech Republic (Unknown Photographer)
Chateau Veltrusy's Laudon Pavilion, Veltrusy, Bohemia, Czech Republic (Picasa)
Chavanon Viaduct, Merlines, Corrèze, France (Fabien 1309)
Chengyang Yongji (Wind and Rain) Bridge, Ma'an, Sanjian, Guangxi, China (Zhangzhugang)
Chillon Viaduct, Veytaux, Vaud, Switzerland (HGolaszewska)
Choluteca Bridges, Choluteca, Choluteca Province, Honduras (Intimaralem85)
Chords Bridge, Jerusalem, Israel (Petdad)
Cidi M'Cid Bridges, Constantine, Constantine Province, Algeria (Yelles)
Citadel Bridge, Aleppo, Syria (Bernard Gagnon)
Cize-Bolozon Viaduct, Daranche, Tarn, France (Kabelleger / David Gubler)
Confederation Bridge, Borden-Carleton, Prince Edward Island, Canada (Chensiyuan)
Cornish-Windsor Covered Bridge, Cornish, New Hampshire, USA (Eurodog)
Covered Bridge, Lovech, Lovech Province, Bulgaria (Klearchos Kapoutsis)
Craigavon Bridge, Londonderry, County Londonderry, Northern Ireland (Jove)
Cross Bayou Bridge, Shreveport, Louisiana, USA (Archangle0)

Da Vinci Bridge, As, Eastern Norway, Norway (Åsmund Ødegård)
Danilo's Bridge, Rijeka Crnojevica, Montenegro (Darij & Ana)
Deh Cho Bridge, Fort Providence, Northwest Territories, Canada (CambridgeBayWeather)
Delal Bridge, Zakho, Kurdistan, Iraq (Moplayer)
Delaware Aqueduct, Lackawaxen, Pennsylvania, USA (National Park Service)
Detroit-Superior Bridge, Cleveland, Ohio, USA (Michael Barera)
Devil's Bridge, Ardino, Smolyan, Bulgaria (Vassia Atanassova - Spiritia)
Devil's Bridge, Gablenz, Saxony, Germany (A.Landgraf)
Devil's Bridge, Martorell, Barcelona, Spain (Jaume Meneses)
Dodhara Chandani Bridge, Bhimdatta, Kanchanpur, Nepal (Ravijung)
Dongshuimen (Yangtze River) Bridge, Chongqing, Sichuan, China (Nyx Ning)
Dragon Bridge, Da Nang, Hai Chau, Vietnam (Bùi Thụy Đào Nguyên)
Dragon Bridge, Ljubljana, Slovenia (Thomas Ledl)
Drava Footbridge, Osijek, Osijek-Barania, Croatia (Zdenko Brkanić)

ACKNOWLEDGEMENTS

Dromana Bridge, Villierstown, Waterford, Ireland (Liam Hughes)
Dubai Floating Bridges Dubai, United Arab Emirates (Imre Salt – Dubai Construction Update Part 11)
Duck Creek Aqueduct, Metamora, Indiana, USA (Rosenthal, James W.)
Duluth Lift Bridge, Duluth, Minnesota, USA (Alfred Essa from Woodbury)

Eads Bridge, St Louis, Missouri, USA (Mitchell Schultheis)
East Taiheng Glasswalk, Zhangjiajie, Hebei Province, China (HighestBridges)
Ebro Bridge, Osera, Zaragoza, Spain (Unknown Photographer)
El Ferdan Railway Bridge, Ismailia, El Qantara, Egypt (H Nawara)
Elgin Bridge, Singapore (Craig Stanfill)
Elisabeth Bridge, Budapest, Central Hungary, Hungary (Peter Siroki)
Emscher Landscape Park Bridges, Duisberg, North Rhine-Westphalia, Germany (Arnoldius)
Enneüs Heerma Bridge, Amsterdam, North Holland, Netherlands (S Sepp)
Enz Viaduct, Bietigheimer, Baden-Württemberg, Germany (Muskiprozz)
Epirus Bridges, Epirus, Greece (Harrygouvas)
Erasmus Bridge, Rotterdam, South Holland, Netherlands (Martin Falbisoner)
Essinger Footbridge, Essing, Bavaria, Germany (Brego)
Eurymedon Bridge, Aspendos, Antalya Province, Turkey (Adam Franco)

Faidherbe Bridge, Saint-Louis, Senegal (Manu25)
Figurny Bridge, Tsaritsyno Park, Moscow, Russia (A.Savin)
Five Circles Footbridge, Copenhagen, Capital Region, Denmark (Colin)
Flisa Bridge, Flisa, Innlandet County, Norway (Jan-Tore Egge)
Foreshore Freeway Bridge, Capetown, Western Cape, South Africa (Paul Mannix)
Franjo Tudjman Bridge, Kantafig, Dubrovnik-Neretva, Croatia (Wolfgang Pehlemann)
Friendship Bridge, Ciudad del Este, Alto Paraná Department, Paraguay (Leonard.inc)

Galata Bridge, Istanbul, Marmara Region, Turkey (Erol Gülsen)
Ganter Bridge, Brig, Valais, Switzerland (rosmary)
Gazela Bridge, Belgrade, Serbia (Miomir Magdevski)
General Rafael Urdaneta Bridge, Maracaibo, Zulia, Venezuela (Wilfredor)
Gignac Bridge, Gignac, Hérault, France (Emeraude)
Gladesville Bridge, Sydney, New South Wales, Australia (Nick-D)
Glienicke Bridge, Potsdam, Brandenberg, Germany (Loewenflausch)
Golden Bridge, Da Nang, Hai Chau, Vietnam (xiquinhosilva from Cacau)
Golden Gate Bridge, San Francisco, California, USA (Rich Niewiroski Jr)
Goodwill Bridge, Brisbane, Queensland, Australia (Margaret R Donald, Sydney)
Gothic Bridge, Vilomara, Barcelona, Spain (David Gaya)
Gouritz River Bridge, Albertinia, Western Cape, South Africa (Damien du Toit)
Grand Canyon Skywalk, Peach Springs, Arizona, USA (Complexsimplellc)
Grandfey Viaduct, Fribourg, Fribourg Canton, Switzerland (ClearFrost)
Great Belt East Bridge, Korsør, Zealand Region, Denmark (Martin Falbisoner)
Guadiana International Bridges, Ayamonte, Huelva, Spain (Roger W. Haworth)
Guangji / Xiangzi (Great Charity) Bridge, Chaozhou, Guangdong, China (Agatha Qiu)
Guyue Bridge, Yiwu, Zhejiang Province, China (Alex Needham)

Ha'penny Bridge, Dublin, County Dublin, Ireland (Bob Collowan)
Haoshang Bridge, Leshan, Sichuan Province, China (Ure Aranas)
Hartland Covered Bridge, Saint-John, New Brunswick, Canada (Dennis Jarvis)
Håverud Aqueduct, Håverud, Vastra Götaland Country, Sweden (Sebbe)
Hawkesworth Suspension Bridge, San Ignacio, Western Belize, Belize (Kaldari)

Helix Bridge, Singapore (Zairon)
Hell Gate Bridge, New York, New York, USA (Dave Frieder)
Henderson Waves Bridge, Singapore (Matthew Hine)
Hercilio Luz Bridge, Florianopolis, Santa Catarina, Brazil (Rodrigo Soldon)
High Bridge, Amsterdam, North Holland, Netherlands (Alain Rouiller)
High Level Bridge, Edmonton, Alberta, Canada (WinderforceMedia)
Hiyoshi Dam Footbridge, Nantan, Kyoto, Japan (Unknown Photographer)
Hohenzollern Bridge, Cologne, North Rhine-Westphalia, Germany (Raimond Spekking
Hongcun Moon Bridge, Huangshan City, Anhui Province, China (Zhangzhugang)
Howrah Bridge, Kolkota (Calcutta), West Bengal, India (Manuel Menal)
Hradecky Bridge, Ljubljana, Slovenia (ModriDirkac)
Hureai Bridge, Nantan, Kyoto, Japan (エコ殿様)

Inca Bridge, Machu Pichu, Cuzco, Peru (Martin St-Amant)
International Railroad Bridge, Sault Ste Marie, Ontario, Canada (Tony Webster)
Irene Hixon Whitney Bridge, Minneapolis, Minnesota, USA (AlexiusHoratius)

Jaargalant Wooden Bridge, Jaargalant, Khövsgöl Province, Mongolia (Torbenbrinker)
Jacques Chaban-Delmas Lifting Bridge, Bordeaux, Nouvelle-Aquitaine, France (A. Delesse (Prométhée))
Japanese Covered Bridge, Hoi An, Quang Nam, Vietnam (Benjamin Vander Steen)
John A. Roebling Suspension Bridge, Covington, Kentucky, USA (EEJCC)
John T. Alsop Jr Bridge, Jacksonville, Florida, USA (Michael Curi)
Juba Nile (Freedom) Bridge, Juba, Central Equatoria, South Sudan (Demosh)
Jules Wijdenbosch Bridge, Paramiribo, Suriname (Santiago Fernandez)
Jungfern Footbridge, Berlin, Brandenburg, Germany (I, DorisAntony)
Juscelino Kubitschek Bridge, Brasilia, Central-West Region, Brazil (Mario Roberto Duran Ortiz)

Kalte Rinne Viaduct, Breitenstein, Lower Austria, Austria (Hiroki Ogawa)
Kassuende Bridge, Tete City, Tete Province, Mozambique (J. Ventura)
Kattwyk Bridges, Hamburg, Lower Saxony, Germany (Gunnar Ruis)
Kawarau Bridge, Queenstown, Otago, New Zealand (Steve & Jem Copley)
Kawazu-Nanadaru Loop Bridge, Shimoda, Shizuoka, Japan (Captain 76)
Kazarma Bridge, Arkadiko, Peloponnese, Greece (Jean Housen)
Kazurabashi Vine Bridge, Shikoku, Ehime Prefecture, Japan (Kimon Berlin)
Keshwa Chaca Bridge, Huinchiri, Cuzco, Peru (Rutahsa Adventures)
Khudafarin Bridges, Jabrayil, Hadrut, Azerbaijan (Abdossamad Talebpour)
Kiel-Hörn Folding Bridge, Kiel, Schleswig-Holstein, Germany (Karlehorn/CC)
King Fahd Causeway Bridge, Al Khobar, Eastern Province, Saudi Arabia (Mohamed Ghuloom)
Kingdom Centre Skywalk, Riyadh, Riyadh Region, Saudi Arabia (BroadArrow)
Kintai Bridge, Iwakuno, Yamaguchi Prefecture, Japan (Sailko)
Kinzua Viaduct, Westline, Pennsylvania, USA (Niagora)
Koh Pen Bamboo Bridge, Kampong Cham Province, Cambodia (James Antribus)
Kohukohu Bridge, Panguru, Northland, New Zealand (Moriori)
Koornbrug, Leiden, South Holland, Netherlands (Michielverbeek)
Krasnoluzhsky Railway Bridge, Moscow, Russia (Denghu)
Kuldiga Bridge, Kurzeme, Courland, Latvia (Staffan Vilcans)
Kurilpa Bridge, Brisbane, Queensland, Australia (Paulguard)

La Cassagne Bridge, Planès, Pyrenees-Orientales, France (Koh Pen Bamboo)
Lacey V. Murrow Memorial Bridge, Seattle, Washington, USA (Tradnor)
Laguna Garzón Bridge, José Ignacio, Rocha, Uruguay (Jimmy Baikovicius)

Lake Shore Drive Bridge, Chicago, Illinois, USA (Marcin Wichary)
Lake Vranov Footbridge, Znojmo, South Moravia, Slovakia (Harold)
Lambézellec Viaduct, Brest, Finistère, France (Henri Moreau)
Land Bridge, Banff National Park, Alberta, Canada (Coolcaesar)
Langeais Bridge, Langeais, Indre-et-Loire, France (ThomasPusch)
Langebrug, Haarlem, North Holland, Netherlands (Bogdan Migulski)
Langkawi Sky Bridge, Langkawi Island, Kedar, Malaysia ("The Dilly Lama")
Langlois Bridge, Arles, Bouches-du-Rhône, France (Guido)
Lego Bridge, Wuppertal, North Rhine-Westphalia, Germany (Morty)
Lejonströmsbron Bridge, Skellefteå, Västerbotten, Sweden (Mattias Hedström)
Lekki-Ikoyi Bridge, Lagos, South West, Nigeria (Chippla)
Leonardo Da Vinci Bridge, Ås, Viken County, Norway (Åsmund Ødegård)
Lerez River Bridge, Pontevedra, Galicia, Spain (innoxiuss)
Lethbridge Viaduct, Lethbridge, Alberta, Canada (Richard Peat)
Lézardrieux Bridge, Lézardrieux, Côtes-d'Armor, France (MiklGds)
Lima Bridge, Ponte de Lima, Norte, Portugal (bob)
Lionel Viera Bridge, La Barra, Maldonado, Uruguay (Axe)
Llinars Bridges, Sant Celoni, Catalonia, Spain (Usuaris)
Lokke Bridge, Sandvika, Greater Oslo, Norway (Patrick Poculan)
Long Biên Bridge, Hanoi, Red River Delta, Vietnam (Pierre De Hanscutter)
Longeray Viaduct, Leaz, Ain, France (Christopher Delaere)
Longten Bridge, Sanyi Township, Miaoli County, Taiwan (王崎)
Lotus Lake Footbridges, Kaohsiung Taiwan (AngMoKio)
Loyang (Ten Thousand Peace) Bridge, Chuanchow, Fujian, China (Vmenkov)
Lucan Bridge, Dublin, County Dublin, Ireland (JP)
Lucano Bridge, Villanova, Padua, Italy (antique print)
Lucky Knot Bridge, Changsa, Hunan Province, China (nextarchitects)
Lumberjack's Candle Bridge, Rovaniemi, Lapland, Finland (themadpenguin)
Lupu Bridge, Shanghai, China (Jurgenlison)
Lusitania Bridge, Mérida, Badajoz, Spain (Luis Javier Gala Orgaz)

Ma Kham Bridge, Kanchanaburi, Thailand (cesar.ruiz)
Mackinac Straits Bridge, Mackinaw City, Michigan, USA (Jeffness)
Magdalena Footbridge, Pamplona, Navarre, Spain (de:Benutzer:Bautsch)
Magdeburg Water Bridge, Magdeburg, Saxony-Anhalt, Germany (Olivier Cleynen)
Main Viaduct, Nantenbach, Bavaria, Germany (Heidas)
Mala Rijeka Viaduct, Podgorica, Central Region, Montenegro (Marcin Konsek)
Malabadi Bridge, Silvan, Diyarbakir, Turkey (Dyrt)
Malan Bridge, Herat, Herat Province, Afghanistan (Artacoana)
Malleco Railway Viaduct, Collipulli, Araucania, Chile (Marcelo Reston)
Marble Bridge, Tsarkoye Selo Park, Saint Petersburg, Northwestern, Russia (Sailko)
Marcelo Fernan Bridge, Cebu City, Central Visayas, Philippines
Marco Polo (Lugou) Bridge, Beijing, China (Fanghong)
Margaret Bridge, Budapest, Central Hungary, Hungary (VinceB)
Marina Cristina Bridge, San Sebastian, Gipuzkoa, Spain (M @ X)
Mary McAleese Boyne Valley Bridge, Drogheda, County Louth, Ireland (Peter Gerken)
Maslenica Bridges, Zadar, Zadar County, Croatia (Ex13)
Matadi Bridge, Matadi, Central Province, Democratic Republic of Congo (3nigma)
Mayfly Footbridge, Szolnok, Jasz-Nagykun-Szolnok, Hungary (Derzsi Elekes Andor)
Medjez-el-Bab Bridge, Béja, Tunisia (Noomen9)
Megyeri Bridge, Budapest, Central Hungary, Hungary (Civertan Grafikai Stúdió)

Mehmed-Pasha Sokolovic Bridge, Visegrad, Republika Srnka, Bosnia-Herzegovina (Branevgd)
Menai Straits Bridges, Menai Bridge, Isle of Anglesey, Wales (Mick Knapton)
Merchants Bridge, Erfurt, Thuringia, Germany (Dr Bernd Gross)
Mes Bridge, Shkodër, Shkodër County, Albania (Diego Delso)
Meuse Footbridge, Chooz, Ardennes, France (MOSSOT)
Miho Museum Bridge, Shigoraki, Shiga, Japan (663highland)
Millau Viaduct, Millau, Aveyron, France (Foto Wolfgang Pehlemann)
Millennium Bridge, Ourense, Spain (Rubén Iglesias)
Millennium Bridge, Podgorica, Central Region, Montenegro (Bjørn Christian Tørrissen)
Mizen Head Bridge, Crookhaven, Munster, Ireland (FUBSAN)
Mohamed VI Bridge, Rabat, Rabat-Salé-Kénitra, Morocco (عدنان حليم)
Monorail Suspension Bridge, Putrajaya, Malaysia (GuillaumeG)
Montaubon Bridge, Montauban, Tarn-et-Garonne, France (Didier Descouens)
Montjean-sur-Loire Bridge, Montjean-sur-Loire, Maine-et-Loire, France (Touriste – Karya sendiri)
Moresnet Viaduct, Plombières, Liège, Belgium (Alupus)
Movable Bridges, Buenos Aires, Buenos Aires Province, Argentina (Roberto Fiadone)
Moving Bridges, Chicago, Illinois, USA (JeremyA)
Mudeirej Bridge, Sawfar, Mount Lebanon, Lebanon (worldmapz)
Murinsel Bridge, Graz, Bundesland, Austria (Walter W.)
Muskauer Park Bridges, Bad Muskau, Saxony, Germany (SchiDD)
Musmeci Bridge, Bad Muskau, Saxony, Germany (Unknown Photographer)

Nanjing Yangtze River Bridge, Nanjing, Jiangsu, China (Juan Gutierrez Andres)
National Post Office Bridge, Guatemala City, Guatemala Department, Guatemala (Infrogmation)
National Unity Bridge, San Pedro Garza Garcia, Nuevo León, Mexico (Supaman89)
Nesenbachtal Bridge, Stuttgart, Baden-Württemberg, Germany (pjt56)
New Bridge, Bratislava, Slovakia (Pudelek (Marcin Szala))
New River Gorge Bridge, Fayetteville, West Virginia, USA (Jaga)
Niagara Railway Arch Bridge, Niagara Falls, New York State, USA (Balcer)
Nine Arch Bridge, Demodara, Uva, Sri Lanka (AntanO)
Nine-Hole Bridge, Hortobágy, Hajdú-Bihar, Hungary (Lily15)
Noorabad Bridge, Marena, Gwalior, Pakistan (Anindo Dey)
Normandie Bridge, Le Havre, Seine-Maritime, France (François Roche)

Obelisk Bridge, Drogheda, North Leinster, Ireland (Jonathan Billinger)
Oberbaum Bridge, Berlin, Brandenburg, Germany (Sarah Jane)
Octavio Frias de Oliveira Bridge, Sao Paulo, Southeast Region, Brazil (Marcosleal)
Ojuela Bridge, Mapimi, Durango, Mexico (Fenerty)
Öland Bridge, Kalmar, Småland, Sweden (Proton)
Old Bridge (Karl Theodor Bridge), Heidelberg, Baden-Württemberg, Germany (BishkekRocks)
Old Bridge, Albi, Tarn, France (Marion Schneider & Christoph Aistleitner)
Old Bridge, Sospel, Alpes-Maritimes, France (Rundvald)
Old Main Bridge, Würzburg, Bavaria, Germany (Christian Horvat)
Old Rhine Bridge, Vaduz, Oberland, Liechtenstein (Rainer Ebert)
Old Suspension Bridge, Mallemort, Bouches-du-Rhône, France (Boerkevitz)
Olstgracht Bridge, Almere, Flevoland, Netherlands (archiexpo)
Orb Aqueduct, Béziers, Hérault, France (Christian Ferrer)
Øresund Bridge, Malmö, Øresund Region, Denmark (Amjad Sheikh)
Osten Transporter Bridge, Osten, Lower Saxony, Germany (Benutzer:Carroy)

Padre Templeque Aqueduct, Hidalgo, Eastern, Mexico (Carmelita Thierry)
Pamban Bridge, Mandapam, Tamil Nadu State, India (IM3847)

Panam Bridge, Sonargaon, Dhaka, Bangladesh (Nasir Khan Saikat)
Paradise Island Hotel Bridge, Nassau, Grand Bahamas, Bahamas (Drumguy8800)
Paris Bridge, Andorra la Vella, Andorra, Andorra (Jordiferrer)
Passerelle Marguerite, La Foa, South Province, New Caledonia (Thomas ballandras)
Peace Bridge, Calgary, Alberta, Canada (Ryan Quan)
Peace Bridge, Plauen, Saxony, Germany (Maler Plauen)
Peacock Island Palace Bridge, Potsdam, Brandenburg, Germany (RThiele)
Pedro e Inês Footbridge, Coimbra, Coimbra, Portugal (CorreiaPM)
Pentele Bridge, Dunaijvaros, Fejer, Hungary (Drobotka)
Petronas Twin Towers Sky Bridge, Kuala Lumpur, Selangor, Malaysia (Wolfgang Holzem)
Pira Delal, Zakho, Kurdistan, Iraq (Moplayer)
Ploče Gate Footbridges, Dubrovnik, Dubrovnik-Neretva, Croatia (Diego Delso)
Plougastel Bridge, Brest, Finistère, France (Toberne)
Pompey's Bridge, Mtskheta, Mtskheta-Mtianeti, Georgia (Igar08)
Pont de Cassagne, Planès, Eastern Pyrenees, France (Ceveno12)
Pont des Amidonniers, Toulouse, Haute-Garonne, France (Mossot)
Pont des Arts, Paris, Ile-de-France, France (TCY)
Pont des Trous, Tournai, Wallonia, Belgium (Jean-Pol Grandmont)
Pont du Gard, Nîmes, Gard, France (Benh Lieu Song)
Pont Flavien, Saint-Chamas, Bouches-du-Rhône, France (maarjaara)
Pont Julien, Bonnieux, Vaucluse, France (Hawobo)
Pont Levant Notre Dame, Tiurnai, Hainaut, Belgium (Karel Roose)
Pont Saint-Jacques, Parthenay, Deux-Sèvres, France (Chris J Wood)
Pont Vieux, Orthez, yrénées-Atlantiques, France (MOSSOT)
Ponte degli Alpini, Bassano del Grappa, Vicenza, Italy (Patrick Denker)
Pont dei Salti, Lavertezzo, Ticino, Switzerland (Peter Sieling)
Ponte della Maddalena, Borgo a Mozzano, Lucca, Italy (H005)
Ponte Scaligero, Verona, Italy (Manfred Heyde)
Port Mann Bridge, Vancouver, British Columbia, Canada (Dgarte)
Poubara Liana Bridge, Franceville, Haut-Ogooué, Gabon (Vincent.vaquin)
Poughkeepsie Railroad Bridge, Poughkeepsie, New York, USA (Mfwills)
Powerscourt Covered Bridge, Hinchinbrooke, Quebec, Canada (Aqk)
Protville Bridge, Kantarat Binzart, Bizerte, Tunisia (El Golli Mohamed)
Puente Nuevo, Ronda, Malaga, Spain (Judas6000)
Putra Bridge, Putrajaya, Malaysia (Ishan)

Quebec Bridge, Quebec, Quebec, Canada (Martin St-Amant (S23678))
Queen Alexandrine Bridge, Stege, Zealand, Denmark (Thue C. Leibrandt)
Queen Juliana Bridge, Willemstad, Curacao, Venezuela (Nelo Hotsuma from Rockwall)
Querétaro Aqueduct, Santiago de Querétaro, Querétaro, Mexico (Asomarte)

Rainbow Bridge, Niagara Falls, Ontario, Canada (Ad Meskens)
Rama VIII Bridge, Bangkok, Thailand (Stygiangloom)
Ramstor Bridge Nur Sultan, Central Kazakhstan, Kazakhstan (Peretz Partensky)
Red Bridge, Meore Kesalo, Kvemo Kartli, Georgia (Tōnis Valing)
Red Gate Aqueduct, Augsburg, Bavaria, Germany (NEITRAM)
Redzinski Bridge, Wroclaw, Lower Silesia, Poland (Mati.laspalmas)
Regensburg Bridge, Regensburg, Bavaria, Germany (Hytrion)
Rendsburg High Bridge, Rendsburg, Schleswig-Holstein, Germany (Malte Hübner)
Rhaetian Railway Brusio Spiral Viaduct, Filisur, Graubünden, Switzerland (Kabelleger)
Rialto Bridge, Venice, Veneto, Italy (Livioandronico2013)
Richmond Bridge, Richmond, Tasmania, Australia (Noodle snacks)

Rio Cruces Bridge, Torobayo, Valdivia, Chile (Dentren)
Rio Negro Bridge, Manaus, Amazonas, Brazil (Ana Claudia Jatahy)
Ritsurin Garden Nanko Bridge, Takamatsu, Kagawa, Japan (663highland)
River Neva Trinity Bridge, St Petersburg, Leningrad Oblast, Russia (Alex 'Florstein' Fedorov)
Rochefort-Martrou Transporter Bridge, Rochefort-sur-Mer, Charente-Maritime, France (Myrabella)
Rock Island Railroad Bridge, Yakima,Washington, USA (Jon Roanhaus)
Roman Aqueduct, Aspendos, Antalya Province, Turkey (Bernard Gagnon)
Roman Bridge, Mérida, Badajoz, Spain (A stray sheep)
Roman Bridge, Trier, Rhineland Palatinate, Germany (Stefan Kühn)
Roquefavour Aqueduct, Ventabren, Bouches-du-Rhône, France (Florent Ruyssen)
Ross Bridge, Ross, Tasmania, Australia (JJ Harrison)
Russky Bridge, Vladivostok, Primorsky Krai, Russia (Баяков Алексей Александрович)

Sai Van Bridge, Macau, Guangdong, China (Netsonfong)
St John's Bridge, Portland, Oregon, USA (Cacophony)
St Patrick's Bridge, Cork, County Cork, Ireland (William Murphy)
Saint-Clément Aqueduct, Montpellier, Hérault, France (martin_vmorris)
Samuel Beckett Bridge, Dublin, County Dublin, Ireland (William Murphy)
San Juanico Bridge, Tacloban, Eastern Visayas, Philippines (Rabosajr)
Sandö Bridge, Kramfors, Vastermorrland County, Sweden (Keibr)
Sanhao Bridge, Shenyang, Liaoning, China (Lau Phang)
Saracens Bridge, Adrano, Sicily, Italy (archenzo)
Savage Mills Bridge, Savage, Maryland, USA (Kjssws)
Saxon Switzerland National Park Bastei Bridge, Kurort Rathen, Saxony, Germany (Thomas Wolf)
Sciotoville Bridge, Sciotoville, Ohio, USA (Vbofficial)
Sebara Dildiy (Broken Bridge), Motta, Amhara, Ethiopia (Krfrantz)
Séjourné Viaduct, Fontpédrouse, Pyrénées-Orientales, France (Thierry Llansades)
Seonimgyo Bridge, Seogwipo, Jeju Volcanic Islands, South Korea (Kevin Miller OR WSTAY)
Seonyu Footbridge, Seoul, Gyeonggi Province, South Korea (이윤범)
Seri Wawasan Bridge, Putrajaya, Malaysia (notsogoodphotography)
Severan Bridge, Arsameia, Adiyaman Province, Turkey (Damien Halleux Radermecker)
Shadorvan Bridge, Shushtar, Khuzestan Province, Iran (Ali Afgah)
Shaharah Bridge, Shaharah, Yemen (Bernard Gagnon)
Shahi Mughal Bridge, Jaunpur, Utter Pradesh, India (Sayed Mohammed Faiz Haider)
Sheikh Zayed Bridge, Abu Dhabi, United Arab Emirates (Still ePsiLoN)
Shinkyo (Sacred) Bridge, Nikko, Tochigi (Miquel Frontera Lladó)
Shiosai Bridge, Shizuoka, Honshu, Japan (Tawashi2006)
Si-o-se Pol (Allah-Verdi Khan Bridge), Isfahan, Isfahan, Iran (Petr Adam Dohnálek)
Sixth Street Bridge, Pittsburgh, Pennsylvania, USA (Nyttend)
Skopje City Centre Footbridges, Skopje, North Macedonia (Rašo)
Sky Bridge, Vancouver, British Columbia, Canada (JMV)
Slauerhoff Flying Drawbridge, Leeuwarden, Friesland, Netherlands (Leineabstiegsschleuse)
Slender West Lake (Five Pavilions) Bridge, Yangzhou, Jiangsu, China (Gisling)
Smithfield Street Bridge, Pittsburgh, Pennsylvania, USA (David Brissard)
Soca Railway Bridge, Solkan, Gorizia, Slovenia (Miran Hladnik)
Spean Praptos Bridge, Chi Kiaeng, Siem Reap, Cambodia (Travelpleb)
Spectacle Bridge, Lisdoonvarna, County Clare, Ireland (Mark Waters)
Spring Creek Trestle Bridge, Hanover, Montana, USA (Mike Cline)
Stadlec Chain Bridge, Stadlec, South Bohemia, Czech Republic (Michael Ritter)
Stanley Bridge, Alexandria, Alexandria Governorate, Egypt (Ahmad Ali)
Stari Most, Mostar, Herzsegovina-Neretva, Bosnia-Herzegovina (BáthoryPéter)

ACKNOWLEDGEMENTS

ACKNOWLEDGEMENTS

Starrucca Viaduct, Lanesboro, Pennsylvania, USA (Niagara)
State Castle Moat Bridge, Ceský Krumlov, South Bohemia, Czech Republic (VitVit)
St-Léger Viaduct, Saint Chamas, Bouches-du-Rhône, France (Jacques Mossot)
Stone Bridge, Zamora, Castile-León, Spain (Antramir)
Sundial Bridge, Redding, California, USA (Prayitno)
Sungai Kebun Bridge, Bandar Seri Begawan, Brunei-Muara, Brunei (Zulfadli51)
Svilengrad Bridge, Svilengrad, Haskova Province, Bulgaria (Andrey Andreev)
Swandbach Bridge, Hinterfultigen, Bern, Switzerland (Хрюша)
Sydney Harbour Bridge, Sydney, New South Wales, Australia (Adam J.W.C.)
Széchenyi Chain Bridge, Budapest, Central Hungary, Hungary (Wilfredor)

Tachogang Lhakang Bridge, Paro, Paro Province, Bhutan (Sameera Vattikuti)
Tacoma Narrows Bridges, Seattle, Washington State, USA (Lderendi)
Taman Negara Canopy Walkway, Kuala Tahan, Pahang, Malaysia (RoB)
Tancarville Bridge, Le Havre, Seine-Maritime, France (Sergey Prokopenko)
Tarn Bridge, Quézac, Lozère, France (Ancalagon)
Tauber Bridge, Rothenburg ob der Tauber, Bavaria, Germany (Lars Schmitt)
Te Rewa Rewa Bridge, New Plymouth, Taranaki, New Zealand (Andrew Smith)
Three-Arched Bridge, Venice, Veneto, Italy (Didier Descouens)
Three Countries Bridge, Huningue, Haut-Rhin, France (Wladyslaw)
Tiberius Bridge, Rimini, Italy (Heiko Trurnit)
Tjörn Bridges, Tjörn, Bohuslän, (Arild Vågen)
Toupin Viaduct, Saint-Brieuc, Côtes-d'Armor, France (Presko)
Traversina Footbridge, St Moritz, Graubünden, Switzerland (Marco Zanoli)
Treponti, Comacchio, Ferrara, Italy (Francesco Gardini)
Trimiklini Double Bridge, Trimiklini, Limassol District, Cyprus (Xaris333)
Trinity Bridge, River Neva Bascule Bridge, Saint Petersburg Northwestern, Russia (Alex 'Florstein' Fedorov)
Triplets Bridges, La Paz, Pedro Domingo Murillo, Bolivia (Russland345)
Tunca Bridge, Erdine, East Thrace, Turkey (MEH Bergmann)
Tunkhannock Viaduct, Nicholson, Pennsylvania, USA (Chuck Walsh)

Umshiang Root Bridge, Cherrapungi, Meghalaya, India (Arshiya Urveeja Bose)
Ushibuka Haiya Bridge, Kumamoto, Japan (Kenta Mabuchi)

Valentré Bridge, Cahors, Lot, France (Accrochoc)
Vanne Aqueduct, Pont-sur-Yonne, Bourgogne-Franche-Comté, France (Charley Andrews 89)
Vanvitelli Aqueduct, Caserta, Italy (ElfQrin)
Vasco da Gama Bridge, Lisbon, Estremadura, Portugal (Paulo Valdivieso)
Verrazzano Narrows Bridge, New York, USA (Ajay Suresh)
Viaur Viaduct, Tanus, Tarn, France (Хрюша)
Victoria Falls Bridge, Victoria Falls, Matabeleland North, Zimbabwe (Kounosu)
Vihantasalmi Bridge, Mäntiharju, Åland, Finland (Motopark)
Vimy Memorial Bridge, Ottawa, Ontario, Canada (Markwrogers)
Vizcaya Transporter Bridge, Portugalete, Biscay, Spain (Javier Mediavilla Ezquibela)
Vlaardingse Vaart Bridge, Vlaardingen, South Holland, Netherlands (Rob Oo)
Vytautas the Great Bridge, Kaunas, Kaunas County, Lithuania (Creative)

Wadi Laban Bridge, Riyadh, Saudi Arabia (Danyal Saeed Photography)
White Nile (Omdurman) Bridge, Khartoum, Sudan (Bertramz)
Wignacourt Aqueduct, Valetta, South Eastern, Malta (Continentaleurope)
Wye Bridge, Keddie, California, USA (Adrian Studer Fthat [1])

Yangtze River Bridge, Nanking, Jiangsu, China (Fllee)
Yibna Bridge, Yavne, Central District, Israel (Ori~)
Yiheyuan Summer Palace Seventeen Arch Bridge, Beijing, China (Zcm11)
Yongle Bridge, Tianjin, Hebei, China (kele_jb1984)

Zayanderud Si-o-se-Pol Bridge, Isfahan, Isfahan, Iran (Reza Haji-pour)
Zhivopisny Bridge, Moscow, Russia (Daryona)